Properties of Crystalline Materials by X-ray Diffraction Methods and Symmetry Groups

A Practical Approach

John Fernando Zapata Mesa
University Institution of Envigado
Envigado, Colombia

CRC Press
Taylor & Francis Group
Boca Raton London New York

CRC Press is an imprint of the
Taylor & Francis Group, an **informa** business
A SCIENCE PUBLISHERS BOOK

First edition published 2026
by CRC Press
2385 NW Executive Center Drive, Suite 320, Boca Raton FL 33431

and by CRC Press
4 Park Square, Milton Park, Abingdon, Oxon, OX14 4RN

CRC Press is an imprint of Taylor & Francis Group, LLC

Library of Congress Cataloging-in-Publication Data (applied for)

ISBN: 978-1-032-38237-1 (hbk)
ISBN: 978-1-032-38308-8 (pbk)
ISBN: 978-1-003-34443-8 (ebk)

DOI: 10.1201/9781003344438

Typeset in Times New Roman
by Prime Publishing Services

DEDICATED

to my wife Adriana and
to my daughters
Daniela who is in heaven and Maria Jose who is on earth

Foreword

Knowledge of materials is fundamental to the development of science and technology that we have today, allowing us to understand and manipulate the properties of matter with unprecedented precision. Among the most powerful tools for their study is X-ray diffraction, a technology for ordered materials, also called crystalline, which allows the atomic arrangement of crystals to be deciphered with extraordinary accuracy. It is in this context that the book Properties of crystalline materials by X-ray diffraction methods and symmetry groups by Professor John Fernando Zapata Mesa stands as an indispensable reference work for the scientific community.

The author, a physicist and a materials scientist with extensive experience in research and teaching, has masterfully captured the knowledge accumulated over years of study and experimentation. As a full-time professor at the University Institution of Envigado in Colombia, his academic career and his commitment to the training of new generations of scientists have been pillars in the consolidation of this work. His clear and precise writing, together with his analytical rigor, make this book an essential guide for all those who wish to enter the world of crystallography and X-ray diffraction.

The book begins with a detailed presentation of the physical foundations of X-rays, addressing the essential principles that govern their interaction with matter. This knowledge is fundamental for understanding X-ray diffraction and its application in the characterization of crystalline materials. Subsequently, the study of symmetry groups and crystallography is introduced, key tools for classifying crystalline structures and predicting their behavior under X-rays.

A highlight of this work is the inclusion of tensor analysis, an essential mathematical resource for describing physical properties that depend on the orientation of crystals, such as thermal conductivity, electrical conductivity, and elasticity. The theory of elasticity, presented in detail, allows us to understand how materials respond to external forces and how these responses can be modeled in the context of crystal symmetry. In addition, the material properties of symmetry groups are explored, establishing deep links between atomic structure and the physical characteristics of materials.

The book also delves into the determination and refinement of crystal structure using the Rietveld method, a modern technique that allows for accurate analysis of X-ray diffraction data and quantification of crystalline phases. Its implementation has revolutionized structural characterization, providing detailed information on the atomic arrangement within crystals.

To complement the topics covered, the book includes several appendices with practical applications and advanced techniques. These include applications of X-ray diffraction using PANalytical's High Score software, the study of preferential orientation in X-ray diffraction, solutions for the elimination of the KAlpha2 peak, and a comprehensive analysis of the material properties of symmetry groups, covering all 230 space groups.

This work not only represents an essential reference for physicists, chemists, engineers and geologists, but also becomes a source of inspiration for university professors and researchers seeking a deeper understanding of the structure of crystalline materials. The author's clarity of exposition, his didactic approach and his scientific rigor make this book an excellent resource for anyone interested in the structural analysis of materials.

I invite the reader to delve into these pages and discover the fascinating world of crystalline materials through the expert eyes of John Fernando Zapata Mesa. Sincerely,

Henry A. Colorado L.

Full Professor, Universidad de Antioquia.

Director, CCComposites Lab., Universidad de Antioquia, Colombia.

PhD and MS, Materials Science, University of California, Los Angeles (2013). BS and MS, Mechanical (and Materials) Engineering, Universidad Nacional de Colombia (2005)

Fulbright Visitor Scholar, Mechanical and Aerospace Engineering, University of California San Diego, 2023–2024

Associate Editor, Materials Science, Heliyon, Elsevier

Associate Editor, Engineered Science, ES Publisher

Associate Editor, ES Materials & Manufacturing, ES Publisher

Scientific Committee, Advanced Composites and Hybrid Materials, Springer Nature

Preface

The development of materials science, and in particular, the understanding of the properties of crystalline materials, has advanced significantly thanks to the application of analytical techniques based on X-ray diffraction and the study of crystallographic symmetry. The ability to investigate the internal structure of crystals with precision has allowed scientists and engineers not only to better understand the fundamental properties of materials, but also develop technological applications that impact areas as diverse as electronics, structural engineering and chemistry.

This book, *Properties of Crystalline Materials by X-ray Diffraction Methods and Symmetry Groups*, was created with the purpose of offering a clear and systematic approach to understand and apply modern X-ray diffraction techniques, together with symmetry group analysis. Throughout these chapters, the basic physical principles of X-rays are explored, as well as advanced methods such as structural refinement using the Rietveld method.

The content is organized so that the reader can follow a logical progression, starting with the "physical foundations of X-rays" (Chapter 1), continuing with "symmetry groups and crystallography" (Chapter 2), and progressing to more abstract concepts such as "tensor analysis" (Chapter 3) and "elasticity theory" (Chapter 4). The intent is to provide a solid foundation in these topics before exploring how the "material properties of symmetry groups" influence the observed physical behaviors of crystals (Chapter 5). Chapter 6 concludes with a detailed review of the process of "structural determination and refinement using the Rietveld method", which is one of the most powerful and versatile tools in the characterization of crystalline materials.

The appendices are designed to complement the core topics of the book with practical examples and solutions to common problems in X-ray diffraction. "Appendix 1" explores "applications of PANalytical's High Score software" in diffraction analysis, while "Appendix 2" addresses the phenomenon of "preferred orientation". "Appendix 3" provides useful methods for "KAlpha2 peak removal", a crucial technique in improving the quality of diffraction data. Finally, "Appendix 4" offers a detailed analysis of the 230 space groups, which are fundamental for describing structural diversity in crystalline materials.

This book is intended for advanced students, researchers and practitioners in the fields of materials science, physics, and chemistry, who seek a deep understanding of X-ray diffraction methods and their relationship to crystallographic symmetry. The aim is to provide both the theoretical knowledge and the practical tools necessary to tackle complex problems in materials characterization.

I hope that this text will be a useful and enriching guide for those who venture into the study of the crystalline properties of materials. Throughout these pages, I have attempted to combine pedagogical clarity with scientific rigor, in an effort to make accessible a topic that, although complex, is exciting and of great scientific relevance.

It is my hope that this text will serve not only as a source of technical knowledge, but also as an inspiration for future research and innovative applications in the vast field of crystalline materials. I am deeply grateful to my colleagues, students, and collaborators, who, with their questions and contributions, have enriched my understanding of this field and motivated the writing of this book.

John Fernando Zapata Mesa

Acknowledgments

A very special thank you to the University of Envigado that has welcomed me for the last 17 years of my life and made the writing of this book possible.

To the dean of the Faculty of Engineering of the University of Envigado, Jonier Rendon, for his unconditional support.

To my psychiatrist, Dr. Carlos Alberto Herrera Cossio, who has helped me in the last 20 years to deal with my depression and other hells that I have had to go through.

To my parents Jesús and Maria Lucila, for the love they have transferred to me throughout their lives.

To the Taylor & Francis Publishing House for inviting me to write this book, an invitation that gave me motivation and energy to carry out this project.

To my wife Adriana and my daughter Maria Jose, for their constant love and company that give meaning and purpose to my life.

Finally, I thank this universe and its physical laws, because thanks to them I am here and I have been able to experience life, love and appreciate the impressive beauty that exists in it.

Contents

Introduction

The study of crystalline materials has been one of the most dynamic and fundamental areas in materials science, with applications ranging from engineering to quantum technology. Within this field, X-ray diffraction has emerged as one of the most powerful tools for exploring and characterizing the internal structure of crystals. Furthermore, the concept of symmetry, which governs the arrangement of atoms in crystals, has become a key element in understanding the physical and chemical properties of these materials.

The book *Properties of Crystalline Materials by X-ray Diffraction Methods and Symmetry Groups* is designed to provide a comprehensive and detailed introduction to these two fundamental concepts: X-ray diffraction and symmetry group theory. Through an integration of theory and practice, this book aims to equip the reader with the knowledge necessary to analyze and understand the properties of crystalline materials.

Throughout its chapters, the book covers a wide range of topics from the "physical foundations of X-rays" to "advanced analysis of material properties based on crystal symmetry". The reader will be introduced to a solid foundation in the physical principles behind X-rays and how they interact with crystals to produce diffraction patterns. These principles are then expanded upon with a discussion of "crystallography and symmetry groups", which provide a mathematical framework for classifying and describing crystal structures.

As we progress, more complex concepts are introduced, such as tensor analysis and elasticity theory, which are essential to describe how the physical properties of materials depend on their symmetry. These concepts allow the reader to understand how stresses, strains, and other physical phenomena are affected by the atomic structure of the material.

In Chapter 1, the physical foundations of X-rays are presented, covering the basic principles underlying their interaction with matter, which is essential knowledge for understanding diffraction.

Chapter 2 covers symmetry groups and crystallography, which are key to classifying crystal structures and predicting how they interact with X-rays. This section provides a solid foundation for using symmetry as a tool for structural analysis.

In Chapter 3, tensor analysis is introduced, a crucial mathematical concept for describing orientation-dependent physical properties in crystals, such as thermal and electrical conductivity, and elasticity.

Chapter 4 focuses on the theory of elasticity, which describes how materials respond to external forces and is fundamental to understanding deformation in crystals.

In Chapter 5, the material properties of symmetry groups are explored, showing how the physical characteristics of materials are intrinsically related to their crystal symmetry.

Chapter 6 delves into crystal structure determination and refinement by the Rietveld method, a modern X-ray diffraction data analysis technique that allows for detailed information on the atomic arrangement within crystals.

Finally, the appendices complement the topics covered with practical applications and advanced techniques. Appendix 1 covers applications of X-ray diffraction using PANalytical's High Score software, while Appendix 2 discusses preferential orientation in X-ray diffraction. Appendix 3 offers solutions for the elimination of the KAlpha2 peak, and Appendix 4 extends the analysis of material properties of symmetry groups to cover all 230 space groups.

CHAPTER 1

Physical Foundations of X-rays

1.1 The discovery of X-rays

This chapter gives an elementary introduction to the physical nature of X-rays and the origin of their continuous spectrum and lines when they are scattered by a Coulomb field or interact with a target, in accordance with the objectives of this book. For aspects of a more in-depth study of the physical nature of X-rays from quantum field theory and from the quantum theory of scattering and diffraction, as well as the associated wave mechanics, the reader can refer to classic texts on the subject like *X-ray Diffraction* (Seeck and Murphy 2015) and *The Nature of X-Rays and Their Interactions with Matter* (Stöhr 2023).

Wilhelm Conrad Röntgen discovered X-rays while studying the phenomenon of electrical discharges in gases. When he used a cathode ray tube with a high voltage of several tens of kilovolts, he observed that the barium salts fluoresced when he brought them close to the tube, although no visible emission from it was perceived. This effect persisted when he wrapped the tube with black cardboard. Röntgen determined that the cause of this fluorescence originated at the point where the stream of energetic electrons collided with the glass wall of the tube and since he did not know its nature, he gave it the name "X-rays".

Röntgen observed the following properties of X-rays, taking the first steps in identifying their nature:

1. There are several substances transparent to X-rays.
2. X-rays cannot be reflected or refracted and do not show interference effects. (These phenomena were present, but they were too subtle for Röntgen to observe at that time.)
3. X-rays could manifest themselves by darkening wrapped photographic plates, discharging electroscopes, and producing fluorescence in different substances.
4. X-rays cannot be deflected by either electric or magnetic fields.

DOI: 10.1201/9781003344438-1

5. Electron bodies, positively or negatively, are discharged by X-rays.
6. X-rays can penetrate low atomic number materials of considerable thickness, while high atomic number substances are opaque.

Röntgen X-rays were very short-wavelength electromagnetic radiation produced by the collision of high-speed electrons on the glass walls of the glass tube. X-rays have a wavelength range (meters) between $1 * 10^{-11}$ y $5 * 10^{-9}$ and an approximate energy of 12.3 KeV (kilo electronvolt).

The most common sources of X-rays are the Bremsstrahlung effect, which I will talk about later, and electronic transitions for the atomic layers close to the nucleus. Two detection methods can be photography or an ionization chamber.

The improvement in the construction of X-ray tubes was the work of the American physicist N. D. Coolidge, in 1913, in the laboratories of General Electric. Figure 1.1 shows a schematic of a Coolidge tube. A thermionic cathode and an anode are enclosed in a glass tube that has been evacuated to an extraordinarily low pressure, so that the electrons emitted by the cathode can travel directly to the anode with a small probability of collision along the way, reaching the anode with a speed corresponding to the difference in the total potential across the tube. Because of the bombardment of the anode surface by the electron stream, said surface emits X radiation.

The first high-voltage sources, applied to the anode and cathode of X-ray tubes, were provided by an induction coil of various shapes. Although some of those supplies are still in use, they were almost completely replaced by a more efficient high-voltage transformer. The emf generated by these transformers varies between 50.000 and 2.000.000 volts. The normal emf used in surgery is 100.000 volts, while higher voltages are used to treat diseases. The alternating current emf supplied by a transformer is not directly applied to the x-ray tube but is first converted into direct current by rectifier tubes.

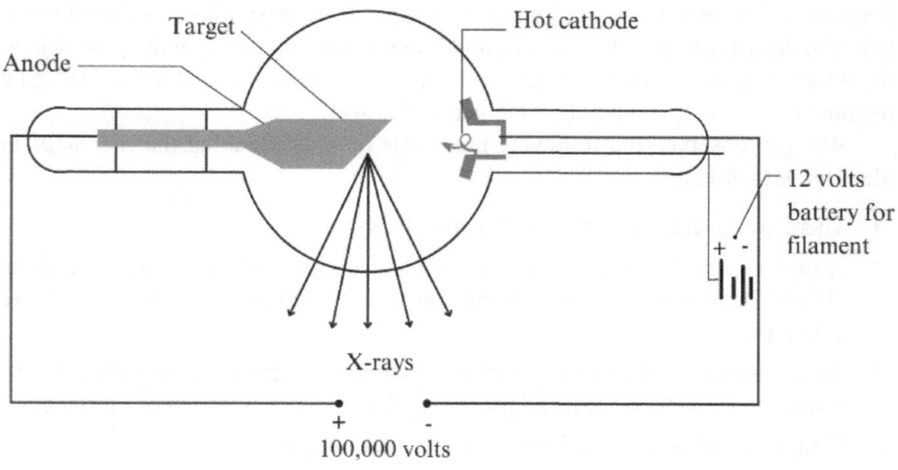

Figure 1.1. Coolidge X-ray tube.

The Coolidge tube is composed by placing a filament of tungsten wire in the center of the cathode, which is heated to incandescence by a low-voltage transformer or accumulator. This filament, which is a copious emitter of electrons, produces a much more intense source of X-rays in the target than was possible with a cold cathode. Most metals melt under the tremendous bombardment of the target with so many electrons. To overcome this difficulty, a metal with a very high melting point, such as Wolfram (tungsten) or molybdenum, is embedded in the face of the copper anode that serves as the target. Copper, which is a good thermal conductor, helps dissipate heat.

The two main types of tubes used in the generation of X-rays are:

1. Gaseous atmosphere tubes
2. Degassed electronic tubes

The first are the most common and simple. In them, the high voltage fractions the gas molecules, converting them into ions and residual electrons. The electric field drives the positive ions against the cathode, so that the electrons are released by bombardment. By bombarding the stream of cathode rays (high-speed electrons), the positive or anti-cathode electrode produces X-rays, the penetrating power of which is determined by the amount of residual gas.

The Coolidge incandescent filament X-ray tube (electron tube) has displaced gas tubes in roentgenological diagnosis and treatment. Furthermore, it is a descendant of the simple Crookes tube and works according to the same principle. It consists of a vial where the most perfect vacuum possible is made and is equipped with electrodes to which a high voltage current can be applied. The negative electrode is called the cathode and is usually a spirally wound tungsten filament. A great advantage of these tubes is the independence of the current in the tube and the voltage. One factor can be altered without affecting the other.

The positive anode is called the target. It is made of a heavy metal, such as tungsten, and soldered to a copper support that helps dissipate the enormous heat generated. The X-ray source is a tube where a high vacuum was generated, and which contains a cathode and an anode inside. The main parts of a large industrial type of pipe are those indicated in Figure 1.1. At one end of the tube is the cathode that emits the electrons, it is formed by a tungsten filament mounted on a cup-shaped metal structure that focuses the electron current on the surface of the anode. The approach is highly precise so that the electrons collide with the anode on a well-defined surface.

The part of the anode that the electrons collide with is the anticathode or focus, a thin disk of pure tungsten. The anticathode is supported by a copper-molybdenum bar. The potential difference between the anode and the cathode for most industrial jobs is between 100,000 and 300,000 volts, that is, between 100 and 300 kilovolts. These high voltages accelerate the electrons, giving them enormous speed, and energy is released at the anticathode at a rate of 50 to 600 watts per cm^2. The anticathode is placed at a high temperature and the heat produced in it must be quickly dissipated. Tungsten does not melt at a temperature below 3300°C, it also does not evaporate to a considerable degree, even at high temperatures and vacuums, it conducts heat well and its characteristics allow a high production of X-rays.

When these electrons emitted at the cathode interact with the coulomb field, as seen in Figure 1.2. They are decelerated and the radiation produced is that predicted by classical electromagnetic theory for an accelerated charge. As the incident electron slows down and loses kinetic energy, the lost energy is used to create a photon with an energy given by

$$hv = K_1 - K_2 \tag{1.1}$$

where the kinetic energy of the heavy nucleus in retreat has been neglected. h is Planck's constant, v is the frequency of the emitted radiation, and K_1 and K_2 are the kinetic energies of the incident electron and of the same electron deflected by the coulomb field of the nucleus.

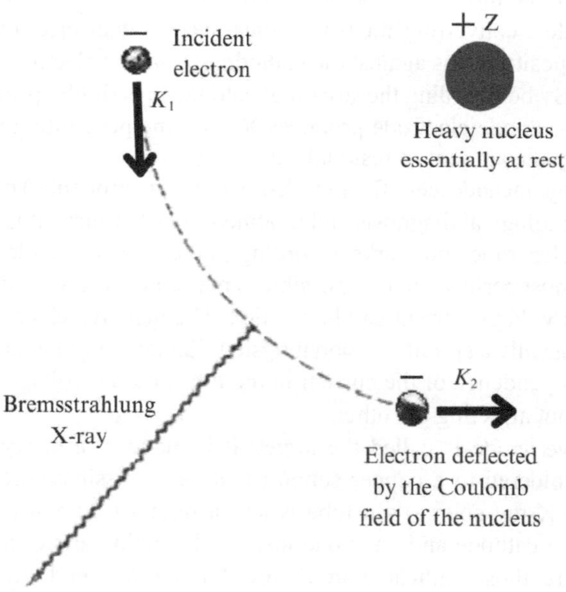

Figure 1.2. Bremsstrahlung produced by the acceleration of an electron in a coulomb field.

The radiation produced by the acceleration of a charged particle is called Bremsstrahlung (German word meaning braking radiation). Bremsstrahlung is also produced in large accelerators, where charged particles are accelerated to high speeds. An X-ray spectrum shows a continuous curve and some peaks, as shown in Figure 1.3.

There is a difference in the way the study of this spectrum is approached, from electromagnetic theory and quantum mechanics: electromagnetic theory predicts continuous radiation from each collision (the continuous curve) while the electron is decelerated, but it does not explain the peaks, the line spectrum, while quantum mechanics predicts the creation of a photon of energy hv, which is, in general,

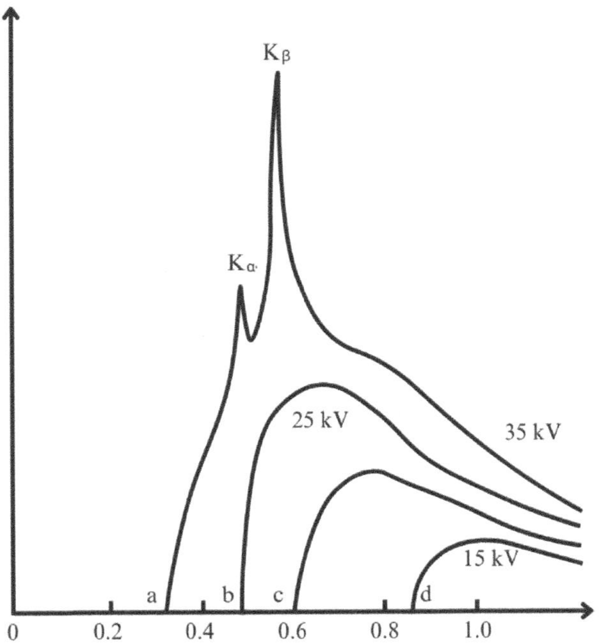

Figure 1.3. The X-ray spectrum of silver shows the Bremsstrahlung and characteristic spectra (peaks k_α and k_β) and the dependence of the limit on the shortest wavelength of the accelerating potential through the tube.

different for each collision (which explains the peaks in the curve). The continuous X-ray spectrum in Figure 1.3, is produced by the Bremsstrahlung.

The peaks of the curve (sharp lines) depend on the material that acts as the target. Although the intensity of the continuous Bremsstrahlung spectrum for a given potential depends on the physical characteristics of the target, the sharp jumps in wavelength at points *a*, *b*, *c*, and *d* in the figure are independent of the target material.

Figure 1.4 shows that when the maximum cutoff frequency is plotted against the accelerator potential *V*, there is a simple linear relationship between these variables.

$$\frac{v_{max}}{V} = \left(\frac{c}{\lambda_{min}}\right)\frac{1}{V} = constant \tag{1.2}$$

where λ_{min} is the minimum wavelength for electron emission and *c* is the speed of light in a vacuum.

Classical theory did not foresee this well-defined cut, which experimental evidence clearly shows.

The trajectory of the incident high-speed electron changes when it is scattered by the Coulomb field of the atoms of the material that acts as a target. According to Planck's quantum hypothesis, the radiation from accelerated electrons must be

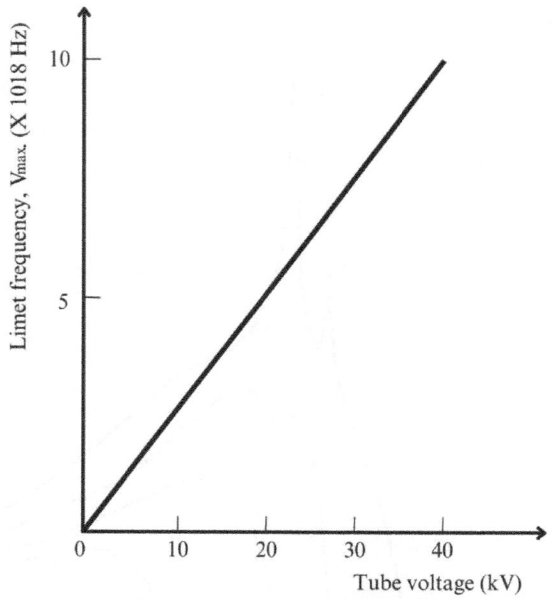

Figure 1.4. Simple linear relationship between maximum cutoff frequency and tube voltage.

emitted in quanta of value hv. However, the maximum energy of any of the energy quanta cannot be greater than the energy of the most energetic incident electron:

$$hv_{max} = K_{max} = eV \tag{1.3}$$

where K_{max} is the maximum kinetic energy that an electron can reach, or it can also be written

$$\frac{hc}{\lambda_{min}} = eV \tag{1.4}$$

where V is the accelerator potential through the X-ray tube, and e is the charge of the proton. The accelerator potential is on the order of several thousand volts, but about 98% of the energy given by the electrons when they collide with the target is transformed into heat energy, and the temperature of the target rises. Because the cutoff frequencies are well-defined and because the wavelengths can be determined exactly, X-ray spectra constitute a good experimental method for determining the value of Planck's constant h. The characteristic spectrum of sharp lines is a function of the target material, although its intensity for a given target depends on the accelerating potential.

The electrons in an atom are arranged in shells around the nucleus, the most tightly bound being in the K shell. Those in the next highest binding position are in the L layer and then the M layer, then the N layer, and so on.

When the highly energetic incident electrons eject an electron from the K shell, an electron in the L shell gives up its energy in the form of X-rays as it passes into the vacancy left in the K shell. This radiation, characteristic of the target material, is called line K_α. The electron in the M shell that fills the vacancy in the K shell gives up energy in the form of another X-ray called the K_β line. These transitions from the L, M, N, etc. layers, to electronic transitions for the layers closest to the nucleus, give rise to the characteristic X-ray spectra.

The K layer gives rise to the series of lines K_α, K_β, K_γ, etc. called the K series. When the incident electrons dislodge electrons from the L shell and the holes are filled by electrons from the remaining M, N, O shells, these transitions give rise to the L series, the first line of which is L_α. The nomenclature for these transitions is illustrated in Figure 1.5. But quantum mechanics establishes that all these known characteristics of energy levels have a close direct relationship with the principal quantum numbers n, l, j, and there are selection rules that establish which transitions are allowed and which are not. These selection rules are given by:

$$\Delta l = \pm 1;\ \Delta j = 0, \pm 1$$

These are the same selection rules that govern the optically active electron in an alkaline atom. The totality of X-rays that are emitted in these transitions, plus others that are emitted infrequently contrary to these rules, constitute the total X-ray spectrum of the atom.

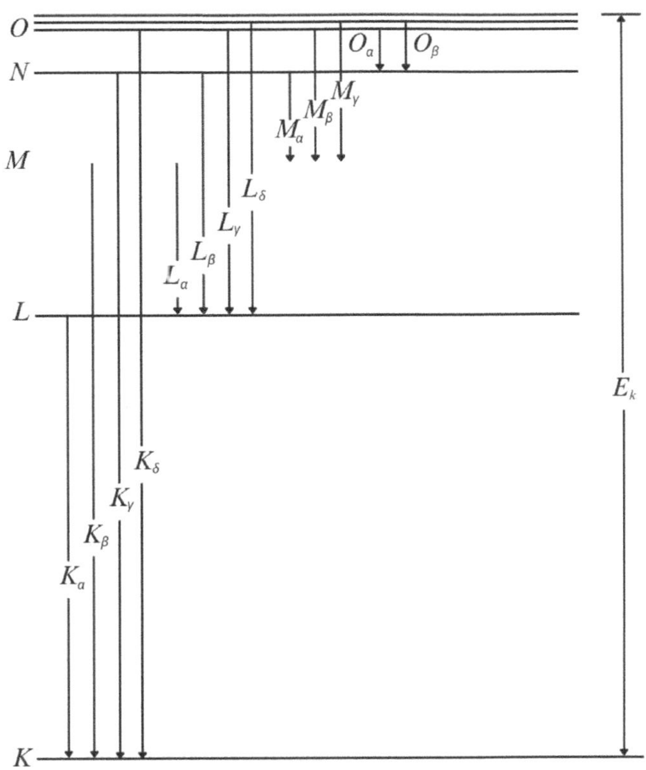

Figure 1.5. Electronic transitions for the layers close to the nucleus, which give rise to the characteristic spectra of X-rays.

As the accelerating voltage in the X-ray tube increases, the Incident electrons produce continuous Bremsstrahlung spectrum until, at a certain critical value of the voltage, the electrons acquire enough energy to dislodge electrons from the inner shells, that is, from the *K*, *L* or *M* shells. Only when the critical potential V_c is reached will the incident electrons be sufficiently energetic for the *K* transitions to occur.

$$eV_c \geq E_k \tag{1.5}$$

where E_k is the energy necessary to release an electron *K* from the atom. The X-rays associated with the more energetic *K* series are called hard X-rays and those associated with the less energetic *L*, *M*, *N* series are called soft X-rays. Hard x-rays are more penetrating.

1.2 Continuous spectrum

The origins of continuous and line spectra are as different as their properties. For this reason, it is advisable to study it separately. In this section, we are going to delve a little deeper into the continuous part of the X-ray spectrum.

The shape of the continuous spectrum depends only on the energy of the electron's incident on the anode and not on their nature. In particular, the short wavelength cutoff point λ_0 (see Figure 1.3) is inversely proportional to the kinetic energy of the incident electron *E*. The cutoff frequency $v_0 = c/\lambda_0$ follows the empirical law.

$$hv_0 = E \tag{1.6}$$

The total energy of the X-rays per incident electron, *w*, which is proportional to the integral over λ of the continuum part of the spectrum, obeys the empirical law

$$w = kZE^2 \tag{1.7}$$

where $k \sim 0.7 * 10^{-4}$ for *w* and *E* given in MeV (mega electron volts) and *Z* is the atomic number of the anode material. The fraction of the electron's kinetic energy that is converted to X-ray energy is:

$$w/E = kZE \tag{1.8}$$

For example, for a typical case of $Z = 90$ and $E = 0.05$ MeV, w/E is only about 0.3%.

The transfer of energy from an incident electron to the electrons in the anode atoms of an X-ray tube is a process that constitutes the most important factor in reducing the speed of the incident electron, and that also leads (in the rare case of a large energy transfer) to the emission of the line spectra of the X-rays; however, It is not the only thing that produces the incident electrons.

Another is the dispersion by Coulomb interaction of these electrons, in collisions close to the nucleus of the atoms that are a part of the anode. In such a collision, the accelerations of an incident electron are very large, and it can be scattered over a large angle. Although the electron does not lose energy, transferring it to the nucleus, because its mass is very large and it does not retreat much, it occasionally loses energy in the form of electromagnetic radiation. This radiation is what forms the continuous X-ray spectrum (Figure 1.2).

Both classical electromagnetic theory and quantum physics predict that a scattered electron emits electromagnetic radiation. According to classical theory, whenever there is scattering, small amounts of radiation are emitted, since it can be deduced from Maxwell's equations that charged particles emit radiation when accelerated.

The same does not happen with quantum theory. In contrast, a scattering process occasionally emits large amounts of radiation (quanta). This situation can be compared with that which exists in the case of the atom, the classical theory predicts the constant emission of radiation by the accelerated electrons that rotate around the nucleus, while the quantum theory predicts the occasional emission of radiation, when the electron changes its quantum states.

In the case of the bound atomic electron, we know that the classical theory makes incorrect predictions, and the quantum theory makes correct predictions, but in the case of the unbound scattered electron it is true that the quantum theory always leads to correct conclusions; However, in this case, classical theory does not lead us to incorrect conclusions.

The classical theory describes quite well certain characteristics of the X-ray production of the continuous spectrum. The contrast between the two cases reflects the fact that the differences between a quantized and a non-quantized theory are not as fundamental in the case of an unbound particle as they are in that of a bound particle.

According to classical electromagnetic theory, a particle of charge Ze, where Z is the atomic number, with an acceleration a, will emit energy in the form of electromagnetic radiation with a speed of:

$$\frac{dE}{dt} = \frac{2}{3}\frac{(Ze)^2}{c^3}a^2 \tag{1.9}$$

When a nucleus of charge Ze scatters this particle, it experiences a Coulomb force that is proportional to zZe^2 and, if the mass of the particle is m, its acceleration is proportional to zZe^2/m. Consequently, the rate of emission of electromagnetic energy can be expressed as:

$$\frac{dE}{dt} = kz^2a^2 = \frac{z^4Z^2}{m^2} \tag{1.10}$$

where k is a proportionality constant. The $1/m^2$ factor predicts that the emission of electromagnetic radiation is more important for electrons than for protons or any other heavy particle. It has been experimentally proven that the energy emitted in the form of X-rays of the continuous spectrum by a heavy charged particle passing through matter is so small that it is almost impossible to measure. The Z^2 factor predicts that high atomic number anodes are the most efficient products of these X-rays.

The dependence on Z^2 also shows that the radiation emitted in the scattering from an atomic electron, that is, for $Z = 1$, is so small that it can be neglected. If the proportionality constant k is evaluated and integrated along the average path of the incident electron, upon reaching rest at the anode, the total energy of the X-rays of the continuous spectrum, w, that is emitted, can be calculated.

The results agree quite well with the empirical equation (1.7). Electromagnetic theory also reasonably explains the polarization of X-rays of the continuous spectrum, predicting that X-rays, emitted in the radial direction, $\vec{r}$, during an acceleration $\vec{a}$, it becomes polarized with the electric vector perpendicular to $\vec{r}$, in the plane defined by the vectors $\vec{r}$ and $\vec{a}$. On the other hand, higher energy X-rays are emitted in frontal crashes in which the acceleration is always approximately antiparallel to the unit vector $\hat{i}$ in the direction of the X axis, taken as the direction of the incident beam of the electrons. Consequently, these X-rays have a large net polarization. In shocks that produce the emission of lower-energy X-rays, the direction of acceleration is not as constrained and, as a result, the net polarization is small.

Classical electromagnetic theory gives an exact prediction of the angular distribution of the emitted X-ray energy in the continuous spectrum. The only aspect of the X-ray emission of the continuous spectrum that classical theory cannot explain is that shape of the spectrum, which can only be explained with quantum mechanics. All the details of the mathematical development of the aspects of electromagnetic theory mentioned in this section and the aspects of the scattering and absorption of scope of this book, the reader can see (Milton and Schwinger 2024) to delve into the given foundations and developments of classical electromagnetic theory for X-rays in the continuum spectrum.

1.3 Line spectrum

In this section we are going to go a little deeper into the part of the X-ray line spectrum. The lengths of the line spectrum or the sharp peaks depend exclusively on the anode material; they depend on its atomic number Z and not on the energy of the incident electrons. However, a line cannot appear in the spectrum unless the kinetic energy of the incident electron, E, has a minimum value, and satisfies the following relationship.

$$E \geq E_T = h\nu = hc/\lambda \tag{1.11}$$

where ν is the frequency and λ is the wavelength of the line. The total energy of the X-rays emitted in a particular line increases with increasing kinetic energy of the incident electron according to the empirical equation:

$$I = k(E - E_T)^n \tag{1.12}$$

where k is another proportionality constant $n \sim 1.5$.

The origin of X-ray line spectra can be understood by analyzing one of the processes by which an energetic electron is forced to slow down as it passes to the anode of an X-ray tube. This process is the transfer of energy from incident electron to the atoms of the anode, through collisions by Coulomb interaction, between the incident electron and the electrons of the atoms.

An energetic electron striking an atom can transfer an appreciable fraction of its kinetic energy to an atomic electron only when the collision is sufficiently head-on. This statement can be justified and made quantitative by analyzing the potential energy of the Coulomb interaction between two electrons:

$$V(r) = \frac{e^2}{r} \tag{1.13}$$

By evaluating using this Coulomb electric potential equation the distance r at which the potential energy is $V(r) \sim 10^2$ eV, we find that $r \sim 10^{-9}$ cm. Since the energy transfer from the incident electron to the atom cannot be greater than the maximum potential energy they experience, we see that energy transfers greater than 100 eV are recorded only in collisions in which the distances of maximum approach between the two electrons. They are smaller than 10^{-9} cm.

This distance is about 10% of an atomic diameter. Collisions as head-on as these are not very frequent, because the atomic electrons in the outermost shells can move away from the path of the incident electron by Coulomb repulsion, and although the atomic electrons in the inner shells cannot do this, since that are firmly bonded to the center of the atom, the probability that the incident electron will approach these electrons is small, because the inner shells are equally small.

Almost all collisions involve an energy transfer of a few electron-volts or, at most, a few tens of electron-volts. In this type of collision, it is energetically possible to excite only one atomic electron from one of the outer shells. The electron can be excited into a bound state or a continuous state, leaving the ionized atom. In both cases, this electron, or any other, will eventually return through the various excited states until reaching the ground state and the atom will emit lines of its spectrum depending on the energy of the incident ray.

One of the most notable differences between the X-ray spectrum and the optical spectrum is that the former vary slightly from one element to another, that is, there are no abrupt changes from one element to the next, contrary to what happens in optical spectra. The reason is that the characteristics of the X-ray spectra basically depend on the binding energies of the electrons in the inner shells. As the atomic number Z increases, these binding energies increase uniformly, due to greater nuclear charge, and are not affected by periodic changes in the number of electrons in the outer shell. Moseley was the first to observe the regularity of X-ray spectra.

1.4 X-ray diffraction

Röntgen's first attempts to experimentally verify the wave nature of X-rays were unsuccessful. The unusually short wavelength of X-rays made interference and diffraction effects very difficult to observe. Early experiments indicated that the wavelength of X-rays was on the order of 10^{-8} cm.

Calculating for a crystal the number of molecules per unit volume, Von Laue found that the average distance between the atoms of a solid was between 10^{-7} and 10^{-8} cm. Von Laue suggested that the regular arrangement of atoms in crystals could be used as a kind of diffraction grating for X-rays, and in 1914, he received the Nobel Prize in physics for establishing the wave character of X-rays by diffracting them by means of crystals. W. H. Bragg and his son Lawrence Bragg received the Nobel Prize the following year for perfecting Von Laue's concepts of ray diffraction by crystals. Von Laue's reasoning ran roughly along the following lines. The macroscopic characteristics of a salt crystal (NaCl) suggest that, on a microscopic scale, the atoms are arranged in a simple regular cubic lattice, taking the molecular weight as a value of 58.45 *g/mol*, also assuming that the electron density is high near the center of the atom, and that the electrons are firmly bonded to it. I take a value for the mass density of 2.165 *g/cc*. This means that one of the molecules in the grain of salt will have a mass of

$$m = \frac{M}{N_A} = \frac{58.45 g/mol}{6.023 * 10^{23} molecule/mol} = 9.705 * 10^{-23} \frac{g}{molecule} \tag{1.14}$$

where N_A is Avogadro's number. The number of molecules per unit volume would be:

$$N = \frac{\rho}{m} = \frac{2.165 g/cc}{9.705 * 10^{-23} \dfrac{g}{molecule}} = 2.24 * 10^{24} molecule/cc \tag{1.15}$$

If each salt molecule contains two atoms (one of *Na* and one of *Cl*), the number of atoms per cubic centimeter is:

$$2N = 4.48 * 10^{22} \; atoms/cc \tag{1.16}$$

If there are *n* atoms along one edge of a 1 *cm* cube of salt, then

$$n^3 = 4.48 * 10^{22} \frac{atoms}{cc} \Rightarrow n = 3.55 * 10^7 \frac{atoms}{cm} \tag{1.17}$$

Thus, the distance between the atoms is:

$$d = \frac{1}{n} = 2.82 * 10^{-8} cm = 2.82 \text{ Å} \tag{1.18}$$

This calculation for the average distance between the atoms of a simple cubic salt structure is summarized by the formula

$$d = \sqrt[3]{\frac{M}{2\rho N_A}} \tag{1.19}$$

Von Laue recognized that the symmetry of a crystal is due to a unit of molecular or atomic size being repeatedly arranged in a regular order. The layers of units are separated by successive uniform distances, which are of the order of the wavelength of the X-rays.

These regularly spaced discontinuities form the basis of a diffraction grating, like that used for visible light, except that the crystal forms a three-dimensional grating. Also, unlike the two-dimensional grating, the three-dimensional grating will not diffract monochromatic light incident on it, at any arbitrary angle. The diffracted X-ray beam will be structurally reinforced only when a wavelength λ finds planes of atoms separated by a distance d and finds them at a certain angle θ.

On the other hand, when an X-ray beam hits on the crystal, when it penetrates the crystal, a part passes through it completely and another part is dispersed between the atoms of the crystal. There are several mechanisms that produce this dispersion. For the moment we are only interested in the one through which X-rays are scattered without changing their wavelength or phase, due to the tightly bound electrons close to the nucleus of the atom. Each atom scatters X-rays through this mechanism at all angles, although the greatest intensity of scattering is in the direction of the incident beam.

In the region outside any atom, the intensity pattern of the scattered X-rays is such that it appears as if all the scattering occurs at its center. For this reason, the scattering made by the crystal can be analyzed in terms of a model crystal, which consists of a group of scattering centers with the same relative locations as the centers of the atoms of the real crystal. Consequently, these scattering centers are located on evenly spaced lattice planes.

Bragg used this idea to analyze the diffraction patterns of X-rays reflected by crystal planes. Figure 1.6 shows an incident beam at an angle θ with respect to an atom-rich plane of a crystal.

When rays 1 and 2 are reflected by points P and Q respectively, the difference in path between the rays is

$$\delta = RQ + QS = 2RQ \tag{1.20}$$

Since $RQ = PQ \sin \theta = d \sin \theta$.

With what you get:

$$\delta = 2d \sin \theta \tag{1.21}$$

Note that, in Bragg reflection, the angles θ of incidence and reflection are the angles that the incident and reflected rays make with the planes of the crystal and not with the normal, as is customary in optics. This angle θ is also called the grazing angle. Thus, the conditions for constructive interference, known as Bragg's laws of reflection, are:

1. The angle of incidence must be equal to the angle of reflection.
2. $\delta = 2d \sin \theta = n\lambda$ for $n = 1, 2, 3, \ldots$ where n is the order of reflection.

Figure 1.6 shows that there are many planes within the crystal, along which rays from various angles of incidence could be reflected to interfere constructively. Depending on the plane chosen, some planes will contain more atoms acting as reflecting agents than others. In general, the rays reflected by these planes will be those of greatest intensity.

A diffraction pattern can also be obtained with X-rays incident on a granulated glass sample. Each crystal fragment within the granules contains atomic planes, which will reflect the X-rays incident on them at a given angle. Since the fragments can be oriented at any azimuthal angle around the incident X-rays, there will be some that reflect light of a given wavelength at a particular angle, to form a circular pattern around the incident beam.

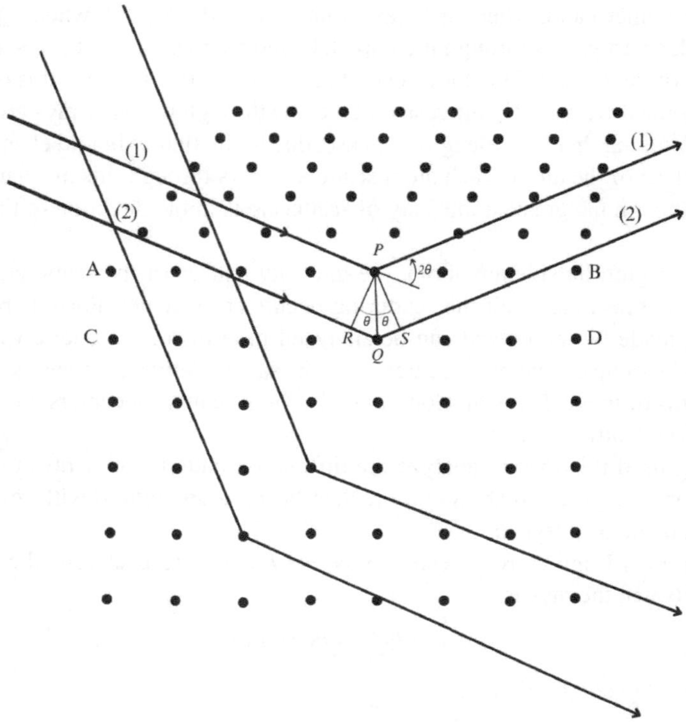

Figure 1.6. Exaggerated view of Bragg reflections through several planes in a crystal.

Those properly oriented at other different angles will form other concentric circles. The resulting diffraction patterns are called Laue patterns. From Figure 1.7 it is observed that the condition for a maximum to occur is:

$$n\lambda = 2d \sin \theta \tag{1.22}$$

where θ is the angle identified in the figure and is in fact the angle of incidence for some of the granulated fragments. Many of the physical properties of crystals can now be investigated with great accuracy with X-ray techniques.

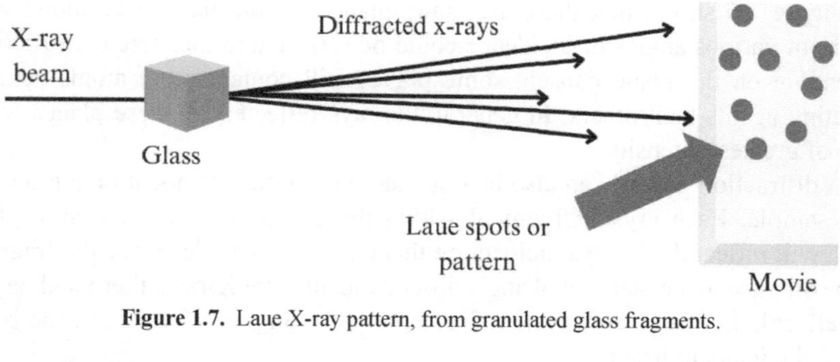

Figure 1.7. Laue X-ray pattern, from granulated glass fragments.

1.5 X-ray diffraction by a diffraction grating

Because the wavelength of X-rays is so short and because they are so penetrating, initial efforts to diffract X-rays with a slit grating were not very successful. It was not mechanically possible to construct slit networks with a slit spacing of 1Å, a typical X-ray wavelength.

Additionally, most mirrors do not reflect X-rays well since they are so penetrating. However, in 1925, Compton and Doan managed to measure the wavelength of X-rays using an ordinary slit grating at very small grazing angles (on the order of 0.001 radians). When rays 1 and 2 in Figure 1.8 are diffracted by the surface of the slotted grating, the path difference between the rays is

$$\delta = CB - AD = d \cos \theta - d \cos (\theta + \phi) \quad (1.23)$$

So, for these rays to interfere constructively, you have to:

$$d \cos \theta - d \cos(\theta + \phi) = n\lambda \quad (1.24)$$

where n is the order of diffraction. Written in terms of an approximation for small angles, and using the cosine series for first terms, $\cos x = 1 - x^2/2 + \cdots$, the above result becomes:

$$n\lambda = d\phi\left(\theta + \frac{\phi}{2}\right) \quad (1.25)$$

It is interesting to know that, when the wavelengths of the After examining the equations, it was found that Avogadro's number in equation (1.14) was not accurate, due to an error in the accepted value of the electronic charge that was used in the determination of N_A.

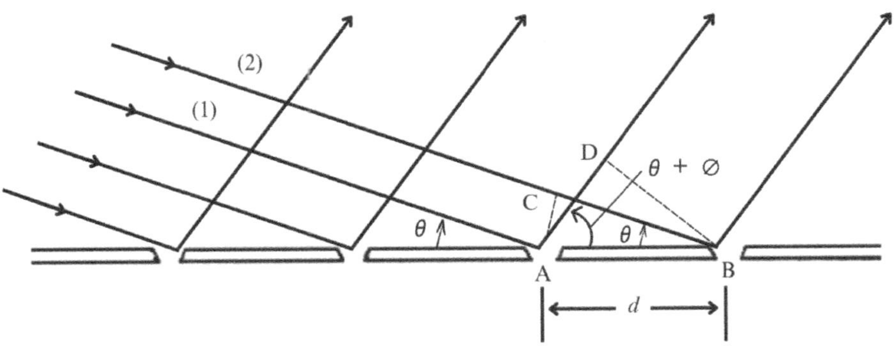

Figure 1.8. Diffraction of incident X-rays at grazing angles on a slotted diffraction grating.

1.6 Compton effect

The first experiments on X-ray scattering showed that the secondary X-rays involved were less penetrating than the primary X-rays. These secondary X-rays were initially thought to consist of the fluorescent radiation characteristic of the radiating element. Although fluorescence is characteristic of heavier elements, later experiments showed a difference in the penetration of secondary X-rays from lighter elements, such as carbon, from which no fluorescent radiation of the type observed can appear.

This led to speculation, by some physicists, that a new radiation had been found, which they called *J* radiation, however, after careful spectroscopic analysis, this idea of *J* radiation could not be supported. A careful experimental analysis determined that the secondary scattered X-rays have the following properties:

1. The scattered radiation consists of two wavelengths, the original wavelength λ_0 and an additional wavelength λ_s that has almost the same value of λ_0.
2. λ_s is always greater than λ_0.
3. λ_s depends on the scattering angle θ and not on the scattering medium.

In 1923 A. H. Compton correctly proposed that X-ray photons, in addition to energy, also have linear momentum, in the same way that a particle has, and that the scattering process is an elastic collision between the photon and an electron.

The change in the wavelength of the photons is due to this elastic scattering and is known as the Compton effect. The energy of an electromagnetic wave is $E = pc$, but the energy of a photon is also $E = hv$, where v is the frequency of the photon, therefore, combining these two expressions we obtain that the linear momentum for the photon can be express as:

$$p = \frac{hv}{c} = \frac{h}{\lambda} \tag{1.26}$$

The X-ray photon in Figure 1.9 is elastically scattered by a stationary free electron. The conservation of linear momentum for this event can be written as:

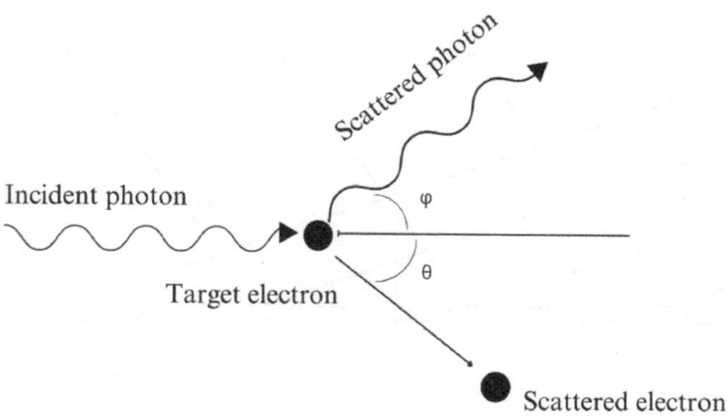

Figure 1.9. Compton scattering of a photon by an electron at rest.

By conserving the linear momentum in the *x*-axis direction, we obtain:

$$p_0 = p_s \cos\varphi + p_e \cos\theta \tag{1.27}$$

By the conservation of linear momentum in the direction of the *y* axis.

$$0 = p_s \sin\varphi - p_e \sin\theta \tag{1.28}$$

where p_s is the linear momentum of the scattered photon and p_e is the linear momentum of the scattered electron. From the law of conservation of energy, we have:

$$E_0 + m_0 c^2 = E_s + m_0 c^2 + K \quad (1.29)$$

where E_0 is the energy of the incident photon, $h\nu_0$; $m_0 c^2$ is the energy of the electron at rest; E_s is the energy of the scattered photon and $m_0 c^2 + K$ is the total energy of the scattered electron. Solving for K we obtain:

$$K = E_0 - E_s = h(\nu_0 - \nu_s) \quad (1.30)$$

Since $E_0 = h\nu_0 = p_0 c$ and $E_s = h\nu_s = p_s c$, we can write equation (1.30) in the form

$$K = (p_0 - p_s)c \quad (1.31)$$

Equation (1.27) can be rewritten as:

$$p_0 - p_s \cos \varphi = p_e \cos \theta \quad (1.32)$$

Squaring this equation on both sides and then adding the result, member by member, with equation (1.28), we obtain that

$$p_0^2 - 2p_0 p_s \cos \varphi + p_s^2 = p_e^2 \quad (1.33)$$

For an electron, from the equations of special relativity, we have to

$$E^2 = (m_0 c^2)^2 + p_e^2 c^2 \quad (1.34)$$

$$E = K + m_0 c^2 \quad (1.35)$$

Eliminating the energy E from equations (1.34) and (1.35), we obtain:

$$\frac{K^2}{c^2} + 2Km_0 = p_e^2 \quad (1.36)$$

Using the expression for K given in equation (1.31) and p_e given in expression (1.33) and multiplying everything by h, in equation (1.36), we arrive at the following result:

$$\frac{h}{p_s} - \frac{h}{p_0} = \frac{h}{m_0 c}(1 - \cos \varphi) \quad (1.37)$$

Replacing $\frac{h}{p_s} = \lambda_s$, which is the scattered wavelength, and $\frac{h}{p_0} = \lambda_0$ into equation (1.37), this equation takes the form:

$$\lambda_s - \lambda_0 = \frac{h}{m_0 c}(1 - \cos \varphi) \quad (1.38)$$

Or equivalently

$$\Delta\lambda = \lambda_c (1 - \cos \varphi) \quad (1.39)$$

where $\lambda_c = \dfrac{h}{m_0 c} = 0.024\text{Å}$ is defined as the Compton wavelength for the electron. In summary, for the explanation based on experimental results of the wavelength shift of the X-rays scattered by an electron, the following can be said:

i. In 1923, Compton experimentally confirmed the results obtained from equation (1.39).
ii. In 1923, Bothe and Wilson observed recoiling electrons.
iii. In 1925, Bothe and Geiger showed that the scattered photon $E_s = h\nu_s$ and the recoiling electron appear simultaneously.
iv. In 1927, Bless experimentally verified the energy of the recoiling electron.

It is important to highlight the following: Some photons are scattered by electrons that are not free but are bound to the atom. Thus, the mass m_0 in equation (1.38) must be replaced by the mass of the entire receding atom. The great value of this mass, compared to that of the electron, will make the variation of the wavelength very small, or in other words the term on the right of equation (1.38) tends to zero, therefore $\lambda_s - \lambda_0 = 0$, which implies that the wavelengths of the incident photon and the scattered photon are almost the same. For this reason, this effect is not observed for X-ray scattering by closely bound electrons.

1.7 X-ray diffraction methods

The information provided by the X-ray diffraction pattern can be seen as two different but complementary aspects: on the one hand, the geometry of the diffraction directions (conditioned by the size and shape of the elemental cell of the crystal) offer us information about the crystalline system. And on the other hand, the intensity of the diffracted rays is closely related to the nature of the atoms and the positions they occupy in the network, such that their measurement constitutes the three-dimensional information necessary to know the internal structure of the crystal.

In general, there are three main methods of X-ray diffraction:

- Laue method
- Rotation or rotating glass methods
- Powder Method

Laue method

Historically it was the first diffraction method. A Polychromatic X-ray is used that hits a fixed glass and perpendicular to this is a flat photographic plate enclosed in a light-proof envelope. The direct beam produces blackening in the center of the film and therefore a small lead disc is placed in front of the film to intercept and absorb it. In his first experiments, he used continuous radiation incident on a stationary crystal. The crystal generated a set of beams that represent the internal symmetry of the crystal. The Laue diagram is simply a stereographic projection of the crystal planes (see Figure 1.7).

There are two variants of this model, depending on the position of the glass with respect to the photographic plate, and they can be:

- *Laue method in transmission mode.* The film is placed behind the glass to record the rays that are transmitted by the glass. One side of the Laue reflection cone is defined by the transmission ray. The film crosses the cone, so the diffraction spots generally lie on an ellipse.
- *Laue method in reflection mode.* The film is placed between the X-ray source and the glass. Rays that are diffracted in a previous direction are recorded. A part of the Laue reflection cone is defined by the transmitted ray. The film crosses the cone, so that the diffraction spots generally lie on a hyperbola.

This method is used to determine symmetry: If a crystal is oriented such that the incident beam is parallel to a symmetry element, the arrangement of the spots in the photograph reveals its symmetry. A photograph by this method of a mineral taken with the incident beam parallel to the binary axis of a monoclinic crystal will show a binary arrangement of spots; if the beam is parallel to the plane of symmetry, the photograph will present a line of symmetry; if it is a rhombic crystal, it will show a double distribution of spots, with two axes of symmetry.

Rotation or rotating glass methods

A single crystal is used. The crystal is oriented in such a way that it can be rotated along one of the main crystallographic axes. The camera is a cylinder of known diameter, coaxial with the axis of rotation of the glass, and has inside it a photographic film protected from light by a black paper cover (see Figure 1.10).

When a rotation photograph is taken, the crystal rotates around one of the main lattice rows, usually a crystallographic axis. This reticular row is perpendicular to the incident beam, and therefore the diffracted rays will always be contained in cones whose axes are common with the axis of rotation of the crystal. This axis is that of the cylindrical film, so the intersection of the cones on the film will be a series of circles,

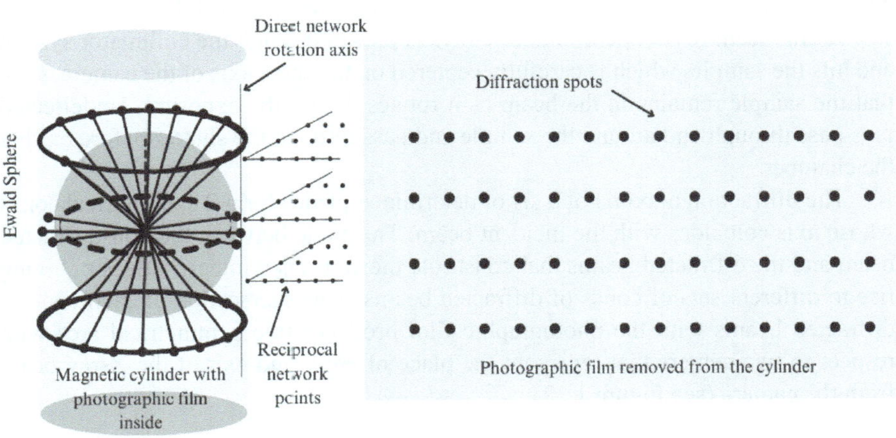

Figure 1.10. Assembly diagram of the rotation method.

which when the film is developed and flattened, will appear as straight parallel lines. Each of them is a layer line, corresponding to a cone of diffracted rays for which n has a certain integer value. In this way, the layer line that includes the incident ray is called the zero layer or equator, the first line is the one that satisfies $n = 1$, the second $n = 2$ and so on. The layer lines are not continuous since the different diffraction spots appear only when the three cones intersect.

The separation of the layer lines is conditioned by the angles of the cones, which in turn depends on the periodicity of the lattice row around which the crystal is rotated. Therefore, knowing the diameter of the cylindrical film, the wavelength of the X-rays and the distance of the n layer above the equator in the film, we can determine the spacing or identity period along the axis of rotation of the crystal.

If in the rotating crystal method rotation photographs are taken with the crystal rotating around each of the three crystallographic axes, we can determine the dimensions of the unit cell, so that the identity periods determined by successively rotating the crystal are the edges of the unit cell, which is true regardless of the symmetry of the crystal.

Powder method

Due to the rarity of truly well-formed crystals and the difficulty of carrying out the precise orientation required by the Laue and spinning crystal methods led to the discovery of the powder method in X-ray diffraction research. In this method, the sample is pulverized as finely as possible and associated with an amorphous material, in the form of an acicular axis of 0.2 to 0.3 mm in diameter. This needle or powder sample is ideally made up of crystalline particles in any orientation; to ensure that the orientation of these small particles is completely random with respect to the incident beam, the sample is usually rotated in the X-ray beam during exposure. The powder chamber is a flat, disc-shaped box with an adjustable needle in the center of it for mounting the sample. The cylindrical wall is cut diametrically by a collimator and a shutter of the beam opposite it. The film is placed inside the camera, with two holes drilled, so that the collimator tube and shutter pass through them once the film adapts properly to the internal surface of the camera (see Figure 1.11).

A thin beam of monochromatic X-rays is passed through the collimator system and hits the sample, which is carefully centered on the short axis of the camera, such that the sample remains in the beam as it rotates during the exposure. Undeflected rays pass through and around the sample and pass through the shutter before exiting the chamber.

The diffraction maxima of a set of determined planes form 2 symmetrical cones whose axis coincides with the incident beam. The angle between the non-diffracted beam and the diffracted beams that constitute the cones is y integer values n, giving rise to different sets of cones of diffracted beams. The intersection of each cone of diffracted beams with the photographic film produces two symmetrical arcs with respect to two centers that represent the place of entry and exit of the X-ray beam from the camera (see Figure 1.12).

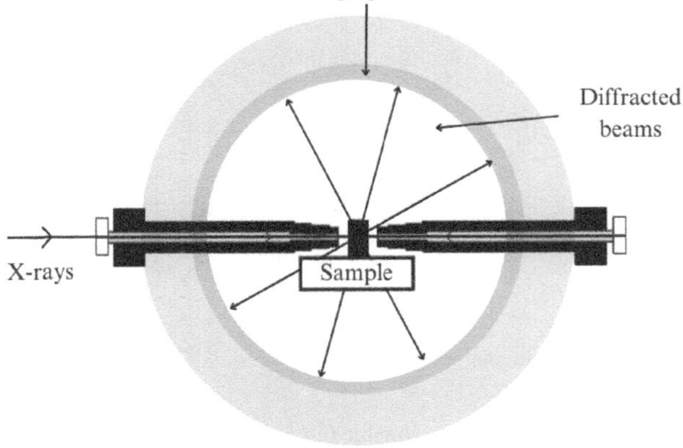

Figure 1.11. Assembly diagram for the powder method.

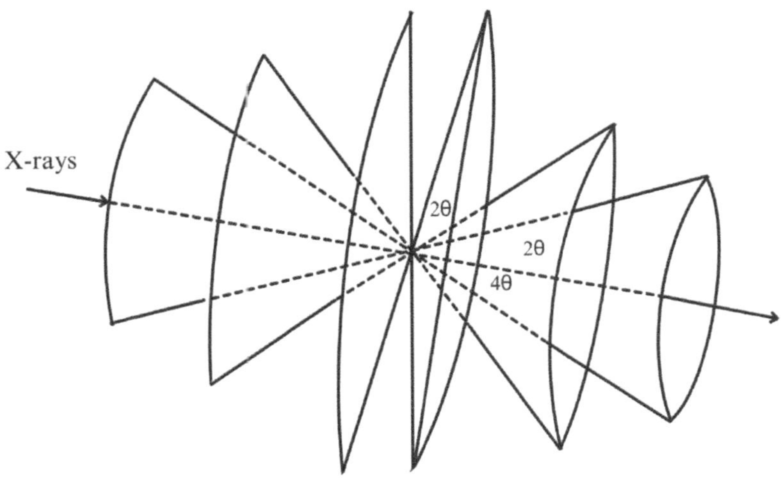

Figure 1.12. Intersecting cones that produce diffraction.

The film is wrapped around the camera (in a circular manner) so that it records the diffraction pattern. Each cone that crosses the film gives the diffraction line, which is seen as arcs. When the film is deployed, a series of concentric and symmetrical arcs are observed with respect to the two holes, see Figure 1.13.

Thus, for example, we will consider the powder model of a crystal from a sample that is known to have a cubic structure, but which one is unknown.

Applying this method, we get the film strip from the Debye camera after exposure, which is then developed and fixed. From the strip, we measure the position of each diffraction line. From the results it is possible to associate the sample with a particular type of cubic structure and determine a value for its lattice parameter.

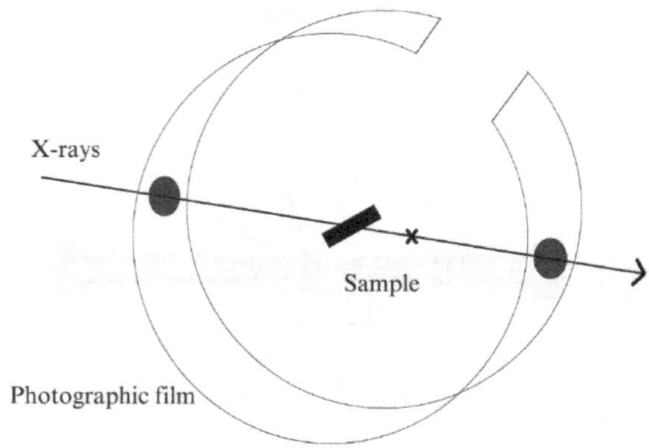

Figure 1.13. Photographic film in the powder method.

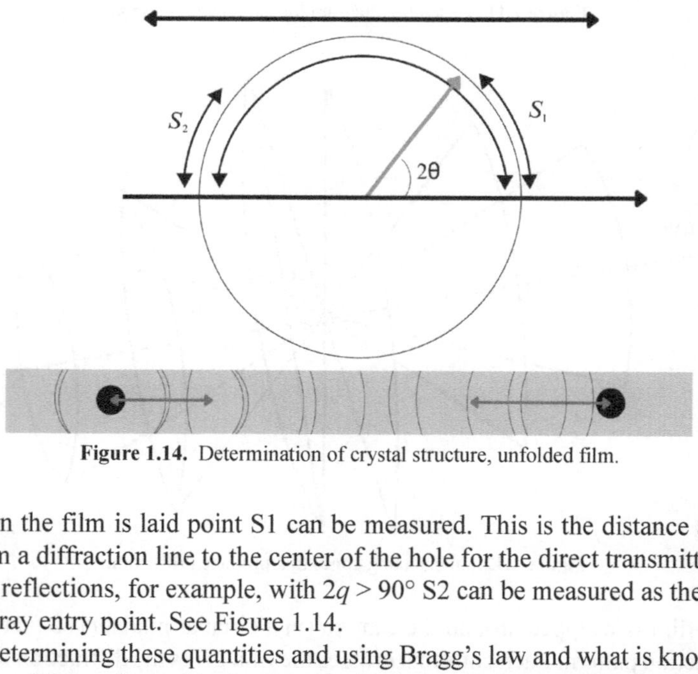

Figure 1.14. Determination of crystal structure, unfolded film.

When the film is laid point S1 can be measured. This is the distance along the film, from a diffraction line to the center of the hole for the direct transmitted beam. For back reflections, for example, with $2q > 90°$ S2 can be measured as the distance from the ray entry point. See Figure 1.14.

By determining these quantities and using Bragg's law and what is known about the planes of the reciprocal lattice, we obtain the type of crystal it is (by analogy).

Bibliography

DUNN, Peter M. Wilhelm Conrad Röentgen (1845–1923), the discovery ofx rays and perinatal diagnosis. Archives of Disease in Childhood-Fetal and neonatal edition, 2001, 84(2): F138–F139.

Kane, S.A. and Gelman, B.A. (2020). Introduction to Physics in Modern Medicine. CRC press

Lee, M. (2017). X-Ray Diffraction for Materials Research: From Fundamentals to Applications. CRC Press

Milton, K. and Schwinger, J. (2024). Classical Electrodynamics. CRC Press

Seeck, O.H. and Murphy, B. (2015). X-ray Diffraction: Modern Experimental Techniques. CRC Press

Stöhr, J. (2023). The Nature of X-Rays and Their Interactions with Matter. Springer.

CHAPTER 2

Symmetry Groups

Crystallography

INTRODUCTION

Chapter 2 covers symmetry groups and crystallography, which are key to classifying crystal structures and predicting how they interact with X-rays. This section provides a solid foundation for using symmetry as a tool for structural analysis.

It also deals with the crystalline structure and its crystalline state as a state of thermodynamic equilibrium under certain conditions of pressure and temperature.

2.1 Geometry of crystals

2.1.1 Crystal structure

The crystalline state of a solid is the state of thermodynamic equilibrium that under certain thermodynamic conditions of pressure and temperature corresponds to a certain crystalline structure.

Crystal structure is the three-dimensional geometry of the periodic and ordered arrangement of atoms, ions, or molecules that constitute a solid in a crystalline state.

The main characteristic of solids in the crystalline state is periodicity. Periodicity implies macroscopic properties such as homogeneity, anisotropy, and symmetry.

Homogeneity: From a macroscopic point of view, it means the invariability of a property over distance, that is, no matter when or how far one advances, the properties of the medium do not depend on the distance to the chosen coordinate system. From the property of homogeneity, at a macroscopic level, we obtain the constancy of the chemical composition and phase state throughout the entire volume of the body in a crystalline state.

This concept of homogeneity allows a body in a crystalline state to be considered as a continuum. This continuum hypothesis will be developed further in Chapter 4 and allows a phenomenological description of many properties of crystalline bodies without the need to consider their discrete atomic structure. When the physical properties of the crystalline body are considered at the macroscopic level, we are

DOI: 10.1201/9781003344438-2

dealing with distances considerably greater than the interplanar distances and with volumes much greater than that of the unit cell.

Anisotropy: A property of a crystalline body is a scalar property when it is independent of the direction in which it is measured, such as, for example, specific weight, heat capacity, etc. Other properties depend on one or more directions, they are said to be vectorial and tensor properties, such as thermal conductivity, dielectric constant, refractive index, etc. These scalar, vectorial, and tensor properties of crystalline bodies will be discussed in depth in Chapters 4 and 6.

If the description of a property is scalar, that is, it is independent of any direction with respect to the chosen reference system, it is said that the crystalline body is isotropic with respect to said property. If, on the contrary, a property in the crystalline solid depends on the direction in which it is measured, it is said that the crystalline body is anisotropic with respect to said property. In any case, a real crystalline body will always be anisotropic with respect to some property, for example, if the configuration of the atoms is different in different directions, this is called structural anisotropy.

Symmetry: The symmetry property refers to symmetry transformations, such as translations, rotations, inversions, among others, are operations that, when applied to a crystalline body, do not change any of its properties, and it is not possible to distinguish between the initial state when applying the transformation, and the final state after applying the symmetry transformation.

According to these characteristics, a body in an ideal crystalline state can be considered at a macroscopic level as a continuous (homogeneous) medium, anisotropic, and symmetrical with respect to some symmetry transformations. However, this is an idealization since a crystalline body is not a static structure, but a lattice of vibrating atoms and these vibrations will depend on the temperature and affect its physical properties.

In many investigations, these vibrations can be omitted or not considered for some given temperatures or subject of interest. On the other hand, the crystalline structure of a real body has imperfections, defects, and impurities, and local variations of its composition, which generates a deviation from the ideal structure and these imperfections, impurities, and defects are of great importance in the thermal behavior and mechanical properties of the materials. However, the working hypothesis of considering a crystalline body with an ideal crystalline structure is very useful in many investigations on physical and chemical properties, since the crystalline structure determines properties such as:

- Density
- Modulus of elasticity
- Thermal expansion
- Melting point
- Thermal conductivity
- Specific heat
- Electrical resistivity
- Magnetic properties, among others.

2.1.2 Crystal lattice

It is a three-dimensional representation of a pattern that is repeated in the crystal structure, as can be seen in Figure 2.1.

Crystalline Lattice = pattern + crystalline structure

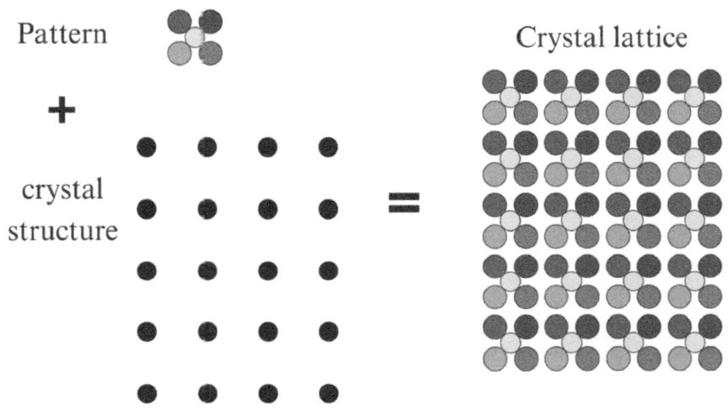

Figure 2.1. Crystalline lattice.

2.1.3 Elements of the crystal lattice

- **Pattern:** It is the set of atoms, molecules, or ions of a crystal that is associated with each knot. Since all knots are geometrically equivalent, the pattern is unique in each crystal lattice.
- **Knot:** It is a point of the crystalline structure equivalent to the operation of translation to all the points of the structure to which a pattern is located or associated, that is, the knots of the different cells are not necessarily occupied by a single atom, in certain materials each knot can have associated a molecule, a group of atoms, or even a group of molecules, that is, by a pattern.
- **Points:** In the crystal lattice, there are points in space, in addition to knots. The knots only occupy a few points in space, where the pattern goes. The points are not equivalent and are located between the knots. The atoms in the crystal structure can occupy the position of a knot or a point, according to the concept of a lattice.
- **Reticular row:** An arrangement of knots along a direction. A lattice direction is therefore a direction containing knots. Each pair of knots in the lattice defines a lattice row.
- **Reticular row symbols:** These are the coordinates *p*, *q*, and *r* of a knot of the lattice, adjacent to another knot taken as origin and are written in brackets, [*pqr*] and fully describe the reticular row.
- **Fundamental rows:** The rows with symbols [100], [010] and [001] are called fundamental rows of the lattice and are parallel respectively to the fundamental

translations $\vec{a}$, $\vec{b}$, and $\vec{c}$. They define the edges of the elementary cell of the lattice.

- **Reticular plane:** It is an arrangement of knots along two directions. Each trio of knots not arranged in the same direction of the lattice defines a plane or two non-parallel reticular rows also define a reticular plane.
- **Symbols of a lattice plane** (Miller indices and Weiss parameters): Consider the plane $ax + by + cz + d = 0$, then $ax + by + cz = -d$, dividing this expression member by member by $-d$, we obtain

$$-\frac{ax}{d}-\frac{by}{d}-\frac{cz}{d}=1 \tag{2.1}$$

Let's do $p = -d/a$, $q = -d/b$, $r = -d/c$. Replacing in equation (2.1) we obtain:

$$\frac{x}{p}+\frac{y}{q}+\frac{z}{r}=1 \tag{2.2}$$

The intersection of the planes with the coordinate axes is as follows:

- With the x-axis is when $y = 0$, $z = 0$, $x = p$. The plane cuts x-axis at point $(p, 0, 0)$.
- With the y-axis is when $x = 0$, $z = 0$, $y = q$. The plane cuts x-axis at point $(0, q, 0)$.
- With the z-axis is when $x = 0$, $y = 0$, $z = r$. The plane cuts z-axis at point $(0, 0, r)$.

That is, p, q, r indicate the intersections of the plane with the coordinate axes and are the so-called Weiss parameters. On the other hand, from equation (2.2), multiplying member by member by pqr, the following expression is obtained:

$$(qr)x + (pr)y + (pq)z = pqr \tag{2.3}$$

Letting $h = qr$, $k = pr$, $l = pq$, the equation becomes:

$$hx + ky + lz = pqr \tag{2.4}$$

These developments show that for each stacking of planes represented by (hkl),

- There is a stacking plane containing a lattice point chosen as the origin.
- There is a stacking plane which makes the intercepts rational: p at lattice points along the x-axis, q at lattice points along the y-axis, and r at lattice points along the z-axis. Assuming that p, q, and r are not common factors, the plane with those intercepts is the first plane from the origin which makes the intercepts rational.
- The stacking plane equation $hx + ky + lz = d$, through the origin, in this case p, q, r are still the Weiss parameters, with $d = pqr$.

On the other hand, multiplying both sides by p the expression $h = qr$, to obtain, $ph = pqr$, as $d = pqr$, implies that $h = d/p$, performing analogous operations we obtain that $k = d/q$ y $l = d/r$, where h, k, l define the Miller indices.

We can now justify the following definition:

The Miller indices are a series of three integers that can be positive or negative and are denoted by the letters *h*, *k*, and *l* in parentheses, (*hkl*), which are obtained from the intersections of a reticular plane with the fundamental translations, and are the inverses of the Weiss parameters, which are denoted by the letters *p*, *q*, and *r* therefore can also be obtained from the intersections of a reticular plane with the fundamental translations.

The number of planes existing between the plane with coordinates *A*, *B*, and *C* and the origin of the lattice is given by the product $H * K * L = N$. The quotients:

$$\frac{N}{H} = h, \ \frac{N}{K} = k, \ \frac{N}{L} = l \tag{2.5}$$

They are integers and would be the Miller indices. The planes that cut two axes and are parallel to the third axis are symbolized by (*hk*0), (*h*0*l*), and (0*kl*), depending on whether they are parallel to the $\vec{c}$ axis, the $\vec{b}$ axis, or the $\vec{a}$ axis, respectively. The simplest planes are: (110), (101), and (011) and are called fundamental planes and define the faces of the elementary cell of the lattice. The planes with the highest density of knots are those with simpler Miller indices and constitute the crystalline faces.

- *Law of rational indices*

  Research on crystals established that for certain faces, the indices could be expressed by simple whole numbers or by zeros; this is the law of rational indices (of Miller and Weiss).

- *Tautozonal planes*

  Set of non-parallel planes that are characterized by having a common edge, that is, they are parallel to a crystallographic direction called the zone axis and whose symbol is [*uvw*]. In Figure 2.2, you can see some families of planes parallel to the $\vec{c}$ axis (perpendicular to the screen).

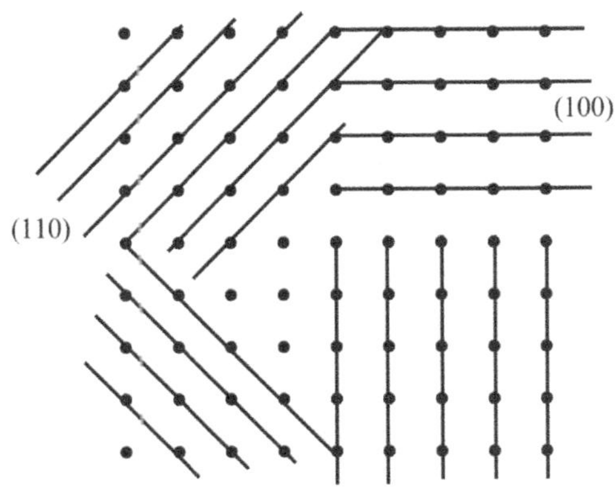

Figure 2.2. Tautozonal planes.

It can also be observed that each set of identical, equally spaced parallel planes has the same symbol and the same lattice density. The condition for a plane (*hkl*) to be parallel to an axis [*uvw*] is given by the Weiss zone law:

$$hu + kv + lw = 0 \tag{2.6}$$

If two planes (*hkl*) and (*h'k'l'*) belong to the same zone or, in other words, are parallel to the same crystallographic direction [*uvw*], the following must be fulfilled:

$$\begin{aligned} hu + kv + lw &= 0 \\ h'u + k'v + l'w &= 0 \end{aligned} \tag{2.7}$$

Solving system (2.7) gives the zone axis symbol:

$$\begin{aligned} u &= hl' - lk' \\ v &= lh' - hl' \\ w &= hk' - kl' \end{aligned} \tag{2.8}$$

- **Crystalline face:** It is the manifestation in the crystal of planes that are characterized by having the highest planar reticular density and simple numbers as Miller indices.
- **Crystal edge:** It is the manifestation in the crystal of a reticular row that is characterized by having the highest linear reticular density and the simplest axis symbol.
- **Reticular density:**
  - **Linear lattice density (λ_{hkl}):** It is the number of knots per unit of length. It is the inverse of the magnitude of the translation vector. The simplest Miller index rows are those with the highest knot density, these are: [100], [010], and [001].

$$\lambda_{hkl} = \frac{1}{d_{hkl}} \tag{2.9}$$

  - **Planar reticular density (σ_{hkl}):** It is the number of knots per unit of surface. It is the inverse of the area, that is:

$$\sigma_{hkl} = \frac{1}{S_{hkl}}, \quad S_{hkl} = \frac{V_{hkl}}{d_{hkl}} \tag{2.10}$$

  The planes with the highest reticular density are those with the largest spacing.
  - **Spatial lattice density (ρ_{hkl}):** It is the number of knots per unit volume. It is the inverse of the volume:

$$\rho_{hkl} = \frac{1}{V_{hkl}} = \frac{1}{d_{hkl} * S_{hkl}} = \frac{\sigma_{hkl}}{d_{hkl}} = \sigma_{hkl} * \lambda_{hkl} \tag{2.11}$$

2.1.4 Types of crystal lattices

Before describing the different types of crystal lattices, it is important to note that in crystal lattice theory, a translation is understood as the distance at which the pattern in the crystal structure is repeated, parallel, and identically along a given direction (see Figure 2.3). This monotonous repetition of the pattern constitutes the fundamental characteristic of a crystal. The crystalline medium can be abstracted from its material content and treated solely in terms of the translations present. The study based on this abstraction is called Crystal Lattice Theory.

There is only one way to arrange the knots of the lattice in one dimension so that each knot has identical environments, that is, a crystalline structure. It is a structure that will have its knots at equal distances, and in one dimension there is no way to form another different crystalline structure, therefore, in one dimension, there is only one lattice structure, so the distances between the knots in different lattices can be different, but here we are talking about the structure, the way in which the atoms can be arranged, independent of their distances, of course, for a single particular lattice the distance between its knots is constant, it would always be the same. In this sense, the statement that in one dimension there can only be one type of crystalline lattice is valid.

In two dimensions something similar happens; there are only five ways to arrange the knots of the lattice to form a crystal structure, so that each knot has identical surroundings. Therefore, in two dimensions, there are only five crystal lattices for different relations of the values of the sides a and b and the angle γ between them (see Figure 2.2). Any other crystal structure different from the five basic ones in two dimensions, is obtained from one of them or from a combination of them. The relations between the lattice parameters are given in Table 2.1.

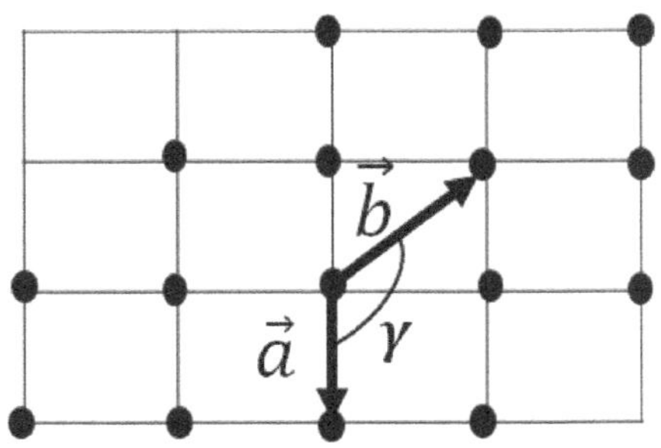

Figure 2.3. Translational symmetry.

Table 2.1. Lattice parameters for the five planar crystal lattices.

Lattice	**Lattice parameters**	
	Sides	Angles
Oblique	$a \neq b$	$\gamma \neq 90°$
Square	$a = b$	$\gamma = 90°$
Rectangular	$a \neq b$	$\gamma = 90°$
Hexagonal	$a = b$	$\gamma = 60°$ or $120°$
Rhombic	$a = b$	$\gamma \neq 90°$, $60°$ or $120°$

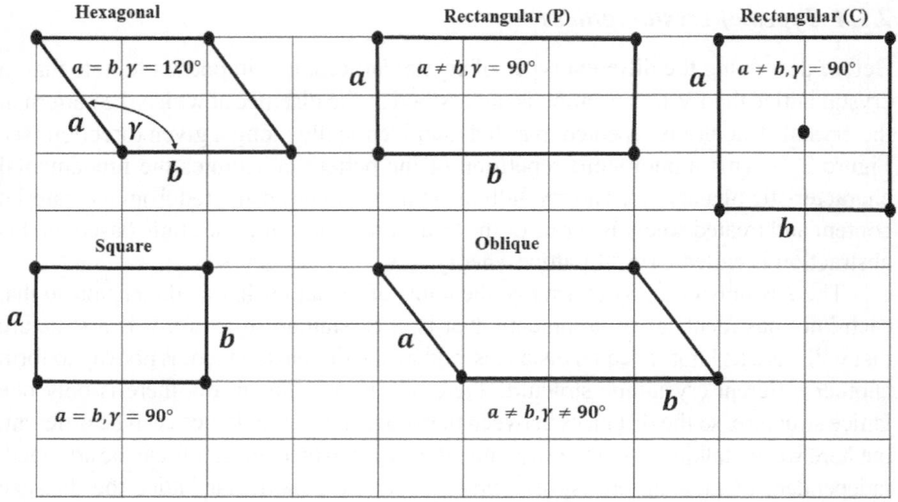

Figure 2.4. Basic structures for two-dimensional lattices.

The respective figures of the five flat crystal lattices can be observed in Figure 2.4.

2.1.4.1 Fundamental translations or unity

These are the smallest translations in the three directions of space. Their magnitudes are respectively represented by *a*, *b*, and *c*, and they are assigned the value of unity. These unitary translations are constants of the crystalline structure, and they are called lattice constants because their values are fixed or constant for a crystal. This trio of vectors defines the reference system of the lattice. In addition to the magnitudes, the reference system is also defined by the angles between these unitary vectors, which here will be called α, β, and γ as shown in Figure 2.5. The sides *a*, *b*, and *c* and their respective angles α, β, and γ are called lattice parameters.

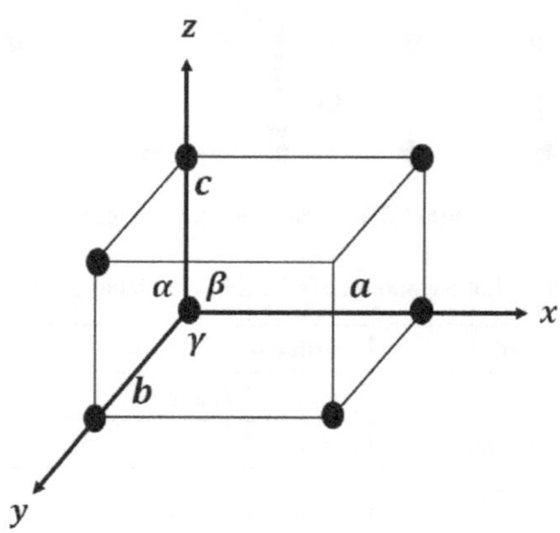

Figure 2.5. Unit cell.

The simplest symmetry that solids have is that of translation; it is also the easiest to understand. The starting point for its study is Bravais' postulate, which we can state as follows:

> "Given a point $\vec{r}$ of the crystalline solid, there exists an infinite, discrete, and unlimited three-dimensional set of points $\vec{r}\,'$ that have the same atomic environment as $\vec{r}$ and with the same orientation."

Points related to each other by Bravais' postulate are called congruent points. It is important to note that this postulate requires that congruent points have an atomic environment with the same orientation. Points in which the atomic environment is the same, but with a different orientation constitute another category, as we will see, and give rise to other symmetry properties.

2.1.4.2 Elementary or unitary cell

It is a parallelepiped delimited by the non-coplanar fundamental translations in a lattice and constitutes the most characteristic part of the crystal, that is, it is the smallest grouping of atoms that preserves the geometry of the crystalline structure and there are two types:

- Primitive elementary cell
- Multiple elementary cell

The primitive elementary cell contains one lattice point, while the multiple elementary cells contain more than one lattice point.

In three dimensions, only 14 crystal lattices can be built from the elementary cells associated with seven crystal systems. These spatial lattices are called Bravais lattices and are obtained by stacking the flat lattices. They can be primitive, which are symbolized by the letter *P*, and multiple, which can be:

- Centered on the bases (either *A*, *B* or *C*)
- Centered on the faces (*F*)
- Centered on the interior (*I*)

They are named according to the relationships between the fundamental translations and the angle between them (see Figure 2.5), as shown in Table 2.2.

And the respective geometries can be observed in Figure 2.6.

It is important to note that, for example, in three dimensions, although there are only 14 Bravais lattices, there can be an indefinite number of crystal lattices because there are an indefinite number of patterns that are observed in nature or that can be synthesized, so many chemical elements have the same crystal structure, but the pattern is made up of different atoms. Copper, for example, has the same face-centered cubic crystal structure as nickel, for which only one atom is associated with a lattices knot, but its constituent atoms are different. In more complex structures, for example, some polymeric, ceramic or biological materials, several atoms, including molecules, can be associated with each lattices point (i.e., the base is greater than one), forming complex unit cells.

Table 2.2. The 14 Bravais Lattices associated with the 7 crystal systems.

Crystalline System	**14 Bravais lattices**	**Parameters**	
Lattice	Type	Sides	Angles
Triclinic|	*P*	$a \neq b \neq c$	$\alpha \neq \beta \neq \gamma \neq 90°$
Monoclinic	*P, A(B,C)*	$a \neq b \neq c$	$\alpha = \gamma = 90° \neq \beta$
Rhombic	*P, I, F, A (B, C)*	$a \neq b \neq c$	$\alpha = \beta = \gamma = 90°$
Tetragonal	*P, I*	$a = b \neq c$	$\alpha = \beta = \gamma = 90°$
Hexagonal	*P*	$a = b \neq c$	$\alpha = \beta = 90° \wedge \gamma = 60°$ or $120°$
Rhombohedral	*P*	$a = b \neq c$	$\alpha = \beta = 90° \wedge \gamma = 60°$ or $120°$
Cubic	*P, I, F*	$a = b = c$	$\alpha = \beta = \gamma = 90°$

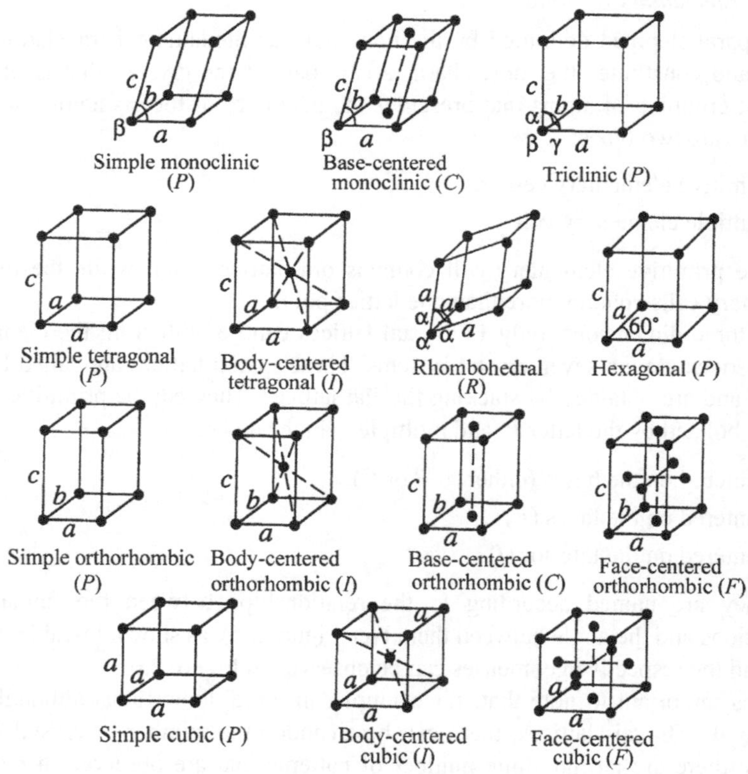

Figure 2.6. The 14 primitive cells in three dimensions or Bravais.

Not all ordered sets of points in space constitute Bravais lattices. For example, in Figure 2.7(a) which represents a two-dimensional lattice with hexagonal symmetry, the neighborhood of point *P* is not identical to the neighborhood of point *P'*, so that both points are not congruent and therefore the lattice is not a Bravais lattice. On the other hand, the lattice in Figure 2.7(b) which also has hexagonal symmetry is a Bravais lattice, since points *Q* and *Q'* are congruent.

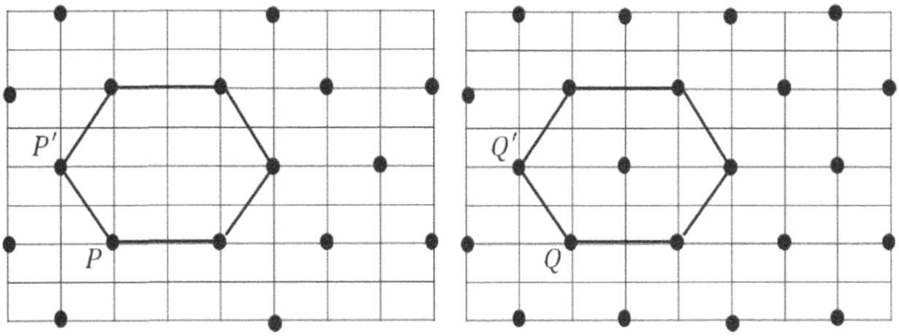

(a) Ordered arrangement of knots that does not constitute a Bravais lattice.

(b) Hexagonal Bravais lattice, which must necessarily be centered.

Figure 2.7. Hexagonal planar lattices related to the Bravais concept.

The points congruent to an arbitrary point $\vec{r}$ can be described by a set of vectors $\vec{t}$ defined by the expression (2.26).

2.1.4.3 The Wigner-Seitz cell

A particularly useful primitive cell is the so-called Wigner-Seitz or proximity cell. The Wigner-Seitz cell of a lattice is defined as the geometric locus of the points in space that are closer to a given knot, arbitrarily taken as the origin, than to any other knot. To construct the Wigner-Seitz cell, the median planes of the segments that join this knot with its first neighbors are drawn from an arbitrary knot.

The intersections of these planes generate a polyhedron that encompasses points that are closer to the knot taken as the origin than to any other knot and, therefore, constitute the proximity cell. By construction, this cell is primitive, since it only contains the origin knot. For example, Figure 2.8a represents the Wigner-Seitz cell of an FCC lattice, which appears displaced by one half-edge in each direction of space (that is, with a knot at the center of the cell and additional knots at the center of each edge) for convenience.

The Wigner–Seitz cell of the FCC lattice is a rhombic dodecahedron in which each of the twelve faces is perpendicular to the line joining the central knot to each

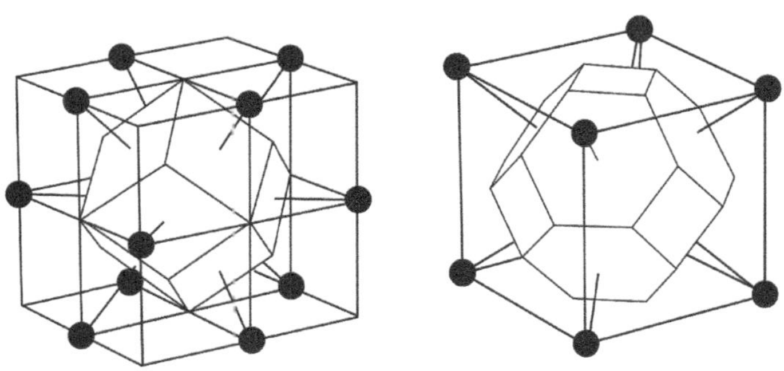

(a) FCC lattice

(b) BCC lattice

Figure 2.8. Wigner-Seitz cells.

knot at the center of the edges. Figure 2.8b shows the Wigner–Seitz cell of a type of cubic cell that has knots at the vertices of the cube and an additional knot at its center; this cell is called a BCC. In this case, the polyhedron is a truncated octahedron in which the hexagonal faces are midplanes of the segment joining the center of the cube to each of the vertices, while the square faces bisect the segments joining the center of the cube to the centers of neighboring cells.

2.1.5 Reciprocal lattice

In what we have seen so far, Bravais' postulate allows us to construct the direct lattice of a solid, which describes its translation symmetry in real space, and which acts as a support for the crystalline structure. But the direct lattice is not the only one that can be associated with a crystal. Indeed, translational invariance forces the functions that describe physical properties to take equal values at congruent points of the lattice. This allows us to construct a second lattice, called the reciprocal lattice, defined in Fourier space. The reciprocal lattice associated with a direct lattice is also a Bravais lattice, and its symmetries have certain correspondences.

The reciprocal lattice was introduced to describe the spacings of families of lattice planes, although it is possible to define it in a formal way. The formal and geometric definitions of the reciprocal lattice are related to each other, as demonstrated by the fundamental theorem of the reciprocal lattice. From the point of view of this book, the concept of reciprocal lattice is useful to describe families of lattice planes and, above all, in the study of X-ray diffraction by crystals. The reciprocal lattice is a fundamental concept in solid state physics, since it is the framework in which the dynamics of crystalline solids is described.

2.1.5.1 Geometric definition, description of reticular planes

Let us see how to associate the reciprocal lattice to solids using a purely geometrical criterion. To do so, let us consider a family of Miller index lattice planes (*hkl*), as represented in Figure 2.9. This family can be completely described by specifying the direction of the normal vector to the planes $\hat{n}_{hkl}$ and the spacing between these planes d_{hkl}.

Ernest Mallard proposed the idea of representing each family of lattice planes by a vector whose direction is that of the normal vector to the planes of the family and whose magnitude is inversely proportional to the spacing between them. Thus, for each family of planes (*hkl*) a vector $\vec{q}_{hkl}$ is defined such that:

$$\vec{q}_{hkl} \parallel \hat{n}_{hkl} \tag{2.12}$$

$$\left|\vec{q}_{hkl}\right| \propto \frac{1}{d_{hkl}} \tag{2.13}$$

The proportionality constant in equation (2.13) is sometimes taken equal to one. Here, for consistency with solid state physics, we will take the constant equal to 2π.

So:

$$\left|\vec{q}_{hkl}\right| = \frac{2\pi}{d_{hkl}} \tag{2.14}$$

Since d_{hkl} is a distance, the vectors $\vec{q}_{hkl}$ are not defined in real space, but in wave vector space (or Fourier space). In the context of solid-state physics, this is called reciprocal space and is represented by the symbol ϵ^*.

Since the set of lattice planes in an ideal crystal is discrete, the set of vectors $\vec{q}_{hkl}$ for all possible values of the indices (*hkl*), is also discrete.

Real space is called direct space in the context of solid-state physics. The set of lattice planes in direct space is infinite, and hence so is the set of vectors $\{\vec{q}_{hkl}\}$.

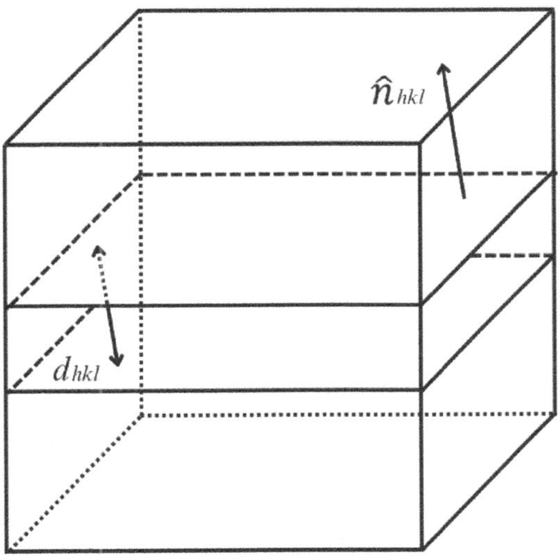

Figure 2.9. Reciprocal lattice.

2.1.5.2 Formal definition of the reciprocal lattice

In addition to the above geometric construction, there is a formal definition of the reciprocal lattice, based on the concept of translational invariance. Indeed, let $f(\vec{r})$ be the function that describes a physical property in a crystal. The translational symmetry of the direct lattice indicates that two points $\vec{r}$ and $\vec{r} + \vec{t}$, are congruent, such that

$$f(\vec{r}) = f(\vec{r} + \vec{t}), \quad \forall \vec{t} \in \mathcal{T} \tag{2.15}$$

where $\mathcal{T}$ is the set of all vectors that preserve translational symmetry, then $f(\vec{r})$ has the translational symmetry of the crystal and therefore, admits a Fourier series expansion of the form.

$$f(\vec{r}) = \sum_{\vec{G}} \tilde{f}(\vec{G}) e^{i\vec{G}\cdot\vec{r}} \tag{2.16}$$

where $\{\vec{G}\}$ are wave vectors, defined in the reciprocal space

$$\tilde{f}(\vec{G}) = \sum_{cell} f(\vec{r}) e^{-i\vec{G}\cdot\vec{r}} \tag{2.17}$$

Expression (2.17) is the Fourier expansion corresponding to a given $\vec{G}$. Considering expression (2.15) for all vectors in the translation group, $\forall \vec{t} \in T$ must hold that:

$$\begin{aligned} f(\vec{r}) = \sum_{\vec{G}} \tilde{f}(\vec{G}) e^{i\vec{G}\cdot\vec{r}} = f(\vec{r}+\vec{t}) = \sum_{\vec{G}} \tilde{f}(\vec{G}) e^{i\vec{G}\cdot(\vec{r}+\vec{t})} \\ = \sum_{cell} \tilde{f}(\vec{G}) e^{i\vec{G}\cdot\vec{r}} e^{i\vec{G}\cdot\vec{t}} \end{aligned} \tag{2.18}$$

Given the orthogonality of plane waves, it forces the vectors $\{\vec{G}\}$ to satisfy $\forall \vec{t} \in T$ that:

$$e^{i\vec{G}\cdot\vec{T}} = 1 \tag{2.19}$$

Or else

$$\vec{G}\cdot\vec{t} = 2\pi n \tag{2.20}$$

where n is an integer. Condition (2.20) implicitly defines the vectors $\{\vec{G}\}$. From condition (2.20), it can be shown that the set $\{\vec{G}\}$ defines a lattice in the reciprocal space. First, this set is closed with respect to addition. Indeed, let us consider an arbitrary vector $\vec{t}$ of the direct lattice and let $\vec{G}_1$ and $\vec{G}_2$ be two vectors of the reciprocal space that satisfy condition (2.19) for that vector $\vec{t}$, therefore:

$$e^{i\vec{G}_1\cdot\vec{T}} = e^{i\vec{G}_2\cdot\vec{T}} = 1 \tag{2.21}$$

In this case the vector $\vec{G}_1 + \vec{G}_2$ also verifies condition (2.19), since

$$e^{i(\vec{G}_1 + \vec{G}_2)\cdot\vec{T}} = e^{i\vec{G}_1\cdot\vec{T}} e^{i\vec{G}_2\cdot\vec{T}} = 1 \tag{2.22}$$

On the other hand, given a triplet of generators of the direct space ($\vec{a}_1$, $\vec{a}_2$, $\vec{a}_3$), we define the vectors

$$\begin{aligned} \vec{a}_1^* &= 2\pi \frac{\vec{a}_2 \times \vec{a}_3}{\vec{a}_1 \cdot (\vec{a}_2 \times \vec{a}_3)} \\ \vec{a}_2^* &= 2\pi \frac{\vec{a}_3 \times \vec{a}_1}{\vec{a}_1 \cdot (\vec{a}_2 \times \vec{a}_3)} \\ \vec{a}_3^* &= 2\pi \frac{\vec{a}_1 \times \vec{a}_2}{\vec{a}_1 \cdot (\vec{a}_2 \times \vec{a}_3)} \end{aligned} \tag{2.23}$$

and they verify that

$$\vec{a}_i \cdot \vec{a}_j^* = 2\pi\delta_{ij} \tag{2.24}$$

where δ_{ij} is the Kronecker delta. The vectors $\vec{a}_1^*$, $\vec{a}_2^*$ and $\vec{a}_3^*$ are non-coplanar, since $\vec{a}_1$, $\vec{a}_2$, and $\vec{a}_3$ are also coplanar. Consequently, the triple ($\vec{a}_1^*$, $\vec{a}_2^*$, $\vec{a}_3^*$) forms a basis of the reciprocal space, such that a vector $\vec{G}$ can be written as:

$$\vec{G} = m_1\,\vec{a}_1^* + m_2\,\vec{a}_2^* + m_3\,\vec{a}_3^* \tag{2.25}$$

For certain coefficients m_1, m_2 and m_3. Let $\vec{t}$ be a direct space translation vector, then it can be written as:

$$\vec{t} = n_1\vec{a}_1 + n_2\vec{a}_2 + n_3\vec{a}_3 \quad (2.26)$$

where the numbers $\{n_i\}$ are integers by construction and the vectors $(\vec{a}_1, \vec{a}_2, \vec{a}_3)$ are the fundamental translation vectors. Taking into account the condition (2.24) and (2.20) implies $\forall\vec{t} \in \mathcal{T}$ that:

$$\vec{G}\cdot\vec{t} = \sum_{ij} m_i n_j \vec{a}_i \cdot \vec{a}_j^* = 2\pi\sum_{ij} m_i n_j \delta_{ij} = 2\pi\sum_i m_i n_i = 2\pi n \quad (2.27)$$

which in turn implies that the coefficients m_1, m_2, and m_3 are also integers. Thus, the vectors $\{\vec{G}\}$ define a Bravais lattice in the reciprocal space, whose generators are related to those of the direct lattice by the relations (2.23). The lattice formed by these vectors is called the reciprocal lattice of the crystal and, like the direct lattice, it appears due to the translational invariance of crystals.

The volume of the primitive cell of the reciprocal space V_c^* is given by the triple product of the basis vectors, that is:

$$V_c^* = \vec{a}_1^* \cdot (\vec{a}_2^* \times \vec{a}_3^*) = \frac{2\pi}{V_c}(\vec{a}_2^* \times \vec{a}_3^*)\cdot(\vec{a}_2^* \times \vec{a}_3^*) = \frac{(2\pi)^3}{V_c}$$

where V_c is the volume of the primitive cell in direct space.

The primitive proximity cell or Wigner-Seitz cell describes the direct lattice. In reciprocal space, the proximity cell is called the first Brillouin zone, and plays a fundamental role in solid state physics.

For example, Figure 2.10(a) depicts the first Brillouin zone of the BCC lattice, which is shaped like a rhombic dodecahedron. Directions of interest in the zone are designated by capital letters. Thus, denotes the center of the zone, P designates the vertex where three faces meet, while H represents the vertex where four faces meet; the letter N is reserved for the center of a face.

Figure 2.10(a) represents the first Brillouin zone for an FCC lattice, which is in the form of a truncated octahedron. Here, K denotes the center of the edge that delimits two hexagonal faces, U the center of the edge between a hexagonal face and a square face, L the center of a hexagonal face, X the center of a square face, and W a vertex of the octahedron.

An interesting thing to mention is that if we calculate the reciprocal lattice of the reciprocal lattice associated with a crystal, we obtain again the original direct lattice. It is then said that both lattices have the involution property. Indeed, since the reciprocal lattice is a Bravais lattice it makes sense to calculate its reciprocal lattice. According to (2.19), the reciprocal lattice of the reciprocal lattice is formed by the vectors $\{\vec{R}\}$ of the direct space such that, $e^{i\vec{G}\cdot\vec{r}} = 1$, comparing this expression with $\vec{G}\cdot\vec{t} = 2\pi n$ (expression (2.20)), implies that the vectors $\vec{t} \in \mathcal{T}$ are among the $\{\vec{R}\}$. Furthermore, these vectors are written in the form

$$\vec{R} = r_1\vec{a}_1 + r_2\vec{a}_2 + r_3\vec{a}_3$$

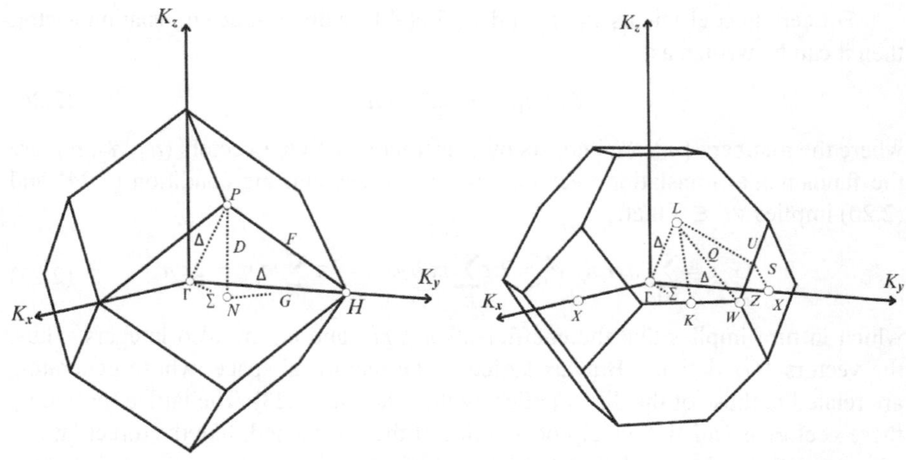

(a) First Brillouin zone of a BCC lattice (b) First Brillouin zone of an FCC lattice

Figure 2.10. First Brillouin zone for BCC and FCC lattices.

Since the vectors $\{\vec{a}_1, \vec{a}_2, \vec{a}_3\}$ form a basis of the direct space, condition (2.20) requires that r_1, r_2, and r_3 be integers, which in turn is equivalent to the vectors $\vec{R}$ being of the form (2.25) and therefore coinciding with the vectors $\vec{t}$ of the translation group of the direct lattice.

2.1.5.3 Fundamental theorem of the reciprocal lattice

The reciprocal lattice we have just formally defined is in fact identical to the lattice generated from the families of planes in the geometric Mallard construction. The fundamental theorem of the reciprocal lattice establishes the correspondence between both definitions. The direct statement of the theorem states that for every family of lattice planes of spacing d_{hkl} there exists an infinite set of reciprocal lattice vectors [of the form (2.25)] perpendicular to said planes, the shortest of which has magnitude $2\pi/d_{hkl}$.

To prove the direct statement of the theorem, consider a family of planes like the one in Figure 2.11, characterized by a normal vector $\hat{n}_{hkl}$ and a spacing d_{hkl} or, alternatively, by a vector $\vec{q}_{hkl}$ given by:

$$\vec{q}_{hkl} = \frac{2\pi}{d_{hkl}} \hat{n}_{hkl} \tag{2.28}$$

Let the position vector be measured from an arbitrary origin O of a generic point in one of the planes of the family; this point need not be a knot of the direct lattice. Then, in the plane considered,

$$\vec{q}_{hkl} \cdot \vec{r} = cte \tag{2.29}$$

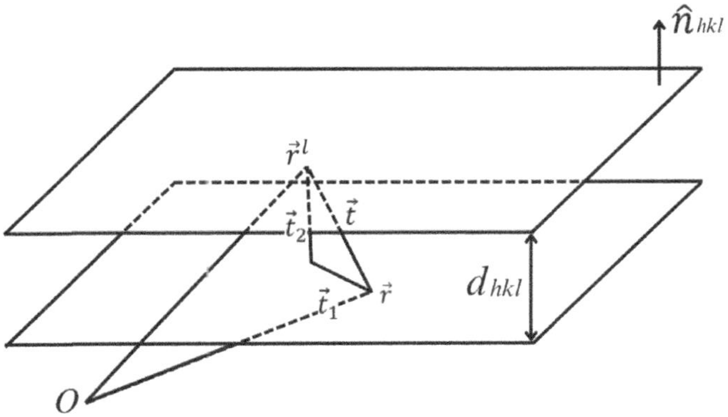

Figure 2.11. Proof of the direct statement of the fundamental theorem of the reciprocal lattice.

Condition that is verified by the knots of the direct lattice that are in the plane. Let us now consider a point $\vec{r}\,'$ that is congruent to $\vec{r}$ in another plane of the family. It is verified that

$$\vec{r}\,' = \vec{r} + \vec{t} \tag{2.30}$$

where $\vec{t}$ is a vector of the translation group which, in turn, can be written as

$$\vec{t} = \vec{t}_1 + \vec{t}_2 \tag{2.31}$$

Being $\vec{t}_1$ a vector in the first plane that we have considered and $\vec{t}_2$ a vector perpendicular to it, of the form

$$\vec{t}_2 = m * d_{hkl} * \hat{n}_{hkl} \tag{2.32}$$

where m is an integer. Therefore, for the point $\vec{r}\,'$ we have:

$$\vec{q}_{hkl} \cdot \vec{r}\,' = \vec{q}_{hkl} \cdot \vec{r} + \vec{q}_{hkl} \cdot \vec{t}_1 + \vec{q}_{hkl} \cdot \vec{t}_2 = \vec{q}_{hkl} \cdot \vec{r} + \vec{q}_{hkl} \cdot \vec{t} = \vec{q}_{hkl} \cdot \vec{r} + 2\pi m \tag{2.33}$$

Relation (2.33) indicates that the product $\vec{q}_{hkl} \cdot \vec{r}$, which is constant in each lattice plane, varies by an amount of $2\pi m$ when passing from one plane of the family to another located at a distance $m * d_{hkl}$ from the first. Thus, between planes of the same family the following is verified:

$$e^{i\vec{q}_{hkl}\vec{r}} = e^{i\vec{q}_{hkl}\vec{r}\,'} = cte \tag{2.34}$$

The constant that appears in (2.34) can be calculated by applying that equality to the plane that contains the origin: Thus, making $\vec{r} = 0$, we get:

$$e^{i\vec{q}_{hkl}\vec{r}} = e^{i\vec{q}_{hkl}\vec{r}\,'} = 1 \tag{2.35}$$

Particularizing now the condition (2.35) to the points of the planes of the family that are knots of the direct lattice $\forall \vec{t} \in T$ we have

$$e^{i\vec{q}_{hkl}\vec{T}} = 1 \tag{2.36}$$

so that $\vec{q}_{hkl}$ is a vector of the reciprocal lattice which, by construction, is perpendicular to the family of index planes (*hkl*) and whose module is inversely proportional to the interplanar distance, which proves the direct statement of the theorem.

It can be shown that there exist infinitely many vectors of the form (2.28) by defining $\vec{q}_{hkl} = p\frac{\pi}{d_{hkl}}\hat{n}_{hkl}$ and $p \in N$. The one with the smallest magnitude corresponds to $p = 1$, which is precisely (2.28).

The converse statement of the fundamental theorem states that given a reciprocal lattice vector $\vec{G}$ there exists a family of direct lattice planes that are normal to it, and whose spacing d_{hkl} is such that $\frac{2}{d_{hkl}}$ is the magnitude of the shortest reciprocal lattice vector that is parallel to $\vec{G}$. To prove the inverse statement, consider a vector $\vec{G}$ of the reciprocal lattice and a point $\vec{r}$ of the direct space, of the form:

$$\vec{G} = m_1\,\vec{a}^*_1 + m_2\,\vec{a}^*_2 + m_3\,\vec{a}^*_3 \tag{2.37}$$

$$\vec{r} = x\vec{a}_1 + y\vec{a}_2 + z\vec{a}_3 \tag{2.38}$$

The geometric locus of the points in direct space at which the scalar product $\vec{G}\,.\,\vec{r} = C$ (where C is a constant) is the plane perpendicular to $\vec{G}$ that passes through the projection of the point $\vec{r}$ onto the vector $\vec{G}$ (point H in Figure 2.12). The equation of this direct space plane is then

$$2\pi(m_1x + m_2y + m_3z) = C \tag{2.39}$$

Now suppose that $\vec{r}$ is one of the translation vectors of the lattice, in which case

$$\vec{r} = n_1\vec{a}_1 + n_2\vec{a}_2 + n_3\vec{a}_3 \tag{2.40}$$

With $\{n_i\}$ integers. The plane in question is a lattice plane of equation

$$m_1n_1 + m_2n_2 + m_3n_3 = \frac{C}{2\pi} \tag{2.41}$$

The constant $C/2\pi$ appearing in (2.41) must be an integer, because all the coefficients on the left are. To each value of the constant $C/2\pi$ one can in fact associate a lattice plane which is normal to the vector $\vec{G}$. The distance of one of

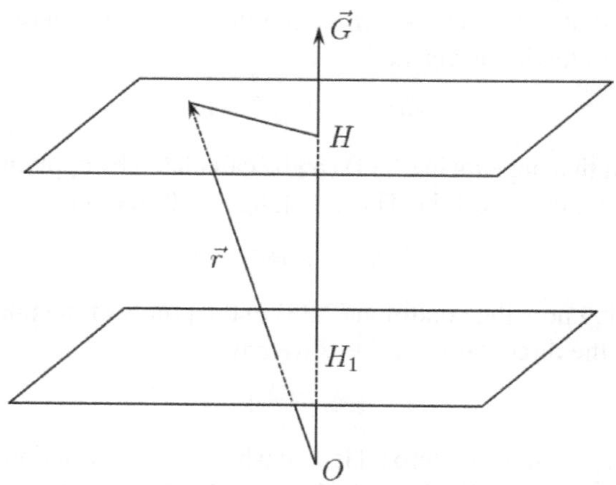

Figure 2.12. Proof of the inverse statement of the fundamental theorem of the reciprocal lattice.

these planes from the origin is given by the length OH which, in turn, is related to the modulus of $\vec{G}$ by

$$OH = \frac{|C|}{|\vec{G}|} \tag{2.42}$$

where C is the constant appearing in (2.41). On the other hand, if OH_1 is the distance from the first plane of the family to the origin, as indicated in Figure 2.12, we can write

$$OH = \frac{|C|}{2\pi} OH_1 = \frac{|C|}{2\pi} d_{hkl} \tag{2.43}$$

Since all planes are equally spaced, applying (2.42) and (2.43) to the plane of the family closest to the origin (i.e., the one with $|C| = 1$) gives:

$$OH_1 = d_{hkl} = \frac{2\pi}{|\vec{G}|} \tag{2.44}$$

Thus, starting from an arbitrary vector $\vec{G}$ we have constructed a family of lattice planes that is perpendicular to it and whose separation is given by $\frac{2}{|G|}$, which proves the inverse statement of the theorem.

2.1.5.4 Relation of a reciprocal lattice vector to Miller indices

The fundamental theorem of the reciprocal lattice states that a family of lattice planes of direct space can be represented by a reciprocal lattice vector of the form (2.37) and a generic point of the family of planes verifies that

$$\vec{G} \cdot \vec{r} = A = Const \tag{2.45}$$

Let us consider a family of lattice planes like the one in Figure 2.13. Among all the planes in the family, let us take the one that contains the origin and, in addition, the one that has the greatest number of intersections of positive coordinates with the coordinate axes; these two planes allow us to define the normal vector to the family and the interplanar spacing. Let the vectors in expression (2.46) be the position vectors of the intersection points of the second plane with the cell axes. Applying condition (2.45) to the first of these points, we obtain (2.47):

$$\begin{aligned} \vec{r}_1 &= x_1 \vec{a} \\ \vec{r}_2 &= x_2 \vec{b} \\ \vec{r}_3 &= x_3 \vec{c} \end{aligned} \tag{2.46}$$

$$m_1 x_1 \vec{a}^* \cdot \vec{a} = 2\pi m_1 x_1 = A \tag{2.47}$$

This implies that

$$x_1 = \frac{A}{2\pi m_1} \tag{2.48}$$

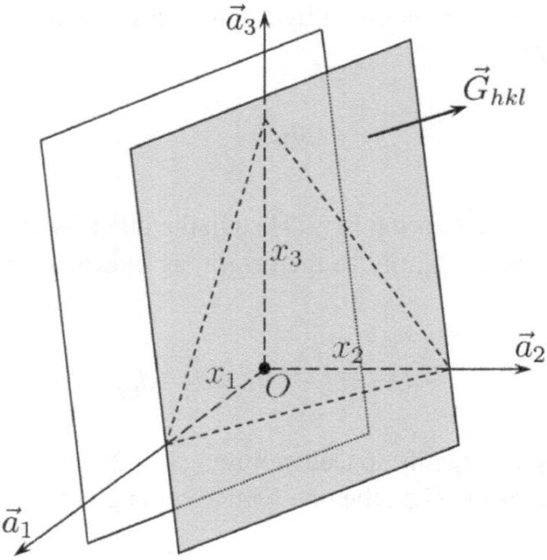

Figure 2.13. Representation of a family of crystalline planes.

And, similarly

$$x_2 = \frac{A}{2\pi m_2} \tag{2.49}$$

$$x_3 = \frac{A}{2\pi m_3}$$

so that the coordinates of the intersections of the plane with the axes are inversely proportional to the components of the reciprocal lattice vector associated with the family of planes:

$$x_1 : x_2 : x_3 = \frac{1}{m_1} : \frac{1}{m_2} : \frac{1}{m_3} \tag{2.50}$$

But these coordinates are also, by definition, inversely proportional to the Miller indices (*hkl*) of the family of planes, which allows them to be given a geometric meaning:

> *"The Miller indices of a family of planes are the components of the reciprocal lattice vector associated with the family of planes in the basis of generators of the reciprocal space."*

Thus, from now on we will denote the vectors of the reciprocal lattice as

$$\vec{G}_{hkl} = h\vec{a}^* + k\vec{b}^* + l\vec{c}^* \tag{2.51}$$

where (*hkl*) are the Miller indices of the family of planes perpendicular to this vector.

2.1.5.5 Crystallographic calculations

Let us return to the zone axes (2.8). Consider two families of planes $\{h_1k_1l_1\}$ and $\{h_2k_2l_2\}$ related by the zone axis:

$$\begin{aligned} u &= k_1l_2 - k_2l_1 \\ v &= l_1h_2 - l_2h_1 \\ w &= h_1k_2 - h_2k_1 \end{aligned} \tag{2.52}$$

since the direction of the zone axis coincides with that of the vector product of the normal vectors to the planes.

On the other hand, planes $\{h_1k_1l_1\}$, $\{h_2k_2l_2\}$, and $\{h_3k_3l_3\}$ belong to the same zone if their normal vectors are coplanar, that is, if

$$\begin{vmatrix} h_1 & k_1 & l_1 \\ h_2 & k_2 & l_2 \\ h_3 & k_3 & l_3 \end{vmatrix} = 0 \tag{2.53}$$

2.1.5.6 Reticular space

It is the distance between the planes of a family of planes. It is measured on the perpendicular to the planes of the family. The spacing of a family of planes (*hkl*) is symbolized by d_{hkl} and is characteristic of each family and can be calculated from the relations (2.14) and (2.51) and the most general form obtained is given by the expression (2.54). These spacings depend on the crystal system that describes the Bravais lattice, since the crystal system determines the orientation of the basis vectors and, therefore, the form of the metric tensor.

$$\begin{aligned} \frac{1}{d_{hkl}^2} = {} & \frac{1}{(1-\cos\alpha^2 - \cos\beta^2 - \cos\gamma^2 - 2\cos\alpha\cos\beta\cos\gamma)} \\ & * \left\{ \left(\frac{h^2}{a^2}\right)\sin\alpha^2 + \left(\frac{k^2}{b^2}\right)\sin\beta^2 + \left(\frac{l^2}{c^2}\right)\sin\gamma^2 \right. \\ & + \left(\frac{2kl}{bc}\right)(\cos\beta\cos\gamma - \cos\alpha) \\ & + \left(\frac{2hl}{ac}\right)(\cos\gamma\cos\alpha - \cos\beta) \\ & \left. + \left(\frac{2hk}{ab}\right)(\cos\alpha\cos\beta - \cos\gamma) \right\} \end{aligned} \tag{2.54}$$

The interplanar spacings for families of planes in the seven crystal systems calculated from expression (2.54) are given in Table 2.3. In it, *a*, *b*, *c*, *α*, *β*, and *γ* are the lattice parameters defined in Figure 2.5.

Table 2.3. Interplanar spacings for families (*hkl*) in the seven crystal systems.

System	$1/d_{hkl}^2$
Cubic P, F, I	$(h^2+k^2+l^2)/a^2$
Tetragonal P, I	$\frac{(h^2+k^2)}{a^2}+\frac{l^2}{c^2}$
Orthorhombic P, I, C, F	$\frac{h^2}{a^2}+\frac{k^2}{b^2}+\frac{l^2}{c^2}$
Hexagonal P	$\frac{4}{3a^3}(h^2+k^2+hk)+\frac{l^2}{c^2}$
Rhombohedral	$\frac{1}{a^2}\left[\frac{(h^2+k^2+l^2)\sin^2\alpha+2(hk+hl+kl)(\cos^2\alpha-\cos\alpha)}{1+2\cos^3\alpha-3\cos^2\alpha}\right]$
Monoclinic P, C	$\frac{h^2}{a^2\sin^2\beta}+\frac{k^2}{b^2}+\frac{l^2}{c^2\sin^2\beta}-\frac{2hl\cos\beta}{ac\sin^2\beta}$
Triclinic P	$(1-\cos^2\alpha-\cos^2\beta-\cos^2\gamma-2\cos\alpha\cos\beta\cos\gamma)^{-1} * \left(\frac{h^2}{a^2}\sin^2\alpha+\frac{k^2}{b^2}\sin^2\beta+\frac{l^2}{c^2}\sin^2\gamma + \frac{2kl}{bc}(\cos\beta\cos\gamma-\cos\alpha) + \frac{2hl}{ca}(\cos\gamma\cos\alpha-\cos\beta) + \frac{2hk}{ab}(\cos\alpha\cos\beta-\cos\gamma)\right)$

In addition to the interplanar distances for families of planes in the seven crystal systems, it is useful to keep in mind the volume of the unit cell, which depends on the geometry, that is, on its lattice parameters, and is given in Table 2.4.

From the dot product of the vector $\vec{q}_{hkl}$ given in (2.14) located in the planes $(h_1k_1l_1)$ and $(h_2k_2l_2)$, respectively and taking into account the expressions for d_hkl corresponding to each crystal system, given in Table 2.3, we can find the angle between the vectors, which is the same angle that exists between the planes $(h_1k_1l_1)$ and $(h_2k_2l_2)$, and are summarized in Table 2.5.

Table 2.4. Unit cell volumes in the seven crystal systems.

Lattice	Unit cell volume
Triclinic	$V = abc\sqrt{1-\cos^2\alpha-\cos^2\beta-\cos^2\gamma+2\cos\alpha\cos\beta\cos\gamma}$
Monoclinic	$abc\sin\beta$
Orthorhombic	$V = abc$
Tetragonal	$V = a^2c$
Hexagonal	$V = \frac{\sqrt{3}}{2}a^2c$
Rhombohedral	$V = a^3\sqrt{1-3*\cos^2\alpha+3*\cos^3\alpha}$
Cubic	$V = a^3$

Table 2.5. Angle θ between planes $(h_1k_1l_1)$ and $(h_2k_2l_2)$.

Lattice	Angle θ between planes $(h_1k_1l_1)$ and $(h_2k_2l_2)$
Triclinic	$\cos\theta = \frac{d_1d_2}{V^2}[S_{11}h_1h_2 + S_{22}k_1k_2 + S_{33}l_1l_2 + S_{23}(k_1l_2+k_2l_1) + S_{12}(h_1k_2+h_2k_1) + S_{13}(l_1h_2+l_2h_1)]$
Monoclinic	$\cos\theta = \frac{d_1d_2}{\sin^2\beta}\left[\frac{h_1h_2}{a^2}+\frac{k_1k_2}{b^2}\sin^2\beta+\frac{l_1l_2}{c^2}-\frac{(l_1h_2+l_2h_1)}{ac}\cos\beta\right]$
Orthorhombic	$\cos\theta = \frac{\frac{h_1h_2+k_1k_2}{a^2}+\frac{l_1l_2}{c^2}}{\sqrt{\left(\frac{h_1^2}{a^2}+\frac{k_1^2}{b^2}+\frac{l_1^2}{c^2}\right)\left(\frac{h_2^2}{a^2}+\frac{k_2^2}{b^2}+\frac{l_2^2}{c^2}\right)}}$
Tetragonal	$\cos\theta = \frac{\frac{h_1h_2+k_1k_2}{a^2}+\frac{l_1l_2}{c^2}}{\sqrt{\left(\frac{h_1^2+k_1^2}{a^2}+\frac{l_1^2}{c^2}\right)\left(\frac{h_2^2+k_2^2}{a^2}+\frac{l_2^2}{c^2}\right)}}$
Hexagonal	$\cos\theta = \frac{h_1h_2+k_1k_2+\frac{1}{2}(h_1k_2+k_1h_2)+\frac{3a^2}{4c^2}l_1l_2}{\sqrt{\left(h_1^2+k_1^2+h_1k_1+\frac{3a^2}{4c^2}l_1^2\right)\left(h_2^2+k_2^2+h_2k_2+\frac{3a^2}{4c^2}l_2^2\right)}}$
Rhombohedral	$\cos\theta = \frac{a^4d_1d_2}{V^2}[\sin^2\alpha(h_1h_2+k_1k_2+l_1l_2) + (\cos^2\alpha-\cos\alpha)(k_1l_2+k_2l_1+l_1h_2+l_2h_1+h_1k_2+h_2k_1)]$
Cubic	$\cos\theta = \frac{h_1h_2+k_1k_2+l_1l_2}{\sqrt{(h_1^2+k_1^2+l_1^2)(h_2^2+k_2^2+l_2^2)}}$

Table 2.5 contd. ...

...Table 2.5 contd.

where

$d_1 = \sqrt{h_1^2 + k_1^2 + l_1^2}$

$d_2 = \sqrt{h_2^2 + k_2^2 + l_2^2}$

$S_{11} = b^2c^2\sin^2\alpha$

$S_{22} = a^2c^2\sin^2\beta$

$S_{33} = a^2b^2\sin^2\gamma$

$S_{12} = abc^2 (\cos\alpha \cos\beta - \cos\gamma)$

$S_{12} = a^2bc(\cos\beta \cos\gamma - \cos\alpha)$

$S_{13} = ab^2c(\cos\gamma \cos\alpha - \cos\beta)$

2.1.5.7 Relationships between direct and reciprocal lattices

We now summarize the relationships established between the direct lattice and the reciprocal lattice, which can be deduced from the previous development. One of the characteristics of the reciprocal lattice is that it has the same symmetry as the direct lattice, when the lattices metric is considered.

- Each direct lattice corresponds to a single reciprocal lattice.
- Every reciprocal lattice has the same symmetry as the direct lattice from which it comes.
- A reciprocal lattice is derived from the direct lattice by a 90° rotation around the knot taken as the origin.
- If the direct lattice is face-centered, its reciprocal will be centered on the interior.
- If the direct lattice is centered on the interior, its reciprocal will be face-centered.
- If the direct lattice is base-centered, its reciprocal is also primitive.
- If the direct lattice is homogeneous, on the contrary, the reciprocal lattice is not. In the direct lattice all knots are equivalent, and the origin can be taken at any knot, while in the reciprocal lattice there is one origin, and the knots are not interchangeable.
- The reciprocal lattice is made up of families of lattice rows, which are perpendicular to planes of the direct lattice, and of families of lattice planes that are perpendicular to lattice rows of the direct lattice.
- The planes of the same family in the reciprocal lattice are not equivalent to each other, unlike what happens in the direct lattice.

2.2 Point symmetry operations

The symmetry operation is a transformation that, when subjected to an object, leads to a configuration indistinguishable from the original. We will return to this concept in section 2.3.1. The symmetry element is an operator that allows the symmetry

Table 2.6. Operations and elements of symmetry.

Symmetry Operations		**Symmetry Element**	
In the plane	**In the space**	**In the plane**	**In the space**
Translation	Translation	Translation vector	Translation vector
Rotation	Rotation	Rotation point	Axis of rotation
Inversion	Inversion	Center	Center
Reflection	Reflection	Line of symmetry	Plane
Reflection-translation (slip)	Rotation - Inversion	Slip line	Rotation-inversion axis
	Rotation - reflection		Rotation axis - reflection
	Reflection - Translation (slip)		Slip plane
			Helical axis
	Rotation-translation		

operation to be performed. Table 2.6 lists the names of the symmetry operations and elements both in the plane and in space. In this section, each of these aspects and their restrictions that give rise to symmetry groups are developed in detail.

So far, we have studied the translation symmetry of crystalline solids. Bravais' postulate naturally leads to the definition of the translation group. In turn, the existence of this group allows us to assign two lattices, the direct and the reciprocal, to each crystal. In particular, the direct lattice describes the "structure" that gives rise to the solid through the repetition of a crystalline structure. But translation symmetry is not the only one that a crystal has. For example, the direct lattice also has inversion symmetry with respect to the origin of the reference system (if this coincides with a knot of the lattice), since both the vector $\vec{t}$ and the vector $-\vec{t}$ belong to the translation group, and additional symmetries may exist.

The crystal structure can also have certain symmetries. This allows the complete crystal to be reconstructed if the symmetry operations it possesses and the maximum unit lacking symmetry, that is, the asymmetric unit, are known.

The symmetries of the lattice and the crystal structure are related. For example, the lattice in Figure 2.14(a) is symmetric under 90° rotations about an axis perpendicular to the plane of the paper through the center of each of the squares defined by four knots. If a crystal is formed from this lattice using an arrow-shaped crystal pattern, as in Figure 2.14(a), the rotational symmetry of the lattice disappears. On the other hand, if the structural pattern is rhombus-shaped, as in Figure 2.14(b), then the crystal has the same 90° rotational symmetry as the lattice. When this occurs, the crystal is said to be holohedral, or to exhibit a holohedric.

The example in Figure 2.14 indicates that, in general, the inclusion of a crystal structure reduces the number of non-translational symmetries in the direct lattice. In other words, a crystal has a few symmetry operations that is less than or equal to that of its direct lattice; the limiting case is the holohedral crystal. On the other hand, the crystal structure of a crystal must have a minimum symmetry and a maximum symmetry. The minimum symmetry must be equal to the symmetry that defines the crystal system to which the crystal belongs. The maximum symmetry must be that of

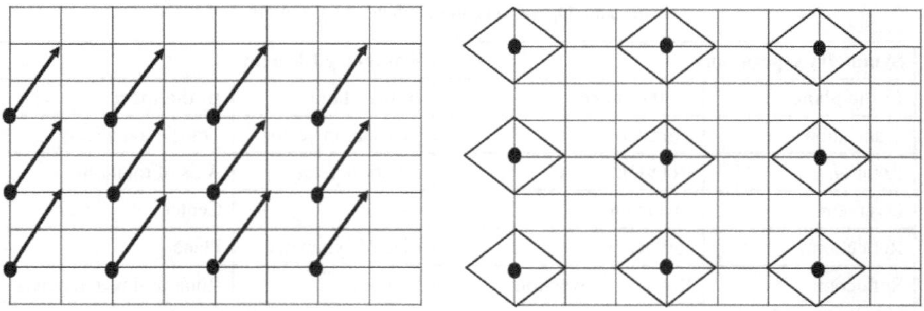

(a) Crystal with less symmetry than the lattice

(b) holohedric

Figure 2.14. Two-dimensional crystals with different symmetries.

the direct lattice. That is, it is not possible for the inclusion of the structural pattern to give rise to symmetries that do not exist in the direct lattice.

The symmetries of a crystal structure are very varied. For example, crystal structures can be invariant under certain rotations or under reflections in certain planes. Symmetry operations of this type are called point-based, because they have at least one point that is invariant. Another example of a point-based operation is inversion, which allows a right-oriented object to be transformed into a left-oriented object. There are also non-point-based symmetry operations, which do not leave any point of the crystal invariant.

In this section of the chapter, we study the symmetry operations (other than translation) that can occur in a crystal and the compatibility relations between them. The systematic study of these operations allows us to justify the existence of the fourteen Bravais lattices classified in the seven crystal systems that we know of.

2.2.1 General definition of symmetry

An essential concept for studying the structure of ideal solids is that of symmetry. According to Evgraf Fedorov, symmetry is a property of bodies that allows them to be brought to coincide with themselves because of certain geometric transformations called symmetry operations. These operations must be isometric; that is, they must preserve the scalar product of two vectors and, therefore, the distances between two points and the angles between two directions.

The isometry condition limits the possible symmetry operations that solids can have to rotations, reflections and, if the solid is allowed to coincide with itself at different points in space, translations.

On the other hand, since chiral molecules exist in Nature (that is, molecules that cannot be superimposed on their mirror images), the operations of interest must also include inversions with respect to a point. Solids related to each other by inversion are called enantiomorphs. Sometimes the word 'symmetry' is also used to refer to the symmetry operations of an object, although strictly speaking they are different concepts.

Isometries verify Chasles' theorem, which states that any isometry can be reduced to a helical motion (i.e., a rotation combined with a translation parallel to the axis of rotation) with or without a center of inversion. If the isometry does not contain any center of inversion, it is called a proper operation; otherwise, it is called an improper operation.

2.2.2 Schoenflies-Fedorov Postulate. Seitz Operators

According to what we have just indicated, in a crystal there may exist points related to each other by symmetries other than the translational symmetries. The inclusion of a crystalline structure may give rise to points of the crystal that are related to each other by symmetries other than the translational symmetries, such that these points are not congruent in the Bravais sense.

For example, consider Figure 2.15, which shows a hypothetical square lattice and a holohedral structural base in the form of four isosceles triangles. Points *A*, *B*, *C*, and *D* are related to each other by a 90° rotation about an axis of rotation perpendicular to the plane of the paper that passes through each node. But these points are not congruent (as *A* and *A'* are, for example) because they do not verify Bravais' postulate. To formally study these symmetries, which are associated with the structural base, it is necessary to adopt an additional principle, the Schnflies-Fedorov postulate:

> *"Given a point $\vec{r}$ of the crystalline solid, there exists an infinite, discrete, and unlimited set of points $\vec{r}\,'$ around which the atomic distribution is identical to that of $\vec{r}$ (with the same orientation or not) or an image of it."*

Two points $\vec{r}$ and $\vec{r}\,'$ related by the Schoenflies-Fedorov postulate are called equivalent points. The postulate states that the atomic neighborhoods at two equivalent points are images of each other; thus, the possibility of enantiomorphic neighborhoods is included. Naturally, the set of equivalent points includes the congruent ones.

Formally, equivalent points are related by a geometric transformation of the form:

$$\vec{r}\,' = S\,\vec{r} + \vec{\tau} \qquad (2.55)$$

where $\vec{\tau}$ is a vector describing the translational part of the isometry (which does not necessarily correspond to a translation of the lattice) and S is a 3 × 3 matrix describing its non-translational part.

Since symmetry operations relating equivalent points are isometries, this matrix must preserve the scalar product of vectors. Let then $\vec{r}_1$ and $\vec{r}_2$ be two arbitrary vectors (which we will represent by column matrices) and let $\vec{r}_1' = S\,\vec{r}_1$ and $\vec{r}_2' = S\,\vec{r}_2$ be respective vectors transformed by the matrix *S*. Then

$$\vec{r}_1' \cdot \vec{r}_2' = \vec{r}_1'^{\,\prime} \cdot \vec{r}_2' = \vec{r}_1^{\,\prime}\, S'S\vec{r}_2 = \vec{r}_1^{\,\prime} \cdot \vec{r}_2 \qquad (2.56)$$

This means that $S'S = I$, where I is the identity matrix, so S must be orthogonal and must verify that the determinant $|S|$ is, $|S| = \pm 1$. The operations for which

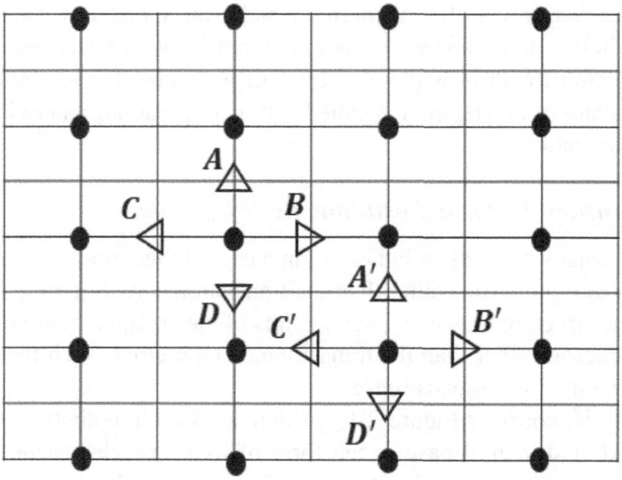

Figure 2.15. Congruent points and equivalent points.

$|S| = 1$ are called direct operations (the rotations) and for which $|S| = -1$, the operation is called inverse or improper rotation (reflections and inversions).

For inverse isometries we can write

$$S = (-S) \cdot (-I) \tag{2.57}$$

where $(-S)$ is a rotation matrix, that is, with determinant +1 and $(-I)$ denotes an inversion. According to (2.57), the inverse operations can always be reduced to the product of a rotation and an inversion.

The transformation (2.55) can be written in matrix form

$$\begin{pmatrix} x' \\ y' \\ z' \\ 1 \end{pmatrix} = \begin{pmatrix} S_{11} & S_{12} & S_{13} & \tau_1 \\ S_{21} & S_{22} & S_{23} & \tau_2 \\ S_{31} & S_{32} & S_{33} & \tau_3 \\ 0 & 0 & 0 & 1 \end{pmatrix} \begin{pmatrix} x \\ y \\ z \\ 1 \end{pmatrix}$$

The rotation matrix is often called the Seitz operator and is denoted as $\{S/\vec{\tau}\}$, that is:

$$\{S/\vec{\tau}\} = \begin{pmatrix} S_{11} & S_{12} & S_{13} & \tau_1 \\ S_{21} & S_{22} & S_{23} & \tau_2 \\ S_{31} & S_{32} & S_{33} & \tau_3 \\ 0 & 0 & 0 & 1 \end{pmatrix}$$

The use of Seitz operators allows us to show that the set of symmetry operations of a crystal has a group structure. First, the composition of symmetry operators is internal. Indeed, the successive application of two symmetry operations of the form (2.55) to an arbitrary point $\vec{r}$ is written:

$$\{S_2/\vec{\tau}_2\}\{S_1/\vec{\tau}_1\}\vec{r} = \{S_2/\vec{\tau}_2\}(S_1\vec{r} + \vec{\tau}_1) = S_2(S_1\vec{r} + \vec{\tau}_1) + \vec{\tau}_2 = S_2S_1\vec{r} + S_2\vec{\tau}_1 + \vec{\tau}_2$$
$$\equiv S\vec{r} + \vec{\tau}$$

where $S = S_2S_1$ and $\vec{\tau} = S_2\vec{\tau}_1 + \vec{\tau}_2$. So, formally

$$\{S_2/\vec{\tau}_2\}\{S_1/\vec{\tau}_1\} = \{S_2S_1/S_2\vec{\tau}_1 + \vec{\tau}_2\} \tag{2.58}$$

which is also a Seitz operator.

On the other hand, the Seitz operator $\{I/\vec{0}\}$ describes an operator that does not modify any point of the crystal and, therefore, corresponds to the neutral element of the set of operations. Using the expression (2.58) it is straightforward to verify that the inverse element with respect to the product of the set of operations is $\{S'/-S'\vec{\tau}\}$ since $\{\{S/\vec{\tau}\}\{S'/-S'\vec{\tau}\} = \{SS'/-SS'\vec{\tau} + \vec{\tau}\} = \{I/\vec{0}\}$. Finally, if $\{S_3/\vec{\tau}_3\},\{S_2/\vec{\tau}_2\},\{S_1/\vec{\tau}_1\}$ are three Seitz operators, then

$$\{S_1/\vec{\tau}_1\}[\{S_2/\vec{\tau}_2\}\{S_3/\vec{\tau}_3\}] = \{S_1/\vec{\tau}_1\}\{S_2S_3/S_2\vec{\tau}_3 + \vec{\tau}_2\} = \{S_1S_2S_3/S_1S_2\vec{\tau}_3 + S_1\vec{\tau}_2 + \vec{\tau}_1\}$$

and

$$[\{S_1/\vec{\tau}_1\}\{S_2/\vec{\tau}_2\}]\{S_3/\vec{\tau}_3\} = \{S_1S_2/S_1\vec{\tau}_2 + \vec{\tau}_1\}\{S_3/\vec{\tau}_3\} = \{S_1S_2S_3/S_1S_2\vec{\tau}_3 + S_1\vec{\tau}_2 + \vec{\tau}_1\}$$

Thus, the Seitz operator is associative. In short, the symmetry operation group of a crystal has group structure, called the symmetry space group of the crystal, and is in general non-commutative. With this notation, the translation group of the crystal is written $\{I/\vec{t}\}$, and can be considered as an invariant subgroup of the symmetry space group.

There is another invariant subgroup formed by operations of the form $\{S/\vec{0}\}$, which contains no translations and therefore leaves at least one point of the crystal invariant. Operations of this type of operations are called point operations; the set of point operations also has group structure and is called the point symmetry group.

2.2.3 Symmetry point operations

2.2.3.1 Classification

2.2.3.1.1 Rotations (axis of Rotation "*n*")

In this section, we describe the possible point operations that can appear in a crystal. The first type is rotation. A crystal is said to have a proper rotation axis, denoted R_α, if the crystal is invariant under a rotation of angle α in the direction of the axis *R*, as shown in Figure 2.16.

By convention, the angle α is taken as positive when the rotation occurs counterclockwise. On the other hand, since each point of the solid must coincide with itself after a certain natural number n of rotations, it must be verified that

$$\frac{2\pi}{\alpha} = n$$

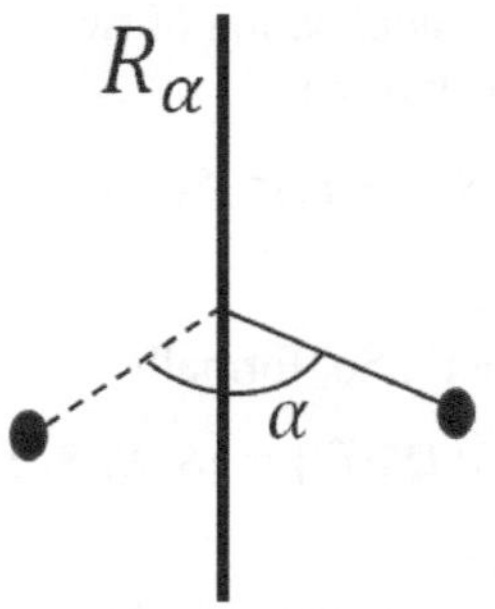

Figure 2.16. Image of a rotation axis—point rotation symmetry.

For some value of n. This property allows to establish a nomenclature for these axes. Thus, for example, an axis $R_{2\pi}$, which makes the crystal coincide with itself after the rotation of an angle 2π, is called a unary axis and is denoted by the symbol "1"; this symmetry operation is the operation, naturally. An axis R_π makes a point of the crystal coincide with itself after two rotations of an angle π; this axis is called a binary axis and is denoted by the symbol "2"; similarly, an axis $R_{2\pi/3}$ makes a point of the crystal coincide with itself after three rotations of $2\pi/3$, is denoted by the symbol "3" and is called a ternary axis, and so on, they are called quaternary and senary axes and are denoted "4" and "6", respectively. Later we will justify that only these five axes of rotation can exist.

The symbols associated with rotation axes perpendicular to the plane of the paper are represented in Figure 2.17. Note the geometric figure located in the center that identifies it.

It is also possible for a binary axis to be parallel to a lattice plane; the position of this axis in the plane is denoted by the symbol "→". In Figure 2.17, we have taken a scalene right triangle as the asymmetric unit. In crystallography, the asymmetric unit is usually represented by a hollow circle.

Rotations are proper operations, and this can be demonstrated only by considering the axis around which the invariant rotation occurs. That is, if $\vec{p}$ is the direction vector of the axis of rotation, it must be verified that:

$$(S - \mathrm{I})\,\vec{p} = \vec{0} \tag{2.59}$$

where I is the identity matrix and S the matrix representing the rotation. According to this,

$$|S - \mathrm{I}| = (1 - tr\,S)(|S| - 1) = 0 \tag{2.60}$$

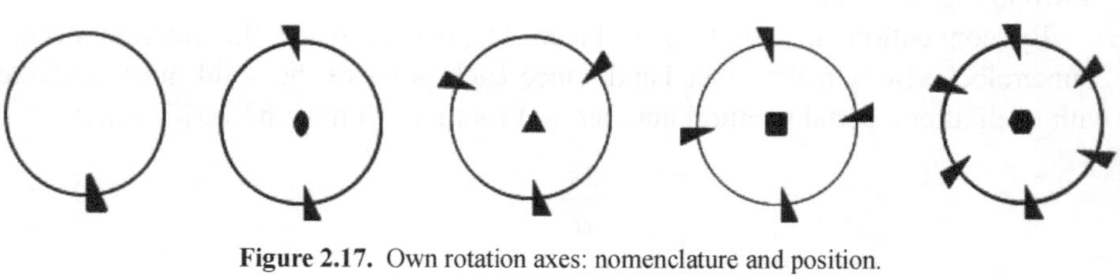

Figure 2.17. Own rotation axes: nomenclature and position.

which implies that

$$|S| = 1 \tag{2.61}$$

As it was intended to be demonstrated.

Indeed, $tr\ S \neq 1$ for rotations is, in fact, for all pointwise operations, so that expression (2.60) implies (2.61) trivially.

2.2.3.1.2 Reflections (m reflection plane)

The second point operation is reflection. The reflection of a point with respect to a plane (called the plane of reflection or mirror plane) produces a point located perpendicular to the plane that passes through the original point and at the same distance from the plane as the latter, but on the other side, as indicated in Figure 2.18; in a certain way, the image of a point by reflection is the one that would be produced if the plane were a mirror.

A crystal is said to have a plane of reflection if it is invariant under this symmetry operation. For example, reflection planes (lines) in a two-dimensional crystal are shown in Figure 2.19. Reflection planes are represented by the letter m, and their

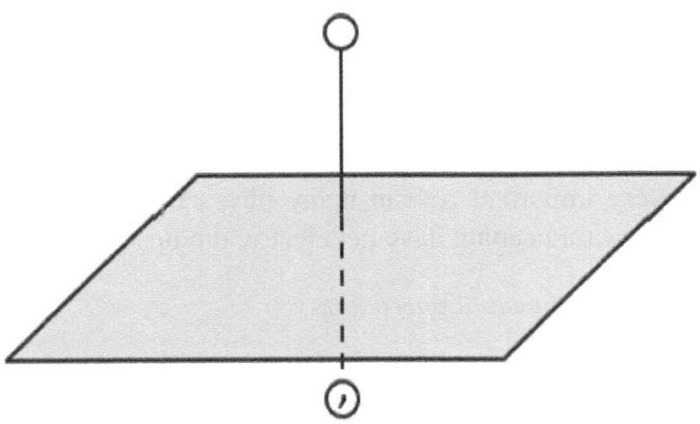

Figure 2.18. Image of a point by a plane of reflection.

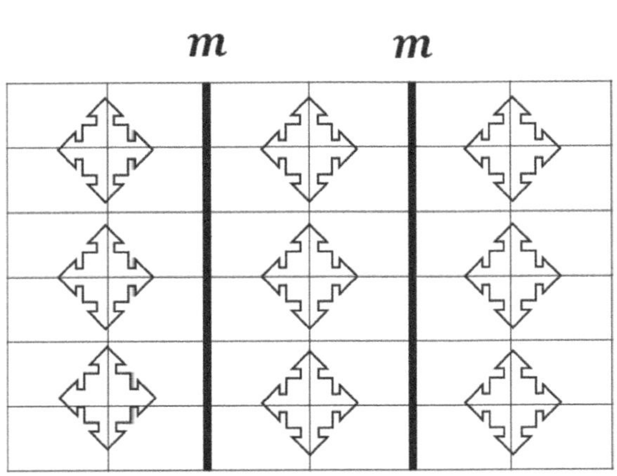

Figure 2.19. Planes (lines) of reflection in a two-dimensional crystal.

position in the unit cell is indicated by a thick line when the planes are perpendicular to the projection plane, or by the symbol "Γ" when it is parallel to it.

On the other hand, the image by a reflection operation of the asymmetric unit is usually represented by the same hollow circle, but with a comma inside. Reflections are inverse operations. However, reflection does not change the orientation of objects and therefore does not allow obtaining a right-oriented object from a left-oriented one or vice versa.

2.2.3.1.3 Inversion (Inversion-Axis $\overline{n}$)

The third point operation is called inversion with respect to a point, the inversion center. Accepting without loss of generality that the inversion center is located at the origin, this operation produces the transformation

$$(x, y, z) \rightarrow (-x, -y, -z)$$

which changes the orientation of the object: if it was on the right, after the inversion it is on the left and vice-versa. A crystal has an inversion center if it remains invariant under inversions. Inversion is only relevant in three-dimensional objects, since in two-dimensional objects it is equivalent to a rotation of angle π with respect to an axis perpendicular to the plane containing the origin. In crystallography, inversion is denoted by the symbol $\hat{\imath}$, or by the symbol $\overline{1}$, and the position of an inversion centre is denoted by an open circle, of the form "○". Inversion with respect to a point is also an inverse operation. When this operation is present in a crystal it is said to be centrosymmetric; otherwise, the crystal is called acentric. The presence of inversion centers plays a very important role in many physical properties; for example, a centrosymmetric structure cannot have net electric dipole moments.

2.2.3.2 Combination of punctual operations

2.2.3.2.1 Rotation: Inversion

The combination of point operations gives rise to other point symmetry operations. For example, suppose that an axis of rotation n contains a center of inversion. In that case, the image of an object is obtained because of a rotation around the axis by an angle $2\pi/n$ followed by an inversion. This combined operation, which is called rotation-inversion, is also point-based and is denoted by the symbol $\overline{n}$. Rotation-inversion is an inverse operation.

Figure 2.20(a) shows the action of a $\overline{4}$ axis on a scalene right triangle initially located in position 1. The images of the triangle are successively positions 2 (below the plane defined by the axis and the inversion center, and inverted with respect to 1), 3, and 4 until finally returning to the starting position.

Figure 2.20(b) shows the projection of positions 1 to 4 onto the plane perpendicular to the axis. In this figure, the solid symbols are located above the plane and the hollow symbols are located below it.

It is important to note that the solid must be invariant under each point operation or combination of operations. If we look at Figure 2.20(a), this implies that if a hypothetical crystal had a triangle like the one in position 1, then it would necessarily also contain 2, 3, and 4, with the same orientation as they appear in the figure.

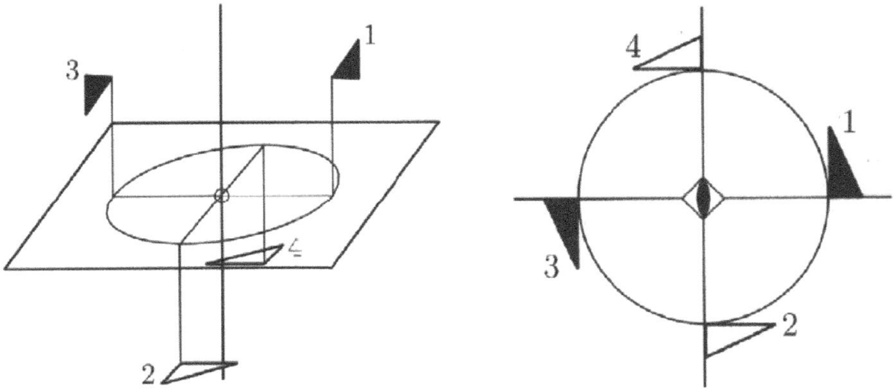

(a) Action of a $\bar{4}$ axis on a triangle located at position 1

(b) Projection on the plane perpendicular to the axis passing through the inversion center

Figure 2.20. Rotation-reversal operation axis.

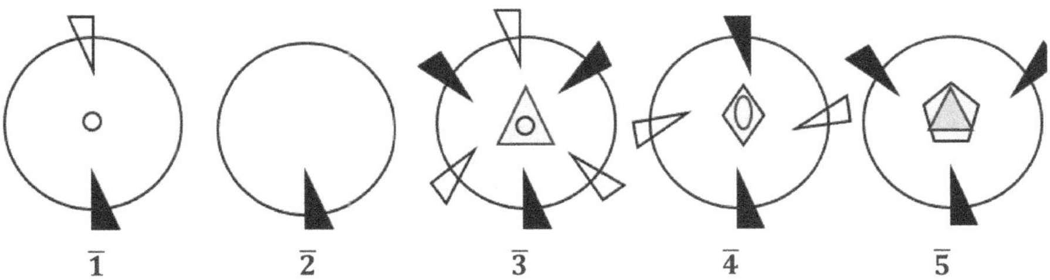

Figure 2.21. Projections of the images of a triangle on its own axes.

Since there are only rotation axes 1, 2, 3, 4, and 6, only the rotation-inversion axes $\bar{1}$, $\bar{2}$, $\bar{3}$, $\bar{4}$, and $\bar{6}$ are relevant. Figure 2.21 shows the projections of the images of a triangle along each of these axes onto a plane perpendicular to them passing through the inversion center; the gray symbols indicate two objects with the same projection, but one above and one below the plane.

According to this figure, axis $\bar{1}$ coincides with the inversion $\hat{\imath}$; hence the two signs for are used for this operation. On the other hand, axis $\bar{2}$ coincides with a reflection around the plane defined by the axis of rotation and the inversion center, so it does not have its own symbol.

2.2.3.2.2 Rotation: Reflection (*n*/*m*)

Now suppose that an axis of rotation *n* is perpendicular to a plane of reflection m and we take the origin of the reference system at the intersection between both. In such a case, the image of an object is obtained by a rotation of angle $2\pi/n$ around the axis followed by a reflection in the mirror plane. This image point is then rotated again by the same angle and reflected in the plane. The process continues until the object is located again in the original (or congruent) position with the same orientation. This combination of operations is called rotation – reflection (or axis of rotation – reflection) and is also a point operation. Rotation – reflection is denoted by the

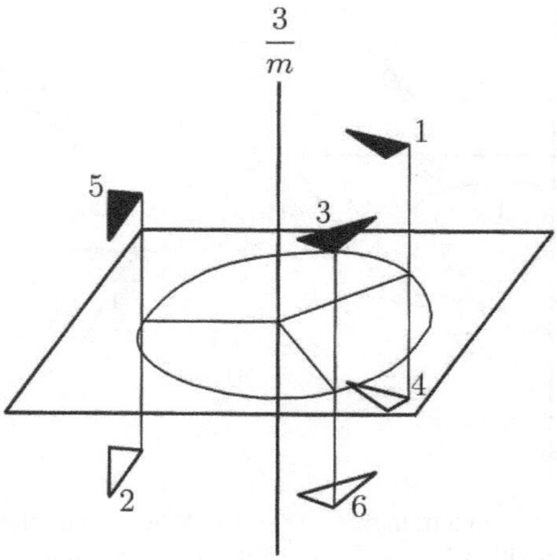

Figure 2.22. Axis of rotation – reflection $3/m$ or $\tilde{3}$.

symbol n/m, or sometimes by the symbol $\tilde{n}$. For example, Figure 2.22 shows the action of an axis of rotation reflection $3/m$ on a scalene right triangle occupying successive positions between 1 and 6.

As in the case of the rotation-inversion axes, only the $\frac{1}{m}, \frac{2}{m}, \frac{3}{m}, \frac{4}{m}$, and $6/m$ reflection rotation axes are relevant. Figure 2.23 represents the projections on the reflection plane of the images by the rotation-reflection axes of the scalene triangle.

By comparison with Figure 2.22, it turns out that the rotation-reflection axes are equivalent to the rotation-inversion axes according to the following correspondences:

$$1/m \equiv \overline{2} \equiv m$$

$$2/m \equiv \overline{1} \equiv \hat{\imath}$$

$$3/m \equiv \overline{6}$$

$$4/m \equiv \overline{4}$$

$$6/m \equiv \overline{3}$$

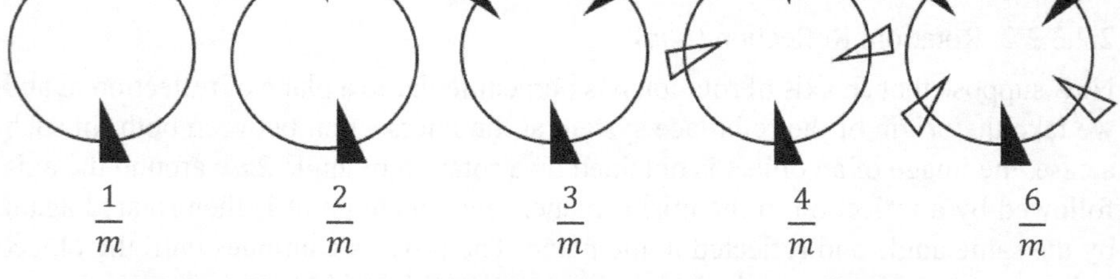

Figure 2.23. Projections of the images of a triangle by rotation-inversion axes.

We can now summarize the relationships between the elements of the lattice and the symmetry operations based on the previous development.

- The number and type of operators that appear in a lattice depend on the metric of the lattice.
- The principle of reticular homogeneity means that every symmetry element that passes through a knot is repeated in parallel and indefinitely in each knot of the lattice.
- Every knot of the lattice is a symmetry center.
- Every axis of symmetry is a reticular row.
- Every plane of symmetry is a reticular plane.
- Perpendicular to every axis of symmetry there is a family of lattice planes.
- Every lattice plane that is a plane of symmetry has a family of lattice rows perpendicular to it and each lattice row in this family is an axis of symmetry.
- Every lattice family that is an axis of even order (2, 4, or 6) has a lattice plane that is a plane of symmetry perpendicular to it.
- A lattice row that is an axis of order 4 or 6, has 4 or 6 families of rows perpendicular to it that are binary axes, and 4 or 6 families of symmetry planes that contain that row.
- The intersection of an axis of even order on an inclined plane of symmetry that is perpendicular to that axis is a center of symmetry.

2.2.4 Constraints imposed by translational invariance

The symmetry operations of a crystal must be compatible with the translation symmetry of its direct lattice, which imposes a few constraints on the way these operations appear. Let us initially consider the rotation operations. Let us take a proper rotation axis of angle α perpendicular to the plane of Figure 2.24 at point A, and let A' be a point congruent to A.

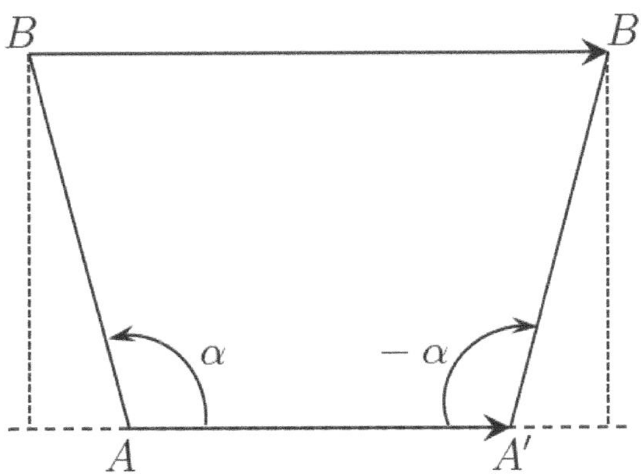

Figure 2.24. Derivation of the order of rotation axes in a crystal.

In such a case, translational symmetry requires that there exists an analogous proper axis perpendicular to the plane at point A'. Let B be the point equivalent to A' by rotation with respect to the axis passing through A and B' the point equivalent to A by rotation about the axis passing through A'. By construction, the distances AA', AB, and BB' must be equal. In addition, points B and B' must be congruent, so that the distance between them must be equal to a natural multiple of the distance AA'. Thus,

$$BB' = AA'[1 + 2\cos(\pi - \alpha)] = m * AA' \tag{2.62}$$

with $m \in N$, so that only values of α are allowed for which

$$1 + 2\cos(\pi - \alpha) = m$$

These angles are as follows:

$$\alpha = 2\pi, \pi, \frac{2\pi}{3}, \frac{\pi}{2}, \frac{\pi}{3}$$

and the corresponding indexes are:

$$n = 1, 2, 3\ 4, 6$$

which define the unary, binary, ternary, quaternary, and senary axes. This proves that compatibility with translational symmetry limits the possible rotation axes to these five axes. This reasoning can be applied analogously to improper rotation axes.

Compactness with translational symmetry also imposes restrictions on reflection planes. Indeed, consider the plane m represented in Figure 2.25 and let $\vec{t}$ be a translation vector perpendicular to the plane.

The image of point A generated by plane m is B, and translational invariance requires the existence of points A' and B' which are congruent, respectively, to A and B. But points A' and B' are also equivalent to B and A by reflection, respectively, so that there must necessarily exist a reflection plane m' parallel to the original plane m and separated from it by a semitranslation parallel to $\vec{t}$ (that is, a distance equal to half the magnitude of $\vec{t}$ in the direction of that vector). As a corollary to this theorem, the net effect of a reflection in two planes parallel to each other is equal to a translation in the direction perpendicular to the planes, the magnitude of which is equal to twice the distance between them.

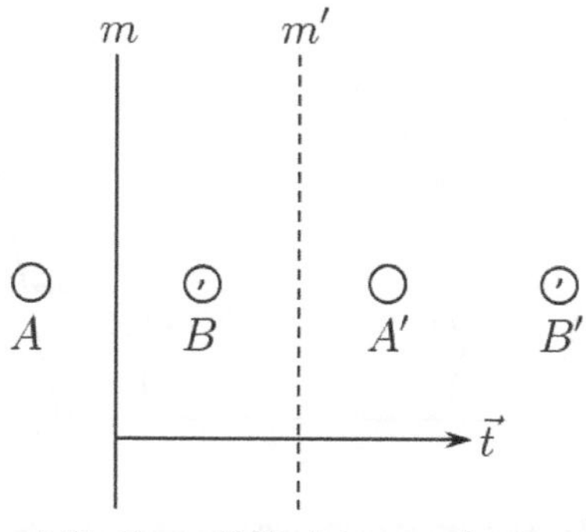

Figure 2.25. Compatibility between reflection planes.

This property is common to all involutive symmetry operations. An operation is involutive if the result of applying it twice is equal to the identity. Thus, in addition to reflections, inversions, rotations by angle π, and some non-pointwise operations that will be defined later are involutive. We can then conclude that the composition of a binary rotation and a translation perpendicular to the axis of rotation is equivalent to a binary rotation with respect to an axis parallel to the first and located a semi-translation with respect to it. Similarly, the composition of an inversion and a translation (in this case arbitrary) is equivalent to an inversion at a point located at the semi-translation with respect to the first.

2.2.5 Compatibility relations between point operations

The symmetry operations of a crystal must also be compatible with each other. Concerning reflections, it is easy to show that the combined reflection in three mutually perpendicular planes is equivalent to an inversion with respect to the point of intersection of the three planes. Symbolically, this is written $m_x m_y m_z = \hat{\imath}$, where x, y, and z denote the three mutually perpendicular axes. Let us now consider two reflection planes m and m' whose normals form a certain angle, $\varphi \neq 0$, as indicated in Figure 2.26.

The composition of the reflections with respect to these two planes is equal to a rotation of angle 2φ around the axis defined by the intersection of the two planes. In fact, let A be a generic point and let A' be the transformation of A by the plane m and A$''$ the transformation of A' by the plane m'. By construction, it is verified

$$\varphi = \theta_1 + \theta_2$$

where θ_1 and θ_2 are defined from Figure 2.26. On the other hand, the distances OA and OA'' must be equal, so that A and A'' are related to each other by a rotation about an axis perpendicular to the plane of Figure 2.26 passing through O. The angle of rotation is

$$2\theta_1 + 2\theta_2 = 2\varphi$$

which proves the theorem.

Two corollaries can be stated to this theorem:

1. The first establishes that the combination of a rotation axis R_α and a reflection plane containing the axis is equivalent to a second reflection plane that belongs to the area defined by the axis and that forms an angle $\alpha/2$ with the first plane.
2. The second corollary states that since the angle of rotation in crystals is restricted to 2π, π, $2\pi/3$, $\pi/2$, $\pi/3$, the reflection planes in a crystal can only form angles the π, $\pi/2$, $\pi/3$, $\pi/4$, $\pi/6$ with each other.

The rotation axes of a crystal must also satisfy certain compatibility relations. Let us consider three rotation axes of angles α, β, γ, parallel to each other and arranged counterclockwise, as in Figure 2.27.

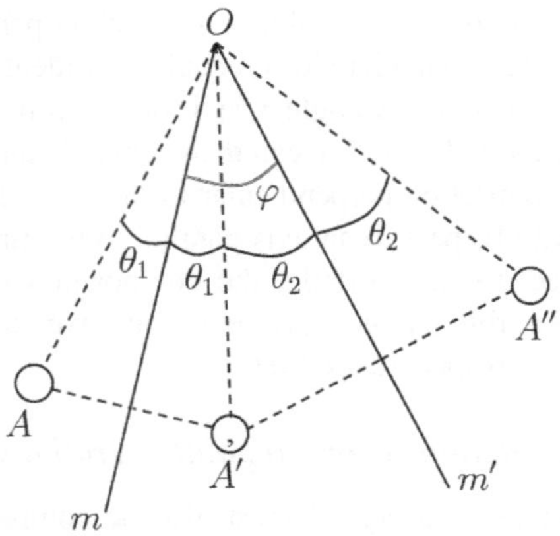

Figure 2.26. Non-parallel reflection planes and equivalent rotation.

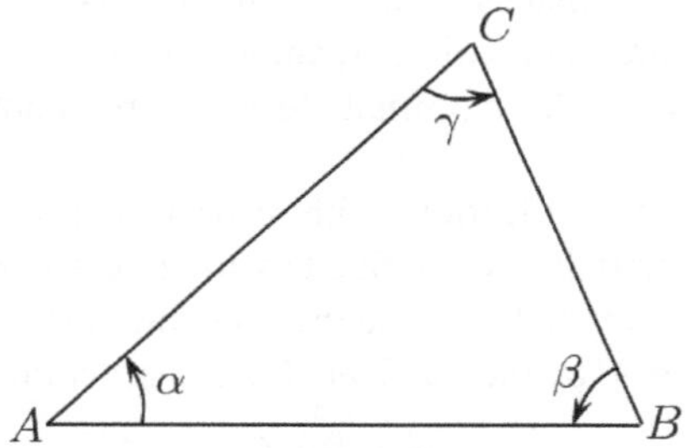

Figure 2.27. Parallel reflection planes and equivalent rotation.

Let *A*, *B*, and *C* be the intersections of these axes with the plane of Figure 2.27. Sylvester's first theorem states that the composition of rotations of angles 2α, 2β, 2γ about *A*, *B*, and *C* is equal to the identity:

$$C_{2\gamma} \cdot B_{2\beta} \cdot A_{2\alpha} = \mathrm{I} \tag{2.63}$$

To prove the theorem, it is enough to note that, as we saw above, the rotation of angle 2α around *A* is equivalent to two successive reflections in the planes *AC* and *AB*. Similarly, the rotation of angle 2β around *B* is equivalent to two successive reflections in the planes *BA* and *BC* and, finally, the rotation of angle 2γ around *C* is equivalent to successive reflections in the planes *CB* and *CA*. Thus, the result is the identity, because all reflections appear in pairs and reflection is an involutive operation. Sylvester's first theorem plays a very important role in the systematic study of space symmetry groups.

A corollary of this theorem is that the composition of two rotations of angles 2α and 2β around parallel axes is equivalent to a rotation of angle $2\alpha + 2\beta$ around a third axis parallel to the previous ones, whose position is determined from the configuration of Figure 2.27. This corollary indicates, for example, that in a crystal there cannot exist a quaternary axis and a senary axis parallel to it, since the combination of both would give rise to an impermissible rotation.

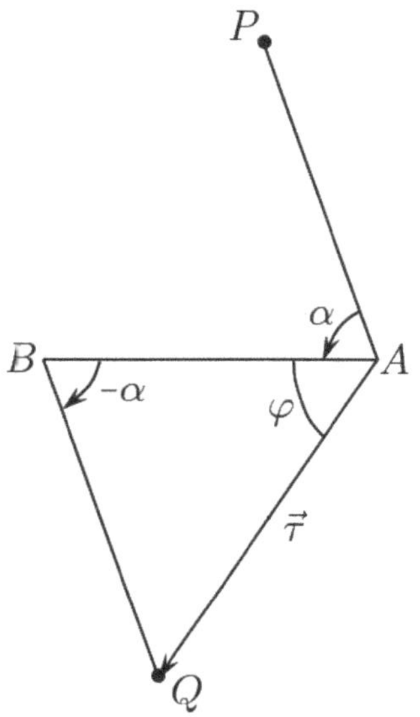

Figure 2.28. Composition of rotations of opposite angles along parallel axes.

Another interesting corollary concerns the situation where $\beta = -\alpha$, in which case the angle γ is zero and the axis C is at infinity. Let us then consider the situation in Figure 2.28.

By rotating an angle α around A, the points AP become aligned along the segment AB, and this becomes the BQ by rotating the angle $-\alpha$ around two parallel axes is equivalent to the translation $\vec{\tau}$ of Figure 2.28, whose module is

$$\tau = 2|AB|\sin\frac{\alpha}{2} \tag{2.64}$$

and whose direction forms an angle $\varphi = (\pi - \alpha)/2$ with the direction AB. This corollary constitutes, in fact, the generalization of that corresponding to binary axes as involutionary operations that has been demonstrated above.

Let us now consider three axes of rotation A, B, and C that intersect at a point O, as represented in Figure 2.29.

If a sphere is drawn around O, the cut of the sphere with the Cartesian planes generates the maximum arcs of Figure 2.29, which in turn determine a spherical triangle with angles α, β, and γ such that

$$\alpha + \beta + \gamma > \pi \tag{2.65}$$

Sylvester's second theorem states that, also in this case, the composition of rotations of angles 2α, 2β, and 2γ around A, B, and C is equal to the identity, identical to expression (2.63), or that the composition of two rotations of angles 2α and 2β around axes A and B that meet at a point (and that are oriented to the right) is equivalent to a rotation of angle 2γ with respect to a third axis C that meets at the same point.

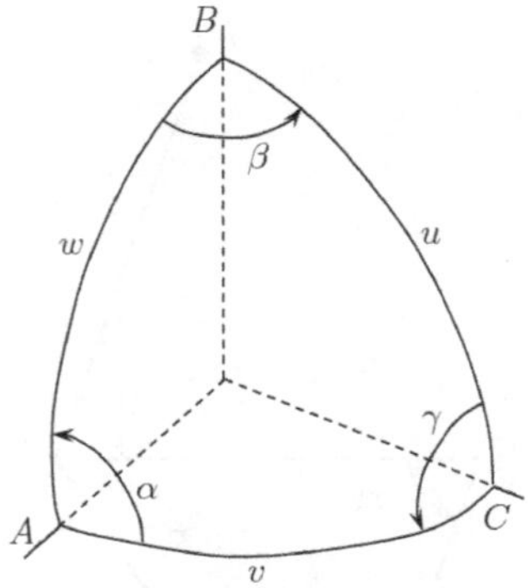

Figure 2.29. Composition of rotations according to axes that intersect at a point.

The position of the third axis C can be determined using the spherical generalization of the cosine theorem. The cosines of the plane angles u, v, and w in Figure 2.29 are given in this case by:

$$\cos u = \frac{\cos\alpha + \cos\beta\cos\gamma}{\sin\beta\sin\gamma} \tag{2.66}$$

$$\cos v = \frac{\cos\beta + \cos\alpha\cos\gamma}{\sin\alpha\sin\lambda} \tag{2.67}$$

$$\cos w = \frac{\cos\gamma + \cos\alpha\cos\beta}{\sin\alpha\sin\beta} \tag{2.68}$$

On the other hand, the angle γ is no longer simply equal to $\alpha + \beta$, as when the axes are parallel. Now each of the angle rotations 2α, 2β, or 2γ is of the form $2\pi/n$, for a certain natural number n. Consequently, (2.65) allows us to write

$$\frac{2\pi}{m} + \frac{2\pi}{n} + \frac{2\pi}{p} > 2\pi \tag{2.69}$$

where m and n are also natural integers, or:

$$\frac{1}{m} + \frac{1}{n} + \frac{1}{p} > 1 \tag{2.70}$$

This is called Euler's relation. The relation (2.70) indicates, for example, that there cannot exist three quaternary rotation axes (nor three senary rotation axes) that converge at the same point. In fact, only certain combinations of rotation axes that converge at a point are possible. Obviously, the unary axes, these are:

1. Combination of the form $n22$, where $n = 2,3,4$, or 6. In these cases, $\alpha = \beta = \pi/2$, $\gamma = \pi/n$. According to (2.66)–(2.68), the two binary axes are perpendicular to the n axis. The axes A and B can then be taken without loss of generality to the equatorial circle with any orientation, in which case the C axis lies at a pole of the sphere. Symmetry groups containing this combination are called *dihedral*

groups and are characterized in that there is only one axis of rotation of higher order than the binary n.

2. Combination 233, in which cases $\alpha = \beta = \pi/3$, $\gamma = \pi/2$. The groups in which this combination appears are called cubic groups and contain three mutually binary 70°32′, 54°44′, and 54°44′. This symmetry is compatible with that of a cube (hence the name of the combination) and with that of a tetrahedron, so cubic groups are also called tetrahedral groups.
3. Combination 234, for which $\alpha = \pi/4$, $\beta = \pi/3$, $\gamma = \pi/2$. In this case, u = 54°44′, v = 45°, w = 35°16′. These systems also contain six binary axes, four ternary axes, and three quaternary axes, and their symmetry is consistent with that of the cube and the octahedron. Groups with this combination of axes are called octahedral groups.

Naturally, the existence of three axes intersecting at a point may imply the existence of more axes. For example, the presence of the ternary axis in the combination 322 implies that there must be a third binary axis perpendicular to it, forming an angle of 60° with the two binary axes already present. Similarly, the presence of the quaternary and senary axes implies the existence of other binary axes coplanar with those represented.

When improper axes of rotation exist, additional restrictions arise. For example, consider two axes of rotation I_1 and I_2 that meet at a point and suppose that I_2 is improper. Given an arbitrary point P, its image along the axis I_1, which we will call Q, is directly equivalent to P while the image of Q along the axis I_2, which we will call R, is enantiomorphic with P. In such a case, a third axis that meets at the same point as I_1 and I_2 and that transforms the point P into the point R must be improper. Thus, in a combination of three axes that meet at a point, the improper axes must appear in pairs.

There are other restrictions concerning the relative arrangement of the rotation axes and reflection planes in a crystal. For example, in any rotation (proper or improper) the direction of the axis must be parallel to a lattice row. To demonstrate this, consider a vector $\vec{v}$ parallel to the rotation axis. This vector is proper to the rotation matrix, so that

$$(S - \mathrm{I})\,\vec{v} = \vec{0} \tag{2.71}$$

The expressions (2.71) constitute a system of three equations with three unknowns, which are the components of the vector $\vec{v}$ in the basis that describes the unit cell. These components are, in general, rational numbers that can be expressed as:

$$v_1 = \frac{x_1}{P}$$

$$v_2 = \frac{x_2}{P}$$

$$v_3 = \frac{x_3}{P}$$

where x_1, x_2, and x_3 are integers and P is a common denominator. In this case, the vector $P\vec{v}$, which is also proper of the rotation matrix, has integer components x_1, x_2, x_3 and is therefore a vector of the translation group. Thus, the direction of a rotation axis must be a lattice row.

On the other hand, every axis of rotation (proper or improper) must be perpendicular to a lattice plane. In fact, let $\vec{v}$ be an eigenvector of a rotation axis of angle α; this vector is parallel to the axis. Let $\vec{t}_1$ be a translation vector of the lattice not collinear to $\vec{v}$. The transformation of $\vec{t}_1$ by the rotation axis $\vec{t}_2$ must verify

$$\vec{v} \cdot \vec{t}_2 = \vec{v} \cdot \vec{t}_1$$

Or,

$$\vec{v} \cdot (\vec{t}_2 - \vec{t}_1) = 0,$$

Hence $\vec{t}_2 - \vec{t}_1$, which is a lattice translation vector, is perpendicular to $\vec{v}$. Let $\vec{t}'_1$ be a lattice vector not collinear with either $\vec{v}$ or $\vec{t}_1$, and let $\vec{t}'_2$ be its transformation by the axis of rotation; in this case, the vector $\vec{t}'_2 - \vec{t}'_1$ is also perpendicular to $\vec{v}$; it is then perpendicular to the plane spanned by $\vec{t}_2 - \vec{t}_1$ and $\vec{t}'_2 - \vec{t}'_1$, which is a lattice plane. From these two results it is easy to conclude that every reflection plane m must be parallel to a lattice plane. This property, for example, prevents the $\vec{c}$ axis of a monoclinic crystal from being an axis of rotation. In fact, only the $\vec{b}$ axis can be.

2.2.6 Non-pointwise symmetry operations

2.2.6.1 Classification

In section 2.3.2, we argued that the most general form possible for a symmetry operation on a crystal involves a translation vector that is not necessarily a vector in the translation group. Operations involving translations do not leave any point in the crystal invariant and are called non-pointwise symmetry operations. These operations are also classified as direct or inverse depending on the nature of their pointwise parts. The translations that appear in these operations do not belong to the translation group of the crystal; for this reason, the alternative term non-symmorphic operation is used to denote non-pointwise operations.

2.2.6.1.1 Helical axis

The first non-pointwise operation that we will consider is the helical axis, which consists of the composition of a proper rotation axis of order n with a translation $\vec{\tau}$ parallel to that axis, as represented in Figure 2.30.

A crystal is said to have a helical axis if its structure is invariant under this combination. The rotational components of the helical axis must be the same as those of the proper rotation axis that defines it, since the argument about the invariance of space remains valid. Thus, only helical axes with rotations 1, 2, 3, 4, and 6 can exist.

The translations $\vec{\tau}$ of the helical axis cannot be arbitrary either. Indeed, if $\vec{t}$ is the vector of the shortest translation group in the direction of the axis of rotation, then the application of the translational component of the helical axis n times (where n is the order of the axis of rotation) produces a displacement $n\vec{\tau}$ of the starting point

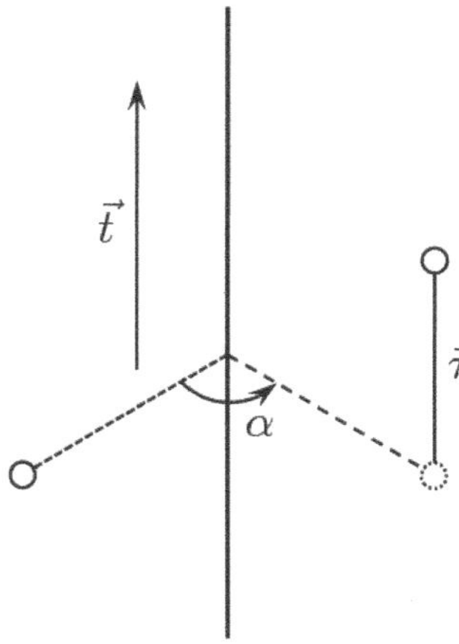

Figure 2.30. Helical axis.

in the direction of the vector $\vec{t}$. Since the crystal must remain invariant under this operation, then it is clear that

$$n\vec{\tau} = p\vec{t}$$

where p is a natural number, such that

$$\vec{\tau} = \frac{p}{n}\vec{t} \tag{2.72}$$

For example, translations parallel to the axis on a helical axis of order 4 are:

$$\vec{0}, \frac{1}{4}\vec{t}, \frac{1}{2}\vec{t}, \frac{3}{4}\vec{t}, \vec{t}, \frac{5}{4}\vec{t}, \cdots,$$

Although only the first four are independent, the fifth and subsequent ones are congruent to them. On the other hand, a helical axis with $\vec{\tau} = 0$, is in fact a proper axis of rotation. Thus, in (2.72) we can obtain

$$1 \leq p \leq n-1$$

The presence of helical axes in a space group is denoted by the symbol n_p. Binary helical axis (2_1), tertiary helical axis (3_1, 3_2), quaternary helical axis (4_1, 4_2, 4_3), senary helical axis (6_1, 6_2, 6_3, 6_4, 6_5).

From the construction itself it can be seen that a 4_2 axis is equivalent to a 2 proper axis; a 6_2 axis is at the same time a 2 proper axis and a 3_2 helical axis, a 6_3 axis is at the same time a 3 axis and a 2_1 and a 6_4 axis is also a 2 proper axis and a 3_1.

Let us now consider the action of a 3_1 axis. If we join the centers of the adjacent points (in height) related by this operation, we obtain a right-handed helix. If we do the same with a 3_2 axis, we also obtain a helix, but this one is left-handed; in fact, both are enantiomorphic. The same occurs with the pairs $4_1 - 4_3$, $6_1 - 6_5$, and $6_2 - 6_4$. Thus, a helical axis behaves like an axis of rotation (or several) or produces a right-handed or left-handed helix, depending on the value of the translation parallel to the axis, according to the following rule:

$$\tau < \frac{1}{2}t \rightarrow \text{Right-handed axis}$$

$$\tau = \frac{1}{2}t \rightarrow \text{Neutral axis}$$

$$\tau > \frac{1}{2}t \rightarrow \text{Left-hand axis}$$

2.2.6.1.2 Slip plane

The second non-pointwise symmetry operation that can occur in crystals is the slip plane, which consists of the composition of a reflection with respect to a mirror plane m and a translation $\vec{\tau}$ parallel to that plane. The crystal has a slip plane if its structure remains invariant under this combination. As for helical axes, the translation must be compatible with the translation symmetry of the lattice, so if a slip plane is applied twice, the net action must be equal to a translation $p\vec{t}$, where $\vec{t}$ is a lattice vector parallel to the plane and p is a natural number. But the net translation of the product of two reflection planes is $2\vec{\tau}$, so it must be verified

$$\vec{\tau} = \frac{p}{2}\vec{t} \tag{2.73}$$

which implies that the possible translations

$$\vec{\tau} = \vec{0}, \frac{1}{2}\vec{t}, \vec{t}, \frac{3}{3}\vec{t}, \cdots,$$

of which only the first two are independent. Moreover, a slip plane with $\vec{\tau} = \vec{0}$ is equivalent to a reflection plane, so that the number p can only take the value of 1. This allows the introduction of a nomenclature for these operations.

- The slip planes in which the translations are $\frac{1}{2}\vec{a}$, $\frac{1}{2}\vec{b}$, and $\frac{1}{2}\vec{c}$ where $\vec{a}$, $\vec{b}$, and $\vec{c}$ are the basis vectors describing the cell, these planes are called axial planes, and are denoted by *a*, *b*, or *c* respectively.
- On the other hand, slip planes in which the translations are of the form $\frac{1}{2}(\vec{a}+\vec{b})$, $\frac{1}{2}(\vec{b}+\vec{c})$, $\frac{1}{2}(\vec{a}+\vec{c})$, or $\frac{1}{2}(\vec{a}+\vec{b}+\vec{c})$, are called diagonal planes, or *n-planes*. Condition (2.73) remains valid when the cell is nonprimitive, although in this case the number p may be rational.
- Slip planes defined in non-primitive cells are labeled with the letter *d* and are called *diamond planes*.

Slip planes perpendicular to the projection plane are represented by a dashed line if the translation is contained in the projection plane; otherwise, they are represented by a dotted line. Diagonal planes are represented by a succession of lines and dots, and diamond planes have their own special nomenclature. When the planes are parallel to the projection plane, the symbol is the same as for mirror planes, but with an arrow indicating the direction of the translation.

2.2.6.1.3 Compatibility of non-point operations

Like point operations, non-point operations must also be geometrically compatible with each other. Let us first consider the compatibility of helical axes. Using Sylvester's second theorem, we can prove that the existence of two orthogonal binary axes that converge at a certain point *O* implies the existence of a third one, perpendicular to the previous ones and which also passes through point *O*. An analysis like that carried out in that case allows us to conclude that:

a) if one of the two initial binary axes is a 2_1 helical axis, then there is a third axis 2 located ¼ from point *O* that orthogonally intersects the helical axis.
b) If there are two perpendicular helical axes that converge at *O*, then there is a binary axis 2 perpendicular to both and that passes to (1/4, 1/4) of *O*.
c) If there exists a pair of axes 2 perpendicular to each other and separated by 1/4 of the translation period, then there exists an axis 2_1 that orthogonally intersects both.
d) if a 2 axis and a 2_1 axis is separated by 1/4 of the lattice period, then there exists a 2_1 axis perpendicular to both of them and which intersects the first 2_1 axis at a distance ¼ of the 2 axes.
e) if two orthogonal binary helical axes are separated by 1/4 of the period, then there exists a third binary helical axis perpendicular to both that passes through (1/4, 1/4)

Let us now consider two slip planes *D* and *D′* which form an angle α as shown in Figure 2.31.

Let *O* be the intersection of the axis of rotation defined by *D* and *D′* with the plane of the paper. Suppose further that the translational components of the planes, *OP* and *OQ*, lie in the plane of the paper, and let *R* be the point of intersection of the perpendicular bisectors of segments *OP* and *OQ*. In Figure 2.31, *X*, *Y*, and *R′* are the images of *R* by *D* and *D′* and the axis through *O*, respectively. The product *DD′* moves point *R* to point *R′* and through *Y* back to *R*. Since *DD′* is a direct motion, it must correspond to an axis of rotation perpendicular to the plane through *R* in Figure 2.31, which is invariant.

Thus, the composition of two slip planes is equivalent to a rotation about an axis parallel to the intersection of the two planes but displaced with respect to it. Moreover, by construction, the angle $RXY = 2\alpha$, which restricts the angles that the slip planes can form with each other, in good agreement with what was discussed for the reflection planes.

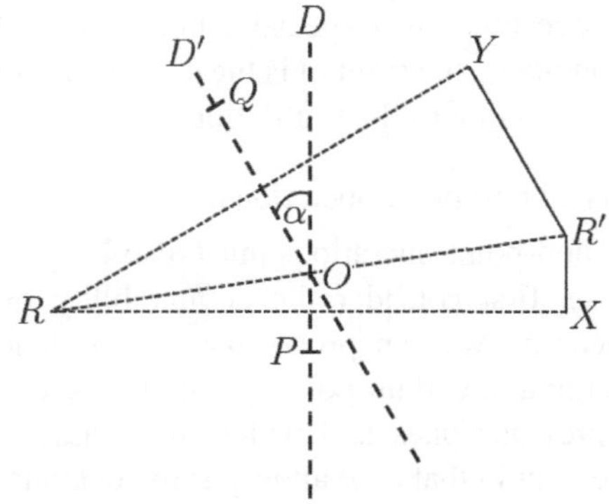

Figure 2.31. Compatibility between sliding planes.

2.3 Point symmetry groups and Laue class

2.3.1 Derivation of point groups: international notation

As we have seen, point symmetry operations generally occur in combination with each other. These combinations must satisfy the compatibility conditions we have just stated, among others. The set of all point symmetry operations constitutes the point group of the crystal. Because there are only three "elementary" point operations and because of the restrictions we have discussed, the number of point symmetry groups that can exist is limited. Johann F. C. Hessel first derived the possible point groups for three-dimensional crystals in 1830 and established their number at 32. On the other hand, the set of crystals that have the same point symmetry group is called a crystal class. Crystal classes are labelled by the name of the point group that corresponds to them, although strictly speaking both concepts are different.

As we saw in an earlier section, the point symmetry group of a crystal is an invariant subgroup of its space group.

Before listing the 32 point symmetry groups, it is useful to define a nomenclature for them. The most widely used, called international notation, assigns each point group a symbol that is constructed according to the following rules:

- Isolated proper rotation axes are denoted by the natural number that defines the axis. When there are several rotation axes that converge at a point, the notation that we defined when studying compatibility relations is used (*n*22,233, or 234).
- Isolated reflection planes are represented by the letter *m*.
- Combinations of axes and planes are denoted differently depending on their relative orientation. Thus, a rotation-reflection axis is denoted by the symbol *n*/*m*, where n is the natural number associated with the axis. On the other hand, a reflection plane parallel to a rotation axis is denoted by *nmm*; the symbol "*mm*" indicates that the angle formed by the reflection planes is half the angle of the rotation axis (i.e., π/n), as we know.

The "*mm*" notation is not necessary in cases $n = 1$ (unary rotation, compatible with a single *m* plane) and $n = 3$ (ternary axis), since in this case the three reflection planes are generated by the action of the rotation axis, as represented in Figure 2.32.

- An improper axis of rotation is denoted by the symbol $\bar{n}$, and the sign $n/\bar{n}$ denotes the presence of both a proper and an improper axis of rotation along the same direction. The combination of improper axes with reflection planes follows the same rules as those just mentioned.

Sometimes a simplified notation is used for point groups. For example, the group 2/m 2/m 2/m is usually denoted as mmm, and the group $\frac{4}{m}\bar{3}\frac{2}{m}$ as $m\bar{3}m$. The reason is that both combinations (complete and simplified) give rise to the same number of point operations.

Let us now begin the systematic study of point groups. The simplest combinations of point operations are a single rotation axis (proper or improper) and a combination of parallel axes. The possibilities are shown in Table 2.7. In constructing this table, we have considered that the combination $1/\bar{1}$ is, in fact, equivalent to a rotation-inversion axis $\bar{1}$ (i.e., an inversion). Similarly, the combination $3/\bar{3}$ is equivalent to an improper axis $\bar{3}$, and should not be considered either. Consequently, the only possible combinations of $n/\bar{n}$ are $2/\bar{2}$, $4/\bar{4}$, and $6/\bar{6}$, which are equivalent to 2/m, 4/m, and 6/m, respectively. Table 2.7 contains 13 point symmetry groups.

Let us now consider the possibility of several rotation axes converging at a single point. According to Euler's relation, if the axes are proper, only combinations of the form $n22$ (with $n = 2,3,4$ and 6), 233, or 234 are possible. If there is a pair of improper axes, then the combinations can be of the form *PII* (where *P* and *I* denote, respectively, proper and improper axis), *IPI* or *IIP*, although there is no independent combination of the latter type. Table 2.8 lists the possible combinations between several rotation axes converging at a point together with the nomenclature of the corresponding groups. In this case, 16 new point groups appear.

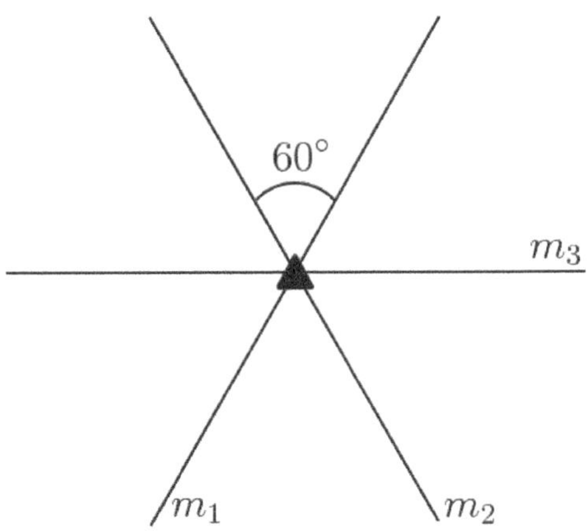

Figure 2.32. Distribution of reflection planes for group 3*m*.

Table 2.7. Point groups with a single axis of rotation.

Own axes	Improper axes	Combinations
1	$\bar{1}$	
2	$\bar{2} \equiv m$	$2/\bar{2} \equiv 2/m$
3	$\bar{3}$	
4	$\bar{4}$	$4/\bar{4} \equiv 4/m$
6	$\bar{6} \equiv 3/m$	$6/\bar{6} \equiv 6/m$

Table 2.8. Point groups with three axes of rotation.

PPP	*PII*	IPI
222	$2mm$	
322	$3m$	$\bar{3}m$
422	$4mm$	$\bar{4}2m$
622	$6mm$	$\bar{6}2\bar{2} \equiv \bar{6}2m$
233	$2\bar{3}\bar{3} \equiv \frac{2}{m}\bar{3}\bar{3} \equiv m\bar{3}$	
432	$4\bar{3}\bar{2} \equiv m\bar{3}m$	$\bar{4}3m$

Finally, there are additional groups that appear when the axes that converge at the same point are simultaneously proper and improper. These point groups are:

$$\frac{2}{\bar{2}}\frac{2}{\bar{2}}\frac{2}{\bar{2}} \equiv \frac{2}{m}\frac{2}{m}\frac{2}{m} \equiv mmm$$

$$\frac{4}{\bar{4}}\frac{2}{\bar{2}}\frac{2}{\bar{2}} \equiv \frac{4}{m}\frac{2}{m}\frac{2}{m} \equiv \frac{4}{m}mm$$

$$\frac{6}{\bar{6}}\frac{2}{\bar{2}}\frac{2}{\bar{2}} \equiv \frac{6}{m}\frac{2}{m}\frac{2}{m} \equiv \frac{6}{m}mm$$

which complete the total of 32 specific groups.

There is an alternative, intuitive, and relatively simple way of deriving point symmetry groups. In this case, we start from the possible proper rotation axes, plus the combinations *n*22, 233, and 432. To obtain all the point groups, these elements are combined with the reflection planes and inversion centers compatible with them until all the possibilities that are consistent with each other are exhausted. In this procedure, improper axes are excluded, because they can be decomposed as a product of a proper rotation and an inverse operation.

The elements associated with the possible improper rotation axes must arise from the general systematic procedure. Table 2.9 contains a derivation of the 32 point groups using this second procedure. The empty cells correspond to combinations that have appeared previously. Let us study as an example some entries in the table.

Table 2.9. Systematic derivation of the 32 point symmetry groups.

Axis	Normal plane	Parallel plane	Diagonal plane	Center of inversion
1	m			$\bar{1}$
2	$\frac{2}{m}$	mmm		
3	$\frac{3}{m}$	$3m$		$\bar{3}$
4	$\frac{4}{m}$	$4mm$		$\bar{4}$
6	$\frac{6}{m}$	$6mm$		
222	mmm		$\bar{4}2m$	
322	$\bar{6}2m$		$\bar{3}m$	
422	$\frac{4}{m}mm$			
622	$\frac{6}{m}mm$			
233	$m\bar{3}$		$\bar{4}3m$	
432	$m\bar{3}m$			

- Combination of a unary axis with a reflection plane. Suppose first that the plane is perpendicular to the axis. Since 1 corresponds to the identity, the plane perpendicular to this axis is the only symmetry operation that exists, so the corresponding entry is a reflection plane m. This reasoning remains valid when the plane is parallel to the axis of rotation, so no entry appears in the corresponding box.
- Combination of a binary axis with a reflection plane parallel to it. Now there must be a second reflection plane that also contains the binary axis and is perpendicular to the first one; the corresponding point group is then $2mm$.
- Combination of the triplet 222 with a reflection plane perpendicular to one of the axes. In this case, the other two binary axes must be included in the reflection plane. Consequently, there must be two reflection planes perpendicular to the first one that contain each of these two binary axes. In short, there is a set of three binary axes, each of which is perpendicular to a reflection plane, such that the corresponding point group is of the form $2/m$ $2/m$ $2/m$ which, as we have seen, is usually represented in an abbreviated form as mmm.

2.3.2 Schnflies notation

In addition to the international notation, point groups can be labeled using so-called Schnflies symbols. The general form of these symbols is

$$A_{nx}$$

where A is a capital letter assigned as follows: "C" (from "cyclic") for uniaxial groups, except for the $\bar{4}$ group; "D" for dihedral groups, "T" for cubic groups, "O" for octahedral groups, and "S" for the $\bar{4}$ group.

The letter x denotes the presence of inverse operations. For $n = 1$, the only index that appears is x, in which case $x = i$ if there is an axis of inversion or $x = s$ (from the German "Spiegel", mirror) if there is a plane of reflection. In other cases, $x = h$, $x = v$, or $x = d$ if there are planes of reflection perpendicular, parallel, or diagonal with respect to the axis of rotation indicated by the index n, respectively. If none of these elements exists, the index is omitted.

The index n denotes the repetition interval of the axis of rotation of maximum symmetry present in the group. This index is omitted in the case of tetrahedral or octahedral groups, because in those cases the axes are 3 and 4, respectively. The index n is also omitted when $x = i$ or $x = s$. The Schnflies s notation is preferred by chemists to describe molecular or crystal symmetry. In crystallography, on the other hand, the international notation is more widely used.

2.3.3 Laue classes and crystal systems

Point groups that contain an inversion center are called centrosymmetric (or centric) groups; of the 32 point groups, 11 are centrosymmetric. These groups are important because, as we will see, physical experiments are invariant under inversions and therefore do not reveal the true symmetry of crystals. In other words, an experiment gives the same result in crystals whose symmetry groups differ only in the existence of an inversion center. The set of point groups that differ only in the presence of an inversion center is called a Laue class and is labeled by the symbol of the centrosymmetric point group it contains.

For example, consider group 2. Combining this with an inversion center gives the group $\bar{2}$, which is formally identical to m. Groups 2 and m therefore belong to the same Laue class. Group $2/m$ also belongs to it. Since the latter is centrosymmetric, the Laue class to which groups 2, m and $2/m$ belong is labeled by this last symbol.

The presence of symmetry elements in a point group imposes certain restrictions on the lattice parameters. It is, therefore, appropriate to group the point groups that give rise to lattice parameters with the same characteristics, so that the corresponding crystal classes can be described by unit cells of the same type. The set of point groups with the same geometric characteristics is called a crystal system. In what follows, we will derive and justify the characteristics of the seven crystal systems that we already know.

Let us start with the point groups 1 and $\bar{1}$. These groups have no rotation axes or reflection planes and therefore do not impose any restrictions on the value of the lengths $\vec{a}$, $\vec{b}$, and $\vec{c}$, nor on the angles α, β, γ. The system in which only the crystal classes 1 and $\bar{1}$ appear is then the triclinic one. The crystal class $2/m$ contains the groups 2, m, and $2/m$, all of which are compatible with a single binary axis. If we take this axis in the direction of the vector $\vec{b}$, then the vectors $\vec{a}$ and $\vec{c}$ must lie in a lattice plane perpendicular to $\vec{b}$. Thus, it must be $\alpha = \gamma = \pi/2$, while β is arbitrary, as are the ratios $\vec{a} : \vec{b} : \vec{c}$. The system to which the crystal classes 2, m and $2/m$ belong is monoclinic.

The crystal classes 222, $2mm$, and mmm contain three mutually perpendicular binary rotation axes. If we take the generating vectors in the directions defined by these axes, then it must be $\alpha = \beta = \gamma = \pi/2$, while the ratios $\vec{a} : \vec{b} : \vec{c}$ are arbitrary. The crystal system corresponding to these classes is orthorhombic.

The seven crystal classes 4, $\bar{4}$, $\frac{4}{m}$, 422, $4mm$, $\bar{4}2m$, and $\frac{4}{m}mm$ are characterized by having a single quaternary axis. If this is taken in the direction of the $\vec{c}$ axis, then the $\vec{a}$ and $\vec{b}$ axes must be equivalent by symmetry and must lie in a plane perpendicular to the $\vec{c}$ axis. Thus, in this case, it must be $\alpha = \beta = \gamma = \pi/2$ and $\vec{a} : \vec{b} : \vec{c} = 1 : 1 : \vec{c}$, with $\vec{c}$ undetermined. The system corresponding to these classes is then the tetragonal one.

For crystal classes with a single ternary axis 3, $\bar{3}$, 32, $3m$, $\bar{3}m$ or senary 6, $\bar{6}$, $6/m$, 622, $6mm$, $\bar{6}2m$, $\frac{6}{m}$ m, the $\vec{c}$ axis parallel to the axis of rotation is usually chosen. These classes belong, respectively, to the trigonal system (characterized by a cell with $\vec{a} : \vec{b} : \vec{c} = 1 : 1 : 1$ and $\alpha = \beta = \gamma$) and to the hexagonal system, characterized by a cell with $\alpha = \beta = \pi/2$ and $\gamma = 2\pi/3$ and ratios $\vec{a} : \vec{b} : \vec{c} = 1 : 1 : \vec{c}$, with $\vec{c}$ indeterminate Finally, crystals with four ternary axes distributed along the diagonals of a cube (classes 23, 432, $\bar{4}3m$, $m\bar{3}$, and $m\bar{3}m$) can be described by a set of three orthogonal vectors parallel to the edges of the cube. The presence of the ternary axes ensures that these three directions are equivalent, which in turn implies that $\alpha = \beta = \gamma = \pi/2$ and $\vec{a} : \vec{b} : \vec{c} = 1 : 1 : 1$. The corresponding system is therefore the cub. All this analysis shows that the existence of seven crystal systems is a consequence of the fact that there are only 32 point groups that give rise to only 11 Laue classes. Table 2.10 shows a classification of the point groups into centrosymmetric and acentric, as well as their distribution into Laue classes and crystal systems.

Table 2.10. Laue classes and classification of point groups.

Symmetry points groups			**Laue class**	**Lattice punctual group**	**Crystalline system**
Acentric		**Centrosymmetric**			
1	1	$\bar{1}$	$\bar{1}$	$\bar{1}$	Triclinic
2	m	$2/m$	$2/m$	$2/m$	Monoclinic
222	$2mm$	mmm	mmm	mmm	Orthorhombic
4	4	$4/m$	$4/m$	$4/mmm$	Tetragonal
422	$4mm,42m$	$4/mmm$	$4/mmm$		
3		3	3	$3m$	Trigonal
322	$3m$	$3m$	$3m$		
6	$3/m$	$6/m$	$6/m$	$6/mmm$	Hexagonal
622	$6mm,62m$	$6/mmm$	$6/mmm$		
233		$m3$	$m3$	$m3m$	Cubic
432	$43m$	$m3m$	$m3m$		

2.4 Group theory

This book is focused on point symmetry groups derived earlier and space symmetry groups which will be discussed in the later section, but it is not focused on group theory itself; however, in this section we will give the definition of groups and some simple examples of how point groups are represented. For the main purpose of this book, which is to study the properties of materials from their crystal structure, it will be sufficient to start from the groups themselves.

2.4.1 Definition of group

Group G is the set of objects or operations (which are called elements of the group) that satisfy the following properties.

1. For this set, a law of internal composition (multiplication) is defined; that is, a law according to which two arbitrary elements A and B of the set G taken in a given order are uniquely matched to a certain element C belonging to the same set; said element is called the product of the elements A and B, and is written $C = AB$.
2. The multiplication operation must satisfy the associative property; that is, for any two elements A, B and D of said set, the equality $(AB)D = A(BD)$ must be verified. The composition law does not have to be commutative; in general, $AB \neq BA$. Groups whose elements commute with each other are called abelian groups.
3. Among the elements of the set there is an element E such that for every element A the equality $AE = EA = A$ is verified. The element E is called the unit element of the group.
4. For every element A belonging to G there exists an element F also belonging to G, such that $AF = E$. This element is called the inverse element of A and is designated by A^{-1}. Thus, the four indicated properties determine a group.

A symmetry element is the group that is determined by a symmetry operation, i.e. the group containing all the mutually inequivalent elements of the given operation. For example, the third-order axis of rotation is the group containing three operations of a 120° rotation, a 240° rotation, ann. These three rotations or operations form a group.

2.4.2 Matrix representations of symmetry operations

For the unambiguous description of crystal directions and planes, the crystallographic coordinate axes X, Y, Z are used. For the description of the physical properties of crystals, as for the analytical representation of their point symmetry groups, the crystallophysical axes X_1, X_2, X_3 are chosen in crystals. The crystallographic and crystallophysical coordinate systems are always straight, and the angles between the positive directions of the corresponding crystallographic and crystallophysical axes are always less than 90°. The positive direction of a crystallographic axis corresponds to a positive direction of the crystallophysical axis. For cubic, tetragonal and

Table 2.11. Direction cosines in relation to two coordinate systems.

Axis	X_1	X_2	X_3
X_1'	C_{11}	C_{12}	C_{13}
X_2'	C_{21}	C_{22}	C_{23}
X_3'	C_{31}	C_{32}	C_{33}

rhombic crystal systems, the directions of the axes coincide in the crystallographic and crystallographic coordinate systems.

Symmetry operations can be written analytically as operations of transformation of the coordinate axes. Let us take the point in space which remains stationary in all symmetric operations of the point group as the origin of the orthogonal crystallophysical coordinate system X_1, X_2, X_3. Every symmetric operation of crystallographic kind will place the axes X_1, X_2, X_3 in new positions X_1', X_2', X_3' .. The angles between the new (X_1', X_2', X_3') and old (X_1, X_2, X_3) axes are determined by Table 2.11.

The first index of the symbol C_{ij} (i, j = 1, 2, 3) refers to the new axes; the second to the old axes. For example, C_{23} is the cosine of the angle between the X_2' axis and the X_3 axis. Thus, every symmetric transformation corresponds to a matrix of direction cosines C_{ij} (in fact, it is these transformation rules that define whether a magnitude is scalar, vector or tensor; this topic will be discussed in depth in Chapter 4), that is, to write it in matrix form. The angle of rotation is considered positive if, observed from the positive end of the axis in the direction of the origin of the coordinates, the rotation from the old axis to the new one is counterclockwise.

Nine coefficients C_{ij} are not independent. Because each row of the matrix (C_{ij}j) represents the direction cosines of the primed axis of coordinates with respect to the non-prime axes X_1, X_2, X_3, in this case we have:

$$C_{ik}\,C_{jk} = 1, \text{ if } i = j \tag{2.74}$$

Since each pair of rows of the matrix (C_{ij}) are direction cosines of two mutually perpendicular directions, therefore

$$C_{ik}\,C_{jk} = 0, \text{ if } i \neq j \tag{2.75}$$

The relations (2.74) and (2.75) are called orthogonality correlations and can be written in the abbreviated form:

$$C_{ik}\,C_{jk} = \delta_{ij} \tag{2.76}$$

where δ_{ij} is the Kronecker delta. The determinant of the matrix of direction cosines is equal to 1, $|C_{ij}| = 1$, for proper symmetry operations and –1, $|C_{ij}| = -1$, for improper ones. The set of all possible inequivalent symmetry operations by which the crystal structure can be transformed into itself forms a mathematical group.

……….The set of matrices (C_{ij}) constructed for all inequivalent symmetry operations that are part of a given symmetry group is called the matrix representation of the symmetry group and the number of inequivalent operations that make up the symmetry group is called the order of the group. A certain system of symmetry operations that make up the given group is called the system of generators if these operations, when multiplied by all modes, give all the operations that make up the group.

..........As an example of a matrix representation of a group, let us write the matrix representation of all the symmetry operations, which are part of the point group *mmm*.

..........The point group *mmm* comprises eight symmetry operations: three reflections in the symmetry coordinate planes, three rotations of 180° around the coordinate axes, also the rotation of 360° and inversion at point O (center). Let us designate the system of coordinate axes, initially obtained by means of the action of the symmetry operations, as follows: X_1', X_2', X_3'. The matrix expression of the reflection operation in the plane X_1X_2 (abbreviated as $m \perp X_3$), is written as follows:

$$C_{ij}(m \perp X_3) = \begin{pmatrix} +1 & 0 & 0 \\ 0 & +1 & 0 \\ 0 & 0 & -1 \end{pmatrix}$$

Similarly, for $(m \perp X_2)$ and $(m \perp X_1)$ we obtain the expressions

$$C_{ij}(m \perp X_2) = \begin{pmatrix} +1 & 0 & 0 \\ 0 & -1 & 0 \\ 0 & 0 & +1 \end{pmatrix}$$

$$C_{ij}(m \perp X_1) = \begin{pmatrix} -1 & 0 & 0 \\ 0 & +1 & 0 \\ 0 & 0 & +1 \end{pmatrix}$$

The determinant $|C_{ij}(m \perp X_i)| = -1$, therefore the reflection symmetry operation in the plane is an improper symmetry operation.

The matrix expression for the 180° rotation operation around X_3 is written (of the rotation matrices in space, this point will be discussed in depth in Chapter 4).

$$C_{ij}(X_3, 180^\circ) = \begin{pmatrix} -1 & 0 & 0 \\ 0 & -1 & 0 \\ 0 & 0 & +1 \end{pmatrix}$$

Similarly, for (X_2, 180°) and (X_1, 180°) we have:

$$C_{ij}(X_2, 180^\circ) = \begin{pmatrix} -1 & 0 & 0 \\ 0 & +1 & 0 \\ 0 & 0 & -1 \end{pmatrix}$$

$$C_{ij}(X_1, 180^\circ) = \begin{pmatrix} +1 & 0 & 0 \\ 0 & -1 & 0 \\ 0 & 0 & -1 \end{pmatrix}$$

The rotation matrix for 360° around any direction in the crystal, for example, the X_3 (X_3, 360°) axis, is written:

$$C_{ij}(X_3, 360^\circ) = \begin{pmatrix} +1 & 0 & 0 \\ 0 & -1 & 0 \\ 0 & 0 & -1 \end{pmatrix}$$

The determinant of $|C_{ij}(X_i, 180^\circ)| = +1$, like the determinant $|C_{ij}(X_i, 360^\circ)| = 1$, i.e., the 180° and 360° rotation operations, are proper symmetry operations. The inversion symmetry operation ($\bar{1}$) is represented in matrix form as follows:

$$C_{ij}(\bar{1}) = \begin{pmatrix} -1 & 0 & 0 \\ 0 & -1 & 0 \\ 0 & 0 & -1 \end{pmatrix}$$

$|C_{ij}(\bar{1})| = -1$, that is, the inversion symmetry operation or transformation is an improper symmetry operation.

..........The set of these 8 matrices that represent the *mmm* group form a group.

2.5 Symmetry Space Groups

2.5.1 Nomenclature

The set of all point and non-point symmetry operations of a crystal constitutes its symmetry space group. There are only 230 symmetry space groups, which were derived independently by Evgraf Fedorov in 1890 and by Arthur Schnflies in 1891.

..........The combination of the 32 point groups with the 14 Bravais cells generates 73 space groups called symmorphic groups. The rest of the space groups are obtained by replacing the rotation axes (proper or improper) by the helical axes of the same order, and the reflection planes by slip planes, considering the mutual compactness conditions that these operations must have with each other. We will not make a systematic derivation of the 230 space groups here because it is beyond the scope of this book, but we will give a description of their nomenclature below.

..........In international notation, space groups are labeled by a letter denoting the type of unit cell that describes the direct lattice, followed by a series of characters describing the minimal set of symmetry operations that generates all the symmetry operations in the group. These characters are different for different crystal systems.

2.5.1.1 Triclinical system

In the triclinic system there can only exist unary axes and inversion centers, corresponding to the space groups $P1$ and $P\bar{1}$, respectively.

2.5.1.2 Monoclinic system

In this case, the binary axis must coincide with the direction of the $\vec{b}$ axis, so after the capital letter it is only necessary to specify an additional symbol indicating which symmetry operations exist in the direction of that axis. For example, the symbol P2/m indicates that the cell is primitive and that there is a reflection plane perpendicular to the binary axis.

2.5.1.3 Orthorhombic system

In the orthorhombic system, the set of symbols consists of three-letter combinations indicating symmetry operations along each of the binary axes. For example, the symbol $Pca2_1$ indicates that the cell is primitive, that there is a slip plane c perpendicular to the $\vec{a}$ axis, that there is a slip plane a perpendicular to the $\vec{b}$ axis, and that there is a helical axis 2_1 parallel to the $\vec{c}$ axis.

2.5.1.4 Tetragonal system

Following the letter for each cell is specified what type of quaternary axis (proper, improper, or helical) exists in the $\vec{c}$-axis direction. Next is indicated what type of binary axis exists in the $\vec{a}$-axis direction, and finally what type of binary axis exists in the [110] direction. Because of quaternary symmetry, it is not necessary to specify symmetry operations along the $\vec{b}$-axis. For example, the symbol $I\frac{4_2}{n}bc$ indicates that the cell is interior-centered, that there is a 4_2 helical axis in the $\vec{c}$-axis direction that is perpendicular to a diagonal slip plane, and that there are also two axial planes *b* and *c* perpendicular to the $\vec{a}$-axis (and $\vec{b}$-axis) and [110] directions, respectively.

2.5.1.5 Trigonal and hexagonal system

In these systems the type of ternary or senary axis along $\vec{c}$ is indicated first, followed by the type of binary axis along $\vec{a}$ and along the $[1\bar{1}0]$, direction. For example, the group $P6_3mc$ describes a primitive cell with a 6_3 helical axis in the $\vec{c}$ direction, a reflection plane perpendicular to the $\vec{a}$ direction, and a slip plane *c* normal to $[1\bar{1}0]$.

2.5.1.6 Cubic System

The symmetries along the $\vec{a}$ axis (equivalently $\vec{b}$ and $\vec{c}$) and the [111] and [110] directions are now specified in this order. Table 2.12 lists the 230 space groups, classified according to the crystal system in which they appear. In this list, point groups that lack reflection planes or inversion centers are underlined. In addition, space groups that can be obtained directly from the symmetry of the corresponding diffraction pattern and missing systems are shown in bold.

……….From the space group of a crystal, the corresponding point group can be determined by simply omitting the symbol associated with the type of lattice and replacing the helical axes and slip planes by the corresponding proper axes and reflection planes. For example, the space groups $P4/m$, $P4_2/m$, $P4/n$, $P4_2/n$, $I4/m$ and $I4_1/a$ correspond to the same point group $4/m$.

2.5.2 General and special positions

Consider an arbitrary point (x, y, z) in the unit cell of a crystal which is not located on any of its symmetry elements. Due to the action of various group operations, there are a certain number of positions in the cell which are equivalent to (x, y, z). This number is called the multiplicity of the space group, and such a position is called

Table 2.12. The 230 symmetry space groups classified by crystal systems

System	Punctual group	Space group
Triclinic	1	$P1$
	$\bar{1}$	$P\bar{1}$
Monoclinic	2	$P2, P2_1, C2$
	m	Pm, Pc, Cm, Cc
	$2/m$	$\frac{P2}{m}, \frac{P2_1}{m} \frac{C2}{m}, \frac{P2}{c}, \boldsymbol{\frac{P2_1}{c}}, \frac{C2}{c}$
Orthorhombic	222	$P222, \mathbf{P222_1}, \mathbf{P2_12_12}, \mathbf{P2_12_12_1}, \mathbf{C222_1}, C222, F222, I222, I2_12_12_1$
	$2mm$	$Pmm2, Pmc2_1, Pcc2, Pma2_1, Pca2_1, Pnc2_1, Pmn2_1, Pba2, Pna2_1, Pnn2, Cmm2, Cmc2_1, Ccc2, Amm2, Abm2, Ama2, Aba2, Fmm2, \mathbf{Fdd2}, Imm2, Iba2, Ima2$
	mmm	$Pmmm, \mathbf{Pnnn}, Pccm, \mathbf{Pban}, Pmma, \mathbf{Pnna}, Pmna, \mathbf{Pcca}, Pbam, \mathbf{Pccn}, Pbcm, Pnnm, Pmmn, \mathbf{Pbcn}, \mathbf{Pbca}, Pnma, Cmcm, Ccma, Cmmm, Cccm, Cmma, \mathbf{Ccca}, Fmmm, \mathbf{Fddd}, Immm, Ibam, \mathbf{Ibca}; Imma$
Tetragonal	4	$P4, \mathbf{P4_1}, P4_2, \mathbf{P4_3}, I4, \mathbf{I4_1}$
	$\bar{4}$	$P\bar{4}, I\bar{4}$
	$4/m$	$\frac{P4}{m}, \frac{P4_2}{m}, \boldsymbol{\frac{P4}{n}}, \boldsymbol{\frac{P4_2}{n}}, I\frac{4}{m}, I\frac{4_1}{a}$
	422	$P422, \mathbf{P42_12}, P4_122, \mathbf{P4_12_12}, \mathbf{P4_222}, \mathbf{P4_22_12}, \mathbf{P4_322}, \mathbf{P4_32_12}, I422, \mathbf{I4_122}$
	$4mm$	$P4mm, P4bm, P4_2cm, P4_2nm, P4cc, P4nc, P4_2mc, P4_2bc, I4mm, I4cm, I4_1md, \mathbf{I4_1cd}$
	$\bar{4}m$	$P\bar{4}2m, P\bar{4}2c, P\bar{4}2_1m, \mathbf{P\bar{4}2_1c}, P4m2, P\bar{4}c2, P\bar{4}b2, P\bar{4}n2, I\bar{4}m2, I\bar{4}c2, I\bar{4}2m, I\bar{4}2d$
	$\frac{4}{m}mm$	$P4/mmm, P4/mcc, \mathbf{P4/nbm}, \mathbf{P4/nnc}, P4/mbm, P4/mnc, \mathbf{P4/nmm}, \mathbf{P4/ncc}, P4_2/mmc, P4/2mcm, \mathbf{P4_2/nbc}, \mathbf{P4_2/nnm}, P4_2/mbc, P4_2/mnm, \mathbf{P4_2/nmc}, \mathbf{P4_2/ncm}, I4/mmm, I4/mcm, \mathbf{I4_1/amd}, \mathbf{I4_1/acd}$
Trigonal	3	$P3, \mathbf{P3_1}, \mathbf{P3_2}, R3$
	$\bar{3}$	$P\bar{3}, R\bar{3}$
	32	$P312, P321, \mathbf{P3_112}, \mathbf{P3_121}, \mathbf{P3_212}, \mathbf{P3_221}, R32$
	$3m$	$P3m1, P31m, P3c1, P31c, R3m, R3c$
	$\bar{3}m$	$P\bar{3}1m, P\bar{3}1c, P\bar{3}m1, P\bar{3}c1, R\bar{3}m, R\bar{3}c$
Hexagonal	6	$P6, \mathbf{P6_1}, \mathbf{P6_2}, P6_3, \mathbf{P6_4}, \mathbf{P6_5}$
	$\bar{6}$	$P\bar{6}$
	$6/m$	$P6/m, P6_3/m$
	622	$P622, \mathbf{P6_122}, \mathbf{P6_222}, \mathbf{P6_322}, \mathbf{P6_422}, \mathbf{P6_522}$
	$6mm$	$P6mm, P6cc, P6_3cm, P6_3mc$
	$\bar{6}m$	$P\bar{6}m2, P\bar{6}c2, P\bar{6}2m, P\bar{6}2c$
	$6/mmm$	$P6/mmm, P6/mcc, P6_3/mcm, P6_3/mmc$
Cubic	233	$P23, F23, I23, \mathbf{P2_13}, I2_13$
	$m\bar{3}$	$Pm\bar{3}, \mathbf{Pn\bar{3}}, Fm\bar{3}, \mathbf{Fd\bar{3}}, Im\bar{3}, \mathbf{Pa\bar{3}}, \mathbf{Ia\bar{3}}$
	432	$P432, \mathbf{P4_232}, F432, \mathbf{F4_132}, I432, \mathbf{P4_332}, \mathbf{P4_132}, \mathbf{I41_32}$
	$\bar{4}3m$	$P\bar{4}3m, F43m, I\bar{4}3m, P43n, F\bar{4}3c, \mathbf{I\bar{4}3d}$
	$m\bar{3}m$	$Pm\bar{3}m, \mathbf{Pn\bar{3}n}, Pm\bar{3}n, \mathbf{Pn\bar{3}m}, Fm\bar{3}m, Fm\bar{3}c, \mathbf{Fd\bar{3}m}, \mathbf{Fd\bar{3}c}, Im\bar{3}m, \mathbf{Ia\bar{3}d}$

a general position. If (x, y, z) is located on one of the point symmetry elements of the crystal (for example, if it is one of the points on an axis of rotation or a plane of reflection), then some symmetry operations leave that point invariant, and the number of positions equivalent to it is less than the multiplicity of the space group. Such a position is called a space position, and its multiplicity depends on the number of symmetry elements sharing the point.

To understand this concept, let us consider the space group P2/m of the monoclinic system, which contains a rotation axis 2 parallel to the axis $\vec{b}$ and a plane m perpendicular to it, plus an inversion center at the intersection point of both. The equivalent positions for this group are:

$$(x, y, z), (\bar{x}, y, \bar{z}), (x, \bar{y}, z), (\bar{x}, \bar{y}, \bar{z}) \tag{2.77}$$

Which means that if an atom exists at an arbitrary position (x, y, z), then an atom of the same type must also exist at the other three positions. The multiplicity of this group is 4. Now consider a point situated on the axis of rotation in the form (0,y,0). The equivalent points of (0,y,0) according to (2.77), are:

$$(0,y,0),\ (0,\bar{y},0)$$

Thus, for the points on the axis of rotation there are only two equivalent positions, not four; they are therefore special positions. The points on the m plane are also special positions, with two equivalent positions. Since both operations (axis 2 and plane *m*) are involutive, the points on the binary axes and the planes in each semi-translation are also special positions, that is, the positions:

$$\left(\frac{1}{2}, y, 0\right), \left(0, y, \frac{1}{2}\right), \left(\frac{1}{2}, y, \frac{1}{2}\right), \left(x, \frac{1}{2}, z\right) \tag{2.78}$$

Let us now consider the inversion center, whose coordinates are (0,0,0). In this case, the four positions of (2.77) give the same result, so the inversion center is also a special position of multiplicity 1. Since inversion is an involution operation, the following are also special positions of multiplicity one:

$$\left(0, \frac{1}{2}, 0\right), \left(0, 0, \frac{1}{2}\right), \left(\frac{1}{2}, 0, 0\right), \left(\frac{1}{2}, \frac{1}{2}, 0\right), \left(0, \frac{1}{2}, \frac{1}{2}\right), \left(\frac{1}{2}, 0, \frac{1}{2}\right), \left(\frac{1}{2}, \frac{1}{2}, \frac{1}{2}\right) \tag{2.79}$$

The equivalent positions for each symmetry space group are tabulated (for example, in the International Tables of Crystallography) and are usually described by a notation due to Ralph Wyckoff. To properly define the Wyckoff positions, it is necessary to know elements of group theory that are beyond the scope of this book. For practical purposes, it is sufficient to indicate here that, in this notation, each set of equivalent positions by point symmetry is associated with a number equal to the multiplicity of the position and a lowercase letter: the position with the lowest symmetry is assigned the letter *a*, and the subsequent ones are assigned the letters *b*, *c*, etc.

The atomic positions of the *P*2/*m* group are shown in Table 2.13.

Table 2.13. Equivalent and special positions for the space group $P2/m$.

Multiplicity	Wyckoff symbol	Coordinates
4	4o	$(x, y, z), (\bar{x}, y, \bar{z}), (x, \bar{y}, z), (\bar{x}, \bar{y}, \bar{z})$
2	2n	$\left(x, \frac{1}{2}, z\right), \left(\bar{x}, \frac{1}{2}, \bar{z}\right)$
	2m	$(x, 0, z), (\bar{x}, 0, \bar{z})$
	2l	$\left(\frac{1}{2}, y, \frac{1}{2}\right), \left(\frac{1}{2}, \bar{y}, \frac{1}{2}\right)$
	2k	$\left(0, y, \frac{1}{2}\right), \left(0, \bar{y}, \frac{1}{2}\right)$
	2j	$\left(\frac{1}{2}, y, 0\right), \left(\frac{1}{2}, \bar{y}, 0\right)$
	2i	$(0,y,0), (0,\bar{y},0)$
1	1h	$\left(\frac{1}{2}, \frac{1}{2}, \frac{1}{2}\right)$
	1g	$\left(\frac{1}{2}, 0, \frac{1}{2}\right)$
	1f	$\left(0, \frac{1}{2}, \frac{1}{2}\right)$
	1e	$\left(\frac{1}{2}, \frac{1}{2}, 0\right)$
	1d	$\left(\frac{1}{2}, 0, 0\right)$
	1c	$\left(0, 0, \frac{1}{2}\right)$
	1b	$\left(0, \frac{1}{2}, 0\right)$
	1a	$\left(\frac{1}{2}, \frac{1}{2}, \frac{1}{2}\right)$

On the other hand, it is worth mentioning that many substances of biological interest (which are, in general, enantiomorphic) crystallize in spatial groups that lack inversion centers and mirror planes, of which there are 63; these are the so-called acentric groups.

Bibliography

De Graef, M. and McHenry, M.E. (2012). Structure of Materials: An Introduction to Crystallography, Diffraction and Symmetry. Cambridge University Press.

Helliwell, J.R. (2017). Science of crystal structures: highlights in crystallography.

Müller, U. (2014). Symmetry of crystals and molecules.

Nespolo, M. (2012). Symmetry in Crystallography, by Paolo Radaelli.

Szwacki, N.G. and Szwacka, T. (2010). Basic Elements of Crystallography. Pan Stanford Singapore.

CHAPTER 3
Introduction to Tensor Analysis

INTRODUCTION

Tensors are mathematical objects of great importance in many areas of physics such as general relativity, electromagnetic theory, materials science, elasticity theory, thermodynamics, fluid mechanics, computational sciences, among others. Tensors constitute a mathematical language that is used to model properties (elastic, optical, magnetic, electrical, thermal, mechanical, chemical in materials with different symmetry properties, whether anisotropic, orthotropic, isotropic, homogeneous, or inhomogeneous) and phenomena such as transport (mass, charge, energy, heat, among others), transformation of the geometry of solid materials and the changes in their properties implied by these transformations.

Tensors are used to model the relationship between crystalline or molecular structure, its symmetries and physical and chemical properties in a direct mathematical way. However, the crystalline structure is defined by the type of atomic bond between its atoms and this type of bond depends on the number of electrons in the atom. In other words, ultimately, a relatively clearer path is seen that goes from the electronic configuration to the mechanical properties of the material. But establishing this connection path from the atomic structure to the properties of the material is an arduous and complex task that requires both experimental and theoretical development and computational implementation. Of course, it is necessary to restrict the type of material, its state and operating temperature to make an analysis of this nature.

A relatively well-defined path has been established for solid and crystalline materials, mainly in metals and some ceramics, as we will see in other sections of this text.

Not everything that has magnitude, direction, and sense is a vector, it can be a tensor without being a vector, such as the refractive index and the elastic constants in anisotropic crystals, which have magnitude, direction, and sense, but do not transform as a vector. A vector is an element of a vector space, but there are other ways to define it more operationally, such as by a linear transformation that follows a certain rule or

DOI: 10.1201/9781003344438-3

property. Tensors are a generalization of vector magnitudes and in this chapter, we will define them from the coordinate transformation, since this approach allows us to explicitly understand precisely their differences with vector magnitudes and in their explicit language of components allows a more direct computational implementation. Another approach that will be developed to define and understand what a tensor is, is the physical approach, to see explicitly how the tensor components connect the causes and effects of the properties in the materials of study, this approach will be developed in later chapters.

3.1 Notation and tensor transformations

3.1.1 Vector concept

Let us first see what a vector is from the point of view of coordinate transformations, and, more precisely, define the vector in terms of the behavior under rotation of the coordinate system, and then generalize these transformations and address the concept of tensor.

We will assume for the moment that our space is isotropic, that is, that all directions are equivalent or it is the same as saying there is no preferred direction or that the properties do not differ by the orientation of the coordinate system chosen, we also assume that space is homogeneous, both in space and in time, that is, no matter how far we go, the properties of space do not change and no matter how much time is considered, the properties of space are preserved and for simplicity let us assume the case of 2 dimensions and then generalize to n dimensions.

Figure 3.1 shows the geometric aspects to be considered when rotating the axes of the coordinate system. Let OX and OY be the primitive axes and $O\overline{X}$ and $O\overline{Y}$ the new axes that arise when rotating the primitive axes through an angle θ.

Let (x, y) be the coordinates of a point P of the plane with respect to the primitive axes and $(\overline{x}, \overline{y})$ the coordinates of the same point P with respect to the new axes. To

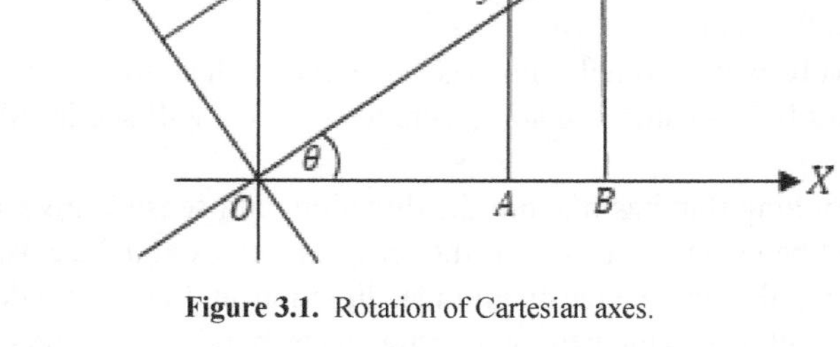

Figure 3.1. Rotation of Cartesian axes.

determine the expressions of the coordinates (x, y) in terms of the new coordinates $(\bar{x}, \bar{y})$, it is observed from Figure 3.1 that:

$$\begin{aligned} x &= OA = OB - AB = \bar{x}\cos\theta - \bar{y}\sin\theta \\ y &= AP = A\bar{A} + \bar{A}P = \bar{x}\sin\theta + \bar{y}\cos\theta \end{aligned} \tag{3.1}$$

Therefore, the equations for the transformation of coordinates are:

$$\begin{aligned} x &= \bar{x}\cos\theta - \bar{y}\sin\theta \\ y &= \bar{x}\sin\theta + \bar{y}\cos\theta \end{aligned} \tag{3.2}$$

The transformation equations (3.2) can be written in matrix form as follows:

$$\begin{pmatrix} x \\ y \end{pmatrix} = \begin{pmatrix} \cos\theta & -\sin\theta \\ \sin\theta & \cos\theta \end{pmatrix} \begin{pmatrix} \bar{x} \\ \bar{y} \end{pmatrix} \tag{3.3}$$

From the transformation equations (3.1), we can obtain the equations for inverse transformation, assuming the equations as a 2*2 system of equations whose unknowns are the new coordinates $(\bar{x}, \bar{y})$, solving the system we obtain the expressions:

$$\begin{aligned} \bar{x} &= x\cos\theta + y\sin\theta \\ \bar{y} &= -x\sin\theta + y\cos\theta \end{aligned} \tag{3.4}$$

Or simply from the matrix form (3.3) it is multiplied on both sides by the inverse matrix and the transformation equations (3.4) are obtained again in matrix form as follows:

$$\begin{pmatrix} \bar{x} \\ \bar{y} \end{pmatrix} = \begin{pmatrix} \cos\theta & \sin\theta \\ -\sin\theta & \cos\theta \end{pmatrix} \begin{pmatrix} x \\ y \end{pmatrix} \tag{3.5}$$

A vector is represented by the coordinates of a point, in this case 2 dimensional, the components (x, y) represent a vector, although we will define it later, therefore, any pair of quantities (A_x, A_y) in the XY coordinate system is transformed into the components $(\bar{A}_x, \bar{A}_y)$ due to the rotation of axes and the transformations (3.2) and (3.4) as follows:

$$\begin{aligned} A_x &= \bar{A}_x\cos\theta - \bar{A}_y\sin\theta \\ A_y &= \bar{A}_x\sin\theta + \bar{A}_y\cos\theta \end{aligned} \tag{3.6}$$

$$\begin{aligned} \bar{A}_x &= A_x\cos\theta + A_y\sin\theta \\ \bar{A}_y &= -A_x\sin\theta + A_y\cos\theta \end{aligned} \tag{3.7}$$

matrix form

$$\begin{pmatrix} A_x \\ A_y \end{pmatrix} = \begin{pmatrix} \cos\theta & -\sin\theta \\ \sin\theta & \cos\theta \end{pmatrix} \begin{pmatrix} \bar{A}_x \\ \bar{A}_y \end{pmatrix} \tag{3.8}$$

Considering the inverse matrix to invert the coordinate transformation (3.8), we obtain

$$\begin{pmatrix} \bar{A}_x \\ \bar{A}_y \end{pmatrix} = \begin{pmatrix} \cos\theta & \sin\theta \\ -\sin\theta & \cos\theta \end{pmatrix} \begin{pmatrix} A_x \\ A_y \end{pmatrix} \tag{3.9}$$

The matrices (3.10) depending on the angle of rotation are called rotation matrices.

$$\begin{pmatrix} \cos\theta & \sin\theta \\ -\sin\theta & \cos\theta \end{pmatrix}, \begin{pmatrix} \cos\theta & -\sin\theta \\ \sin\theta & \cos\theta \end{pmatrix} \tag{3.10}$$

3.1.1.1 A first definition of vector

The components (A_x, A_y) that are transformed into components $(\bar{A}_x, \bar{A}_y)$ according to the coordinate transformation equations (3.8) define a vector or are components of a vector that we will call $\vec{A}$. In this way, a vector is defined by the way its components transform under a rotation of axes in space. Rotating the reference system and finding the coordinates in the new rotated system is what is called coordinate transformation in orthogonal reference systems.

But let us look at several aspects involved in this definition. Note the following, in the *XY* coordinate system of Figure 3.1, the squared magnitude of the vector $\vec{A}$ is given by

$$\|\vec{A}\|^2 = A_x^2 + A_y^2 \tag{3.11}$$

If we replace the transformation equations given in expression (3.6) by expression (3.11), it can be observed, after developing a short and simple algebra, that

$$\|\vec{A}\|^2 = \bar{A}_x^2 + \bar{A}_y^2 = \|\vec{\bar{A}}\|^2 \tag{3.12}$$

That is, the vector preserves its norm under this rotation transformation, a result that was expected because the rotation matrices given in expression (3.10) are orthogonal; from linear algebra, it can be shown that orthogonal matrices preserve the norm. The vector also does not change its direction and sense under rotation transformations, it changes the value of its components, obviously, whose values are given by the transformation equations (3.6) and (3.7), but not its magnitude, direction, and sense. The fact that the vector does not change its magnitude due to the rotation of the axes is a fundamental characteristic of what properly defines a vector, which is inherited from the orthogonality of the rotation matrix and this in turn carries the properties of the configuration space. The same does not necessarily happen with a tensor. A tensor can expand or contract when rotated, so if we want to preserve its norm it is necessary to make an additional transformation. We will return to this point later to obtain more clarity on what is meant. Given a pair of quantities $(-y, x)$, let us see how it behaves under coordinate transformations. Using the same vector notation, we set $A_x = -y$ and $A_y = x$, and applying the coordinate transformation given in expression (3.9), we obtain:

$$\bar{A}_x = -y\cos\theta + x\sin\theta$$

$$\bar{A}_y = -y \sin\theta + x \cos\theta$$

Replacing the x and y components by the transformations given in expressions (3.2), we obtain after a short and simple algebra that

$$\bar{A}_x = -\bar{y}$$

$$\bar{A}_y = \bar{x}$$

That is, $(-y, x) = (-\bar{y}, \bar{x})$ or equivalently $(A_x, A_y) = (\bar{A}_x, \bar{A}_y)$, which proves that this structure generates components of a vector.

Let us now consider the components $\vec{V} = x\hat{\imath} - y\hat{\jmath} = (x, -y)$, making an analogous development using the coordinate transformations, it can be seen that $(x, -y) \neq (\bar{x}, -\bar{y})$.

Therefore, the components $(-y, x)$ are not components of a vector since they do not transform like the components of a vector.

Let us now consider the transformations (3.4) or their equivalent (3.5) and make the following association to make a change of notation that allows us to generalize these results to more dimensions and establish more general results.

$$x \to x_1$$

$$y \to x_2$$

$$\bar{x} \to \bar{x}_1$$

$$\bar{y} \to \bar{x}_2$$

$$a_{11} = \cos\theta;\ a_{12} = \sin\theta;\ a_{21} = -\sin\theta;\ a_{22} = \cos\theta$$

Then the transformations (3.4) can be written

$$\begin{aligned} \bar{x}_1 &= a_{11}x_1 + a_{12}x_2 \\ \bar{x}_2 &= a_{21}x_1 + a_{22}x_2 \end{aligned} \tag{3.13}$$

The coefficients a_{ij} can be interpreted as cosines of direction between the $\bar{x}_i$ and x_j axes, that is,

$$a_{12} = \cos(\bar{x}_1, x_2) = \sin\theta$$

$$a_{21} = \cos(\bar{x}_2, x_1) = \cos(\theta + \pi/2) = -\sin\theta$$

$$a_{11} = \cos(\bar{x}_1, x_1) = \cos\theta$$

$$a_{22} = \cos(\bar{x}_2, x_2) = \cos\theta$$

$$a_{12} = \cos(\bar{x}_1, x_2) = \sin\theta$$

$$a_{21} = \cos(\bar{x}_2, x_1) = \cos(\theta + \pi/2) = -\sin\theta$$

$$a_{11} = \cos(\bar{x}_1, x_1) = \cos\theta$$

$$a_{22} = \cos(\bar{x}_2, x_2) = \cos\theta$$

On the other hand, from the transformations (3.13) we can observe that:

$$\frac{\partial \overline{x}_1}{\partial x_1} = a_{11}; \frac{\partial \overline{x}_1}{\partial x_2} = a_{12}; \frac{\partial \overline{x}_2}{\partial x_1} = a_{21}; \frac{\partial \overline{x}_2}{\partial x_2} = a_{22}$$

That is, $\frac{\partial \overline{x}_i}{\partial x_j} = \frac{\partial x_j}{\partial \overline{x}_i} = a_{ij}$

Therefore, we can write the transformations (3.13) as:

$$\overline{x}_i = \sum_{j=1}^{2} \frac{\partial \overline{x}_i}{\partial x_j} x_j; \; i = 1,2$$

This expression can be generalized in dimensions as follows.

3.1.1.2 A second definition of vector

The set of N quantities, A_j, is said to correspond to the components of an N-dimensional vector when the transformation of its components under a rotation operation of the coordinate axes is given by:

$$\overline{A}_i = \sum_{j=1}^{N} \frac{\partial \overline{x}_i}{\partial x_j} A_j = \sum_{j=1}^{N} \frac{\partial x_j}{\partial \overline{x}_i} A_j \tag{3.14}$$

Einstein's addition convention establishes that whenever we see repeated indices, it is understood that it is a sum, applying this convention we can write

$$\overline{A}_i = \frac{\partial \overline{x}_i}{\partial x_j} A_j = \frac{\partial x_j}{\partial \overline{x}_i} A_j$$

It can be proven that the direction cosines a_{ij} satisfy an orthogonality condition as follows:

$$\sum_i a_{ij} a_{jk} = \delta_{jk} \text{ or } \sum_i a_{ji} a_{ki} = \delta_{jk} \tag{3.15}$$

where δ_{jk} is the kronecker delta function defined by:

$$\delta_{jk} = \begin{cases} 1 \; ; \; j = k \\ 0 \; ; \; j \neq k \end{cases}$$

To verify the orthogonality of the direction cosines, in general, we use the equivalent form of the expression (3.15):

$$\sum_i \frac{\partial x_j}{\partial \overline{x}_i} \frac{\partial x_k}{\partial \overline{x}_i} = \sum_i \frac{\partial x_j}{\partial \overline{x}_i} \frac{\partial \overline{x}_i}{\partial x_k} = \frac{\partial x_j}{\partial x_k} = \delta_{jk} \tag{3.16}$$

To obtain the expression (3.16), the same property of transformation of the components given in the expression (3.14) has been used and the final result $\frac{\partial x_j}{\partial x_k}$ is equal to δ_{jk}, since x_j and x_k ($j \neq k$) have been assumed to be orthogonal for any number of dimensions. If ($j = k$), obviously the partial derivative is equal to 1.

3.1.1.3 Multiplication between vectors

Now that we have defined and understood what a vector is, we can combine them into two very useful definitions that refer to the way of multiplying two vectors, such as the scalar product and the vector product. Let's analyze the mathematical nature of both definitions.

1. **Point or scalar product of vectors**

   Let $\vec{A}$ and $\vec{B}$ be two vectors with N components each. The dot product or scalar product of two vectors is defined as:

$$\vec{A}\cdot\vec{B}=\sum_{i=1}^{N}A_iB_i$$

For simplicity and without loss of generality, let us take $N = 3$ and see the behavior of this operator under transformation of the components under rotation. For this case, let us take the transformed starting point components, so we have, expanding the summation, the following expression:

$$\vec{\bar{A}}\cdot\vec{\bar{B}}=\sum_{i=1}^{3}\bar{A}_i\bar{B}_i=\bar{A}_x\bar{B}_x+\bar{A}_y\bar{B}_y+\bar{A}_z\bar{B}_z \tag{3.17}$$

Replacing it in expression (3.17), the transformation rule for each component gives:

$$\begin{aligned}&\bar{A}_x\bar{B}_x+\bar{A}_y\bar{B}_y+\bar{A}_z\bar{B}_z\\&=\sum_i a_{xi}A_i\sum_j a_{xj}B_j+\sum_i a_{yi}A_i\sum_j a_{yj}B_j+\sum_i a_{zi}A_i\sum_j a_{zj}B_j\end{aligned} \tag{3.18}$$

Let us use the indices k and l to sum over x, y and z and writing the expression (3.18) in a more compact way, we obtain:

$$\sum_k\bar{A}_k\bar{B}_k=\sum_l\sum_i\sum_j a_{li}A_ia_{lj}B_j \tag{3.19}$$

Expression (3.19) can be reorganized as follows:

$$\sum_k\bar{A}_k\bar{B}_k=\sum_l\sum_i\sum_j(a_{li}a_{lj})A_iB_j \tag{3.20}$$

$$\sum_k\bar{A}_k\bar{B}_k=\sum_i\sum_j\delta_{ij}A_iB_j \tag{3.21}$$

$$\sum_k\bar{A}_k\bar{B}_k=\sum_i A_iB_i \tag{3.22}$$

This proves that the dot product between two vectors is an invariant under a rotation transformation of the coordinate system, that is, a number that does not change under this rotation transformation, in this way:

$$\vec{\bar{A}}\cdot\vec{\bar{B}}=\vec{A}\cdot\vec{B}$$

which shows that the point product is indeed a scalar. To arrive at expression (3.22) the orthogonality of the direction cosines (3.15) was used in expression (3.20), then the definition of the Kronecker delta was used in expression (3.21) to obtain expression (3.22).

2. **Cross or vector product of two vectors**

The cross or vector product of two vectors is defined, for simplicity and without loss of generality, in three dimensions as:

$$\vec{A}\times\vec{B}=\sum_{i=1}^{3}\left(\sum_{j,k=1}^{3}\epsilon_{i\,jk}\;a_j b_k\right)\hat{e}_i$$

or simply $\vec{A}\times\vec{B}=\vec{C}$, with

$$c_i=\sum_{j,k=1}^{3}\epsilon_{i\,jk}\;a_j b_k$$

using Einstein's convention we have:

$$c_i=\epsilon_{ijk}\;a_j\;b_k$$

where ϵ_{ijk} is the Levi-Civita symbol defined as:

$$\epsilon_{i\,jk}=\begin{cases}+1;(i,j,k)\,is\,(1,2,3),(2,3,1)\,or\,(3,1,2)\\-1;(i,j,k)\,is\,(3,2,1),(1,3,2)\,or\,(2,1,3)\\0;i=j\,or\,j=k\,or\,k=1\end{cases}$$

To prove that indeed the vector product of two vectors generates another vector, without loss of generality, let us take the transformed components of the vector product, that is:

$$\bar{C}_i=\bar{A}_j\,\bar{B}_k-\bar{A}_k\,\bar{B}_j \tag{3.23}$$

i, j, k take values in cyclic order. Replacing the transformation of the components (3.14) in the expression (3.23), we obtain:

$$\begin{aligned}\bar{C}_i&=\bar{A}_j\bar{B}_k-\bar{A}_k\bar{B}_j=\sum_i a_{jl}A_l\sum_m a_{km}B_m-\sum_l a_{kl}A_l\sum_m a_{jm}B_m\\\bar{C}_i&=\sum_{l,m}(a_{jl}a_{km}-a_{kl}a_{jm})A_lB_m\end{aligned} \tag{3.24}$$

The direction cosines given in parentheses vanish for $m=l$. Consequently, we have that j and k acquire fixed values, depending on the choice of i, and six combinations of l and m. If $i=3$, then $j=1$, $k=2$ (cyclic order) and we have the following combinations of direction cosines and their negatives.

$$\begin{aligned}a_{11}\,a_{22}-a_{21}\,a_{12}&=a_{33}\\a_{13}\,a_{21}-a_{23}\,a_{11}&=a_{32}\\a_{12}\,a_{23}-a_{22}\,a_{13}&=a_{31}\end{aligned} \tag{3.25}$$

Substituting again in equations (3.24), we obtain, for $i = 3$, the following expression:

$$\bar{C}_3 = a_{33}A_1B_2 + a_{32}A_3B_1 + a_{31}A_2B_3 - a_{33}A_2B_1 - a_{32}A_1B_3 - a_{31}A_3B_2$$
$$\bar{C}_3 = a_{31}C_1 + a_{32}C_2 + a_{33}C_3 \tag{3.26}$$
$$\bar{C}_3 = \sum_n a_{3n}C_n$$

Expression (3.26) shows that indeed the component $\bar{C}_3$ transforms as a vector, which complies with the definition of a vector (3.14). Similarly, by permuting the indices, it is proven that

$$\bar{C}_1 = \sum_n a_{1n}C_n$$

$$\bar{C}_2 = \sum_n a_{2n}C_n$$

which proves that the vector product $\vec{A} \times \vec{B} = \vec{C}$, is a vector.

1. Defining vectors as components that follow a transformation rule brings many advantages: We realize that not everything that has direction, magnitude, and sense is a vector.
2. Properties are demonstrated in a general way, in N dimensions, such as that the scalar product of vectors is a scalar, and the vector product is a vector.
3. Vector equations thus posed will be independent of the coordinate system used; it does not have to be Cartesian.
4. Due to its indexical notation, the computational implementation of algorithms to solve these equations is facilitated.
5. It is directly generalized to the concept of tensor.

3.1.2 Tensor concept

We have already seen that a quantity is a scalar when it is invariant under the transformation of coordinates by rotation, that is, it can be written in the form:

$$\bar{\varphi}(\bar{x}, \bar{y}, \bar{z}) = \varphi(x, y, z) \tag{3.27}$$

In this case we say that φ is a scalar quantity which is also called an invariant. See equations (3.12) and (3.22), where we observe that the dot or scalar product of vectors and the norm of a vector are scalar quantities or operators. This idea is extended to the concept of a scalar field, when a function, say of n dimensions, assigns a real number to each point in space, that is, φ: $R^n \rightarrow R$, and satisfies condition (3.27) then φ is called a scalar field. We also saw that a quantity is a vector if its components transform according to the expression (3.14), that is, assuming Einstein's addition convention, according to the expression:

$$\bar{A}_i = \frac{\partial \bar{x}_i}{\partial x_j} A_j = \frac{\partial x_j}{\partial \bar{x}_i} A_j \tag{3.28}$$

Note that the concept of a vector implies a rotation because there is a derivative $\frac{\partial x_j}{\partial \overline{x}_i}$, and these derivatives define the direction cosines of the rotation transformation, as already explained in previous paragraphs. Note also that a scalar quantity does not have a rotation, since in expression (3.27), there are no derivatives related to the direction cosines, because of this, scalar quantities are called rank zero tensors, vectors are called rank 1 tensors, and therefore it is natural to define rank two tensors as mathematical objects that transform their components from two rotations as we will see in the next section. These arguments justify the following definition of a vector:

Suppose that in a coordinate system $(x^1, x^2, ..., x^N)$ there are N quantities $(A^1, A^2, ..., A^N)$ related to other N quantities $(\overline{A}^1, \overline{A}^2, ..., \overline{A}^N)$ in another coordinate system $(\overline{x}^1, \overline{x}^2, ..., \overline{x}^N)$ by the transformation equations:

$$\overline{A}^i = \sum_{j=1}^{N} \frac{\partial \overline{x}^i}{\partial x^j} A^j, i = 1, ..., N$$

or using the Einstein summation convention:

$$\overline{A}^i = \frac{\partial \overline{x}^i}{\partial x^j} A^j$$

These are then called components of a contravariant vector or contravariant tensor of first rank or first order.

A convention has been introduced here, when deriving rotated coordinates with respect to non-rotated ones, a super index is placed on the components and they are called contravariant components, and likewise, when deriving non-rotated or non-transformed coordinates with respect to transformed coordinates, a subscript is placed on the components and they are called covariant components, as in the equation (3.29)

$$\overline{A}_i = \frac{\partial \overline{x}_i}{\partial x_j} A_j = \frac{\partial x_j}{\partial \overline{x}_i} A_j \tag{3.29}$$

Now in a more formal way the following definition is justified:

Suppose that in a coordinate $(x^1, x^2, ..., x^N)$ there are N quantities $(A_1, A_2, ..., A_N)$ related to other N quantities $(\overline{A}_1, \overline{A}_2, ..., \overline{A}_N)$ in another coordinate system $(\overline{x}^1, \overline{x}^2, ..., \overline{x}^N)$ by the transformation equations.

$$\overline{A}_i = \sum_{j=1}^{N} \frac{\partial x^j}{\partial \overline{x}^i} \overline{A}_j, i = 1, ..., N$$

or using the Einstein summation convention:

$$\overline{A}_i = \frac{\partial x^j}{\partial \overline{x}^i} A_j \tag{3.30}$$

These quantities are called covariant vectors or covariant tensors of the first rank or first order. Instead of saying a tensor whose components are A^i or A_i, the literature often says the tensor A^i or A_i, without any confusion. In a Cartesian coordinate system, in a Euclidean space, there is no difference between contravariant or covariant vectors or tensors. These aspects of expressing a tensor in a basis and in its reciprocal basis become very relevant in the theory of general relativity and in some aspects of the theory of plasticity and viscoelasticity in the study of the mechanical properties of materials.

3.2 Second- and higher order

A generalization to second-order or rank tensors is straightforward and they are defined as follows:

Suppose that in a coordinate system $(x^1, x^2, ..., x^N)$ there are N^2 quantities A^{ij} related to other N^2 quantities $\overline{A}^{pr}$ in another coordinate system $(\overline{x}^1, \overline{x}^2, ..., \overline{x}^N)$ by the transformation equations

$$\overline{A}^{pr} = \sum_{j=1}^{N}\sum_{i=1}^{N} \frac{\partial \overline{x}^p}{\partial x^i}\frac{\partial \overline{x}^r}{\partial x^j} A^{ij};\ p, r = 1, 2, ..., N$$

or also

$$\overline{A}^{pr} = \frac{\partial \overline{x}^p}{\partial x^i}\frac{\partial \overline{x}^r}{\partial x^j} A^{ij} \tag{3.31}$$

According to the adopted conventions, they are called contravariant components of a second-rank tensor or rank two.

Similarly, let us take coordinate system $(x^1, x^2, ..., x^N)$ in which N^2 quantities A_{ij} are related to other N^2 quantities $\overline{A}_{pr}$ in another coordinate system $(\overline{x}^1, \overline{x}^2, ..., \overline{x}^N)$ by means of the transformation equations

$$\overline{A}_{pr} = \sum_{j=1}^{N}\sum_{i=1}^{N} \frac{\partial \overline{x}^p}{\partial x^i}\frac{\partial \overline{x}^r}{\partial x^j} A_{ij};\ p, r = 1, 2, ..., N$$

or also

$$\overline{A}_{pr} = \frac{\partial \overline{x}^p}{\partial x^i}\frac{\partial \overline{x}^r}{\partial x^j} A_{ij} \tag{3.32}$$

They are called covariant components of a second rank or rank two tensors, also often called the component of a covariant tensor of order two.

Similarly in the same coordinate systems if the N^2 quantities A^i_j are called components of a second rank mixed tensor if

$$\overline{A}^p_r = \frac{\partial \overline{x}^p}{\partial x^i}\frac{\partial x^j}{\partial \overline{x}^r} A^i_j \tag{3.33}$$

It is easy to generalize the definition to higher rank tensors in the following way, the expression:

$$S = S^{i_1 i_2 \ldots i_p}_{j_1 j_2 \ldots j_q}$$

It defines a mixed tensor of order $p + q$, contravariant of order p and covariant of order q. For example, in the case of S^{qst}_{kl}, it represents the components of a tensor of rank 5, contravariant of order 3 and covariant of order 2 and they are transformed according to the relations:

$$\bar{A}^{pm}_{i\,j} = \frac{\partial \bar{x}^p}{\partial x^q} \frac{\partial \bar{x}^r}{\partial x^s} \frac{\partial \bar{x}^m}{\partial x^t} \frac{\partial x^k}{\partial \bar{x}^i} \frac{\partial x^l}{\partial \bar{x}^j} A^{qst}_{kl} \tag{3.34}$$

3.3 Tensor algebra

Before entering the consideration of algebraic operations, aspects of calculations, and matrix representation of tensors, let us define a tensor that is very useful called the Kronecker Delta, it is denoted δ^i_j and is defined as:

$$\delta^i_j = \begin{cases} 0, & if\ i \neq j \\ 1, & if\ i = j \end{cases} \tag{3.35}$$

As its notation indicates, it is a second-rank mixed tensor.

Let's see with simple examples how this tensor operates, consider the following example for $j = 1, 2$ in the following summation

$$\begin{aligned} \sum_{i=1}^{2} a_{ij}\delta^i_j &= a_{1j}\delta^1_j + a_{2j}\delta^2_j \\ &= a_{11}\delta^1_j + a_{2j}\delta^2_j \\ &= a_{11}\delta^1_1 = a_{11}, \quad if\ j = 1 \\ &= a_{22}\delta^2_2 = a_{22}, \quad if\ j = 2 \end{aligned}$$

since $\delta^2_1 = \delta^1_2 = 0$ by its definition.
That is,

$$\sum_{i=1}^{2} a_{i\,j}\delta^i_j = a_{jj}$$

for $j = 1$ y $j = 2$.

Note that what the Kronecker delta did was destroy the summation and change the index i to j. So, the Kronecker deltas do is destroy summations and change the contracted or repeated indices to the other index, for example, if it had been

$$\sum_{i=1}^{2} a_{i\,j}\delta^i_k = a_{k\,j}$$

Let's see another example, for $k = 1$, $k = 2$, in the following summation.

Note that it is not summed over k, but over i and j, that is why the sum is taken separately for when $k = 1$ and $k = 2$

$$\begin{aligned}\sum_{i=1}^{2}\sum_{i=1}^{2} a_{i_j}\delta_k^i &= \sum_{i=1}^{2}(a_{i1}\delta_k^i + a_{i2}\delta_k^i) \\ &= a_{11}\delta_k^1 + a_{12}\delta_k^1 + a_{21}\delta_k^2 + a_{22}\delta_k^2 \\ &= a_{11}\delta_1^1 + a_{12}\delta_1^1 = a_{11} + a_{12}, \quad k = 1 \\ &= a_{21}\delta_2^2 + a_{22}\delta_2^2 = a_{21} + a_{22}, \quad k = 2\end{aligned}$$

therefore,

$$\sum_{i=1}^{2}\sum_{i=1}^{2} a_{i\,j}\delta_k^i = a_{11} + a_{12} = \sum_{j=1}^{2} a_{1j}, \quad k = 1$$

$$\sum_{i=1}^{2}\sum_{i=1}^{2} a_{i\,j}\delta_k^i = a_{21} + a_{22} = \sum_{j=1}^{2} a_{2j}, \quad k = 2$$

So, we can write in summary form

$$\sum_{i=1}^{2}\sum_{i=1}^{2} a_{i\,j}\delta_k^i = \sum_{j=1}^{2} a_{k\,j}$$

For $k = 1, k = 2$.

In this way, we observe again that the Kronecker delta destroys a summation sum and changes the repeated index i to k. If we make use of the summation convention, we can write the previous examples in a general form as

$$a_{ij}\,\delta_j^i = a_{jj}$$

$$a_{ij}\,\delta_k^i = a_{kj}$$

This is valid regardless of the order and dimension of the tensors.

The basic algebraic operations between tensors are defined as follows.

Addition (subtraction) of two or more tensors is defined only with tensors of the same rank and type (i.e., the same number of contravariant indices and the same number of covariant indices) and is also a tensor of the same rank and type. So, if we have tensors $S_{j_1 j_2 \cdots j_q}^{i_1 i_2 \ldots i_p}$ and $P_{j_1 j_2 \cdots j_q}^{i_1 i_2 \ldots i_p}$, then

$$T_{j_1 j_2 \cdots j_q}^{i_1 i_2 \ldots i_p} = S_{j_1 j_2 \cdots j_q}^{i_1 i_2 \ldots i_p} \pm P_{j_1 j_2 \cdots j_q}^{i_1 i_2 \ldots i_p} \tag{3.36}$$

It is also a tensor. Tensor addition obeys the properties of commutativity and associativity.

Outer product. The product of two tensors is a tensor whose rank is the sum of the ranks of the given tensors. This product, which involves ordinary multiplication of the tensor components, is called an outer product. Thus, if we have tensors $S_{j_1 j_2 \cdots j_q}^{i_1 i_2 \ldots i_p}$ and $P_{j_1 j_2 \cdots j_q}^{i_1 i_2 \ldots i_p}$, then

$$T_{j_1 j_2 \cdots j_{2q}}^{i_1 i_2 \ldots i_{2p}} = S_{j_1 j_2 \cdots j_q}^{i_1 i_2 \ldots i_p}\, P_{j_1 j_2 \cdots j_q}^{i_1 i_2 \ldots i_p} \tag{3.37}$$

A particular case would be to consider the tensors S^{mp}_{q} and P^{abc}_{d}, then

$$S^{mp}_{q} P^{abc}_{d} = T^{mpabc}_{qd}$$

Note that a rank 3 tensor is multiplied by a rank 4 tensor and a rank 7 tensor is generated. However, not every tensor can be decomposed into the product of two tensors of lower rank, so division between tensors is not always possible.

Contraction. If a contravariant and a covariant index of a tensor are equated, the result indicates that a summation over the equal indices is to be taken according to the summation convention. This resulting sum is a tensor of rank two times smaller than the original tensor. The process is called contraction. For example, in the rank 7 tensor, T^{mpabc}_{qd}, we set $m = q$, so

$$T^{mpabc}_{qd} = T^{mpabc}_{md} = T^{pabc}_{d} \tag{3.38}$$

Thus, the contraction of the m indices transform a rank 7 tensor into a rank 5 tensor.

Inner product. By the process of the outer product of two tensors followed by a contraction, we obtain a new tensor called the inner product of the given tensors. The process is called inner multiplication. For example, given the tensors S^{mp}_{q} and P^{abc}_{d} by multiplying them and making $m = q$ and $d = a$, we obtain another tensor of rank 4 times smaller, for each contraction or what is the same, equalization of two indices, the tensor decreases its rank by two, in this way,

$$S^{mp}_{q} P^{abc}_{d} = S^{mp}_{m} P^{abc}_{a} = S^{p} P^{bc} = T^{pbc} \tag{3.39}$$

Note that the inner product of a rank 3 tensor by a rank 4 tensor produces a rank 3 tensor.

Quotient law. Suppose it is not known whether a quantity X is a tensor or not. If an inner product of X with an arbitrary tensor is a tensor, then X is also a tensor. This is called the quotient law.

3.4 Cartesian tensors

Let x^i represent a rectangular Cartesian coordinate system in a three-dimensional Euclidean space, and φ^i represent any curvilinear or rectangular coordinate system (i.e., cylindrical or spherical coordinates) in the same space. The vector $\vec{x}$ having Cartesian components x^i is called the position vector of the arbitrary point $P(x^1, x^2, x^3)$ relative to the rectangular Cartesian axes. The square of the differential element of the distance between two very close points $P(\vec{x})$ and $Q(\vec{x} + d\vec{x})$ will be

$$(ds)^2 = dx^i\, dx^i \tag{3.40}$$

From the transformation of coordinates

$$x^i = x^i\,(\varphi^1, \varphi^2, \varphi^3) \tag{3.41}$$

Relating the two systems, the differential distance is:

$$dx^i = \frac{\partial x^i}{\partial \varphi^p} d\varphi^p \tag{3.42}$$

And therefore, the expression (3.40) can be written as:

$$(ds)^2 = \frac{\partial x^i}{\partial \varphi^p}\frac{\partial x^i}{\partial \varphi^q} d\varphi^p d\varphi^q = g_{pq} d\varphi^p d\varphi^q \tag{3.43}$$

where

$$g_{pq} = \frac{\partial x^i}{\partial \varphi^p}\frac{\partial x^i}{\partial \varphi^q} \tag{3.44}$$

g_{pq} defined thus is a covariant second order tensor, which is usually called the metric tensor or fundamental tensor of the space. φ^i represents a rectangular Cartesian system, say the system x'^i, then:

$$g_{pq} = \frac{\partial x^i}{\partial x'^p}\frac{\partial x^i}{\partial x'^q} = \delta_{pq} \tag{3.45}$$

where δ_{pq} is the Kronecker delta already defined in (3.35).

Any coordinate system for which the squared distance differential element takes the form of (3.40) is called a homogeneous coordinate system. Coordinate transformations between homogeneous systems are orthogonal transformations, and when we consider such transformations, tensors so defined are called Cartesian tensors and are transformed according to the transformation laws for tensors given in Sections 3.1 and 3.2. This is the case for the transformation laws between rectangular Cartesian coordinate systems with a common origin. As mentioned in Section 3.1, for Cartesian tensors there is no distinction between covariant and contravariant components and therefore it is usual to use only subscripts in expressions representing Cartesian tensors.

3.5 Principal values and directions of second-order symmetric tensors

Let us consider a second-order symmetric Cartesian tensor T_{ij}, i.e., $T_{ij} = T_{ji}$. For each symmetric tensor T_{ij} defined at some point in space, there is a vector given by the inner product that is associated with each direction, specified by a unit normal vector n_i at that point, i.e.:

$$v_i = T_{ij} n_j \tag{3.46}$$

The tensor has a matrix representation; therefore, it can be considered as a linear operator that originates the vector v_i associated with the direction n_i. If the direction is such that v_i is parallel to n_i, the inner product can be expressed as a scalar multiple of n_i. In this case then we can write:

$$T_{ij} n_j = \lambda n_i \tag{3.47}$$

and the direction n_i is called the principal direction or principal axis of T_{ij}. Considering that $n_i = \delta_{ij}\, n_j$, equation (3.47) can be written as:

$$(T_{ij} - \lambda\delta_{ij})\, n_j = 0 \tag{3.48}$$

Expression (3.48) represents a system of three equations for the four unknowns, n_i and λ, associated with each principal direction. The system to be solved explicitly is:

$$\begin{aligned} (T_{11} - \lambda)n_1 + T_{12}\, n_2 + T_{13}\, n_3 &= 0 \\ T_{21}\, n_1 + (T_{22} - \lambda)n_2 + T_{23}\, n_3 &= 0 \\ T_{31}\, n_1 + T_{32}\, n_2 + (T_{33} - \lambda)n_3 &= 0 \end{aligned} \tag{3.49}$$

If the $n_i = 0$, it is the trivial solution, it satisfies the equations for any value of λ, but it is a matter of obtaining non-trivial solutions. To find these non-trivial solutions, the condition that $n_i n_i = 1$ is imposed for the system of equations (3.49), due to the homogeneity of the system, there is no loss of generality by imposing this restriction, which is what generates mathematical consistency, since in this way we would have 4 equations with four unknowns.

On the other hand, for the system of equations (3.49) to have non-trivial solutions, the determinant of the coefficients must be zero, that is:

$$|T_{ij} - \lambda\delta_{ij}| = 0 \tag{3.50}$$

The development of this determinant leads to a cubic polynomial in λ,

$$\lambda^3 - \mathrm{I_T}\,\lambda^2 + \mathrm{II_T}\,\lambda - \mathrm{III_T} = 0 \tag{3.51}$$

Which is known as the characteristic equation of T_{ij} and for which the scalar coefficients are:

$$\mathrm{I_T} = T_{ii} = trT_{ij} \tag{3.52}$$

trT_{ij} means the trace of T_{ij}

$$\mathrm{II_T} = \frac{1}{2}(T_{ii}T_{jj} - T_{ij}T_{ij}) \tag{3.53}$$

$$\mathrm{III_T} = |T_{ij}| = \det\,(T_{ij}) \tag{3.54}$$

The expressions (3.52), (3.53), (3.54) are called respectively, the first, second, and third invariants of T_{ij}. The three roots of (3.51) which we will denote as $\lambda_{(1)}$, $\lambda_{(2)}$, $\lambda_{(3)}$, are called the principal values of T_{ij}. Since we have assumed that T_{ij} is symmetric in real components, the principal values are also real; and if these values are distinct, the principal directions are mutually orthogonal. When referred to their principal axes, both the tensor and its matrix representation appear in the diagonal form as follows:

$$T = \begin{pmatrix} \lambda_{(1)} & 0 & 0 \\ 0 & \lambda_{(2)} & 0 \\ 0 & 0 & \lambda_{(3)} \end{pmatrix} \quad (3.55)$$

If $\lambda_{(1)} = \lambda_{(2)}$, the tensor is independent of the choice of the axes of $\lambda_{(1)}$ and $\lambda_{(2)}$, once the principal axis associated with $\lambda_{(3)}$ has been chosen.

If all principal values are equal, any direction is a principal direction. It is customary to order the principal values in the following way: $\lambda_{(I)} > \lambda_{(II)} > \lambda_{(III)}$.

Let us denote the principal axes as: $O\bar{x}_1\bar{x}_2\bar{x}_3$, the transformation from the Cartesian axes $Ox_1x_2x_3$ is given by the elements of Table 3.1.

Table 3.1. Transformation matrix for the principal axes with respect to the $Ox_1x_2x_3$ system.

	x_1	x_2	x_3
$\bar{x}_1$	$a_{11} = n_1^{(1)}$	$a_{12} = n_2^{(1)}$	$a_{13} = n_3^{(1)}$ (3.56)
$\bar{x}_2$	$a_{21} = n_1^{(2)}$	$a_{22} = n_2^{(2)}$	$a_{23} = n_3^{(2)}$
$\bar{x}_3$	$a_{31} = n_1^{(3)}$	$a_{32} = n_2^{(3)}$	$a_{33} = n_3^{(3)}$

Where the $n_i^{(j)}$ are the cosines of the j-th principal directional.

3.6 Hamilton-Cayley equation

In different applications of semiconductor physics, materials science, optics, elasticity theory, plasticity theory, among others, it is necessary to know the powers of the tensors, to find powers, for example, the square of the tensor T_{ij} is obtained by the inner product of $T_{ik}\,T_{kj}$, the cube by $T_{ik}\,T_{km}\,T_{mj}$ and so on. Therefore, when the tensor T_{ij} is in the diagonal form (3.55), the n-th power can be written:

$$T^n = \begin{pmatrix} (\lambda_{(1)})^n & 0 & 0 \\ 0 & (\lambda_{(2)})^n & 0 \\ 0 & 0 & (\lambda_{(3)})^n \end{pmatrix} \quad (3.57)$$

Expressions (3.57) and (3.55) indicate that T_{ij} and all its integer powers have the same principal axes. Since each principal direction satisfies the polynomial (3.51) and due to the diagonal form of the tensor T^n given by (3.57), the tensor T itself will satisfy the polynomial (3.51), i.e.:

$$T^3 - \mathrm{I_T}\,T^2 + \mathrm{II_T}\,T - \mathrm{III_T}\,J = 0 \quad (3.58)$$

where J is the identity matrix. The polynomial (3.58) is known as the Hamilton-Cayley equation. Multiplying both sides of the polynomial (3.58) by T gives the equation:

$$T^4 = \mathrm{I_T}\,T^3 - \mathrm{II_T}\,T^2 + \mathrm{III_T}\,T \quad (3.59)$$

Substituting T^3 from equation (3.58) into equation (3.59) gives:

$$T^4 = (\mathrm{I}_\mathrm{T}^2 - \mathrm{II}_\mathrm{T})T^2 + (\mathrm{III}_\mathrm{T} - \mathrm{I}_\mathrm{T}\,\mathrm{II}_\mathrm{T})T + \mathrm{I}_\mathrm{T}\,\mathrm{III}_\mathrm{T}\,J \tag{3.60}$$

Continuing the same procedure, positive powers of the tensor T are obtained as linear combinations of T^2, T and J.

3.7 Tensor fields

A scalar field attaches or associates a number to each point in space, a vector field associates with a vector, in general, a tensor field associates with a tensor $T(\vec{x}, t)$ to each point in space for a given time $(\vec{x}, t)$, where the position vector $\vec{x}$ varies in a particular region of space and time t varies in a particular interval of time. A tensor field is continuous and differentiable when all its components $T(\vec{x}, t)$ are continuous or differentiable functions of $\vec{x}$ and t. If the components of the field do not depend on time but only on the position vector, $\vec{x}$ then the tensor field is said to be stationary.

In a rectangular Cartesian coordinate system, the position vector of an arbitrary point can be written as:

$$\vec{x} = x_i \hat{e}_i \tag{3.61}$$

In general, tensor fields of different orders can be written in indexical or symbolic form as shown in Table 3.2.

Table 3.2. Tensor fields of different orders.

Tensor field name	Indexical form	Symbolic form	Tensor field order
Scalar field	$\phi = \phi(x_i, t)$	$\phi = \phi(\vec{x}, t)$	Zero
Vector field	$v_i = v_i(\vec{x}, t)$	$v = v(\vec{x}, t)$	One (3.62)
Tensor field of order two	$T_{ij} = T_{ij}(\vec{x}, t)$	$T = T(\vec{x}, t)$	Two

The differentiation of the components of the tensor field with respect to the coordinates x_i is written as $\partial / \partial x_i$ or with a shorter notation as ∂_i and is represented by the nabla operator ∇ defined as:

$$\nabla = \hat{e}_i \frac{\partial}{\partial x_i} = \hat{e}_i \partial_i \tag{3.63}$$

The partial derivative with respect to the variable x_i is often represented by the so-called subscript convention, with a comma, as shown in Table 3.3.

Table 3.3. Derivatives of tensor fields with subscript notation.

$\frac{\partial \phi}{\partial x_i} = \phi_{,i}$	$\frac{\partial v_i}{\partial x_i} = v_{i,i}$
$\frac{\partial v_i}{\partial x_j} = v_{i,j}$	$\frac{\partial T_{ij}}{\partial x_k} = T_{ij,k}$ (3.64)
$\frac{\partial^2 v_i}{\partial x_j \partial x_k} = v_{i,jk}$	$\frac{\partial^2 T_{ij}}{\partial x_k \partial x_m} = v_{ij,km}$

In these examples, we see that the operator ∂_i generates a higher order tensor if the index i remains as a free index and a lower order tensor if the index i becomes a dummy index in the derivative.

Differential operators are fundamental in all physics and are written in tensor notation as shown in Table 3.4.

Table 3.4. Differential operators in tensor notation.

Name of the operator	**Tensor form**	**Symbolic form**
Gradient	$\partial_i \phi = \phi_{,i}$	$\nabla \phi = \frac{\partial \phi}{\partial x_i} \hat{e}_i$
Divergence	$\partial_i v_i = v_{i,i}$	$div\, \vec{v} = \nabla \cdot \vec{v}$ (3.65)
Rotational	$\epsilon_{ijk}\, \partial_j\, v_k = \epsilon_{ijk}\, v_{k,j}$	$rot\, \vec{v} = \nabla \times \vec{v}$
Laplacian	$\partial_{ii}\, \phi = \phi_{,ii}$	$\nabla^2 \phi = \nabla \cdot \nabla \phi$

3.8 Quadratic transformation forms

Quadratic transformation forms are useful for establishing transformation rules for higher rank tensors from transformation rules for second rank tensors. Assuming these second rank tensors to be symmetric, we can make use of the linear transformation law:

$$S' = ASA' = \begin{pmatrix} a_{1} & a_{12} & a_{13} \\ a_{21} & a_{22} & a_{23} \\ a_{31} & a_{32} & a_{33} \end{pmatrix} \begin{pmatrix} S_{11} & S_{12} & S_{31} \\ S_{12} & S_{22} & S_{23} \\ S_{31} & S_{23} & S_{33} \end{pmatrix} \begin{pmatrix} a_{11} & a_{21} & a_{31} \\ a_{12} & a_{22} & a_{32} \\ a_{13} & a_{23} & a_{33} \end{pmatrix} \qquad (3.66)$$

where A' is the transposed matrix of the matrix A.

Carrying out the indicated multiplication and arranging the components of S' and S into (6 × 1) column matrices permit equation (3.66) to be represented in the quadratic form

$$\begin{bmatrix} S_1' \\ S_2' \\ S_3' \\ S_4' \\ S_5' \\ S_6' \end{bmatrix} = \begin{bmatrix} a_{11}^2 & a_{12}^2 & a_{13}^2 & 2a_{12}a_{13} & 2a_{13}a_{11} & 2a_{11}a_{12} \\ a_{21}^2 & a_{22}^2 & a_{23}^2 & 2a_{22}a_{23} & 2a_{23}a_{21} & 2a_{21}a_{22} \\ a_{31}^2 & a_{32}^2 & a_{33}^2 & a_{32}a_{33} & 2a_{33}a_{31} & 2a_{31}a_{32} \\ a_{21}a_{31} & a_{22}a_{32} & a_{23}a_{33} & a_{22}a_{33}+a_{23}a_{32} & a_{21}a_{33}+a_{23}a_{31} & a_{21}a_{32}+a_{22}a_{31} \\ a_{31}a_{11} & a_{32}a_{12} & a_{33}a_{13} & a_{12}a_{33}+a_{13}a_{32} & a_{11}a_{33}+a_{13}a_{31} & a_{11}a_{32}+a_{12}a_{31} \\ a_{11}a_{21} & a_{12}a_{22} & a_{13}a_{23} & a_{12}a_{23}+a_{13}a_{22} & a_{11}a_{23}+a_{13}a_{21} & a_{11}a_{22}+a_{12}a_{21} \end{bmatrix} \begin{bmatrix} S_1 \\ S_2 \\ S_3 \\ S_4 \\ S_5 \\ S_6 \end{bmatrix} \quad (3.67)$$

Or simply in symbolic matrix notation

$$S' = \gamma S \quad (3.68)$$

where

$$S = \begin{bmatrix} S_{11} \\ S_{22} \\ S_{33} \\ S_{23} \\ S_{31} \\ S_{12} \end{bmatrix} = \begin{bmatrix} S_1 \\ S_2 \\ S_3 \\ 2S_4 \\ 2S_5 \\ 2S_6 \end{bmatrix} = \begin{bmatrix} S_1 \\ S_2 \\ S_3 \\ S_4 \\ S_5 \\ S_6 \end{bmatrix} \quad (3.69)$$

Here, γ is a (6 × 6) matrix called the quadratic transformation matrix for a symmetrical second-rank tensor.

$$\gamma = \begin{bmatrix} a_{11}^2 & a_{12}^2 & a_{13}^2 & 2a_{12}a_{13} & 2a_{13}a_{11} & 2a_{11}a_{12} \\ a_{21}^2 & a_{22}^2 & a_{23}^2 & 2a_{22}a_{23} & 2a_{23}a_{21} & 2a_{21}a_{22} \\ a_{31}^2 & a_{32}^2 & a_{33}^2 & a_{32}a_{33} & 2a_{33}a_{31} & 2a_{31}a_{32} \\ a_{21}a_{31} & a_{22}a_{32} & a_{23}a_{33} & a_{22}a_{33}+a_{23}a_{32} & a_{21}a_{33}+a_{23}a_{31} & a_{21}a_{32}+a_{22}a_{31} \\ a_{31}a_{11} & a_{32}a_{12} & a_{33}a_{13} & a_{12}a_{33}+a_{13}a_{32} & a_{11}a_{33}+a_{13}a_{31} & a_{11}a_{32}+a_{12}a_{31} \\ a_{11}a_{21} & a_{12}a_{22} & a_{13}a_{23} & a_{12}a_{23}+a_{13}a_{22} & a_{11}a_{23}+a_{13}a_{21} & a_{11}a_{22}+a_{12}a_{21} \end{bmatrix}$$

Bibliography

Clayton, J.D. (2019). Applications of tensor analysis in continuum mechanics.
Koranga, B.S. and Padaliya, S.K. (2022). An Introduction to Tensor Analysis. River Publishers.
Young, E.C. (2017). Vector and Tensor Analysis. CRC Press.

CHAPTER 4

Introduction to Elasticity Theory

INTRODUCTION

In materials science, when investigating the behavior of matter against external factors, it has been useful in a wide area not to consider the molecular structure of the material but to think about the material like an everything. With this approach, the Quantum physics in relation to atomic and molecular structure, instead it is assumed that the material is distributed continuously throughout its volume, completely filling the space it occupies, this is called the continuous medium hypothesis.

Even within the limitations that this hypothesis may imply, it has been very useful to investigate the behavior of materials in solid, liquid and gaseous states. The continuum hypothesis implies that the mathematical description of the behavior of a material, field quantities such as stresses, displacements and strains, are expressed as continuous functions for intervals of the coordinates in space and time and that the initial and boundary conditions are well defined. In this chapter, the hypothesis in the stress and strain tensor is deepened, in addition to the conservation laws and Hooke's generalized law for linear elasticity, as well as some mathematical modeling that derives from it.

4.1 Stress tensor

In this section we are going to establish the difference between mass force, surface force and volume or volumetric force and study the main properties and characteristics of the stress tensor. The forces that act on all volume elements of a continuous medium are known as mass forces, for example, gravitational or inertial forces are mass forces. We are going to represent these forces by the symbol B_i (force per unit mass), or as p_i (force per unit volume), which are related through density (ρ) by equation (4.1):

$$\rho B_i = p_i \tag{4.1}$$

Surface forces are those that act on a surface element, whether a portion of the boundary surface of the continuous material or an arbitrary internal surface and are

DOI: 10.1201/9781003344438-4

represented by the symbol f_i (force per unit area). Contact forces between two solids are an example of this type of surface forces.

4.1.1 Cauchy Stress Principle

Cauchy's stress principle states that the change in surface forces on a portion of the surface of the continuous material, denoted as $\Delta f_i/\Delta S_i$, infinitesimally does not diverge in the limit when ΔS_i tends to zero at a point P arbitrary, which is taken as a reference to establish a vector normal to the surface, while at the same time the change in forces, Δf_i, is nullified by taking the limit.

This limit defined as df_i/dS generates a resultant vector (force per unit area) and is called the stress vector and is denoted as $t_i^{(\hat{n})}$, or vectorically as $t^{(\hat{n})}$. This notation highlights the fact that the stress vector at a given point P of the continuum medium depends explicitly on the surface element ΔS chosen and represented by the unit vector n_i or vectorially as $\hat{n}$. For a differently oriented surface element, which has a different unit normal vector, the stress vector associated with P will also be different.

The stress vector that arises from the action of the matter inside the volume through the ΔS considered at point P and on the exterior matter is the vector that is a consequence of Newton's third law of action-rection, this implies that if the force goes outwards it is equivalent if it had been exerted in the negative direction of the unit normal vector to the surface, that is:

$$-t_i^{(\hat{n})} = t_i^{(-\hat{n})} \tag{4.2}$$

At an arbitrary point P of a continuous medium, the Cauchy stress principle associates a stress vector $t_i^{(\hat{n})}$ to each unit normal vector n_i which represents the orientation of an infinitesimal surface element that contains P as an interior point. All possible pairs define the stress state at this point. Fortunately, it is not necessary to specify every pair of vectors, stress and normal to the plane, to fully describe the stress state at a given point. This can be achieved by specifying the stress vector in each of three mutually perpendicular planes that intersect at P. Then, the coordinate transformation equations serve to relate the stress vector of any other plane that passes through the point P, with the three given plans. Each of the three stress vectors associated with the coordinate planes can be written in terms of their Cartesian components as shown in equations (4.3):

$$\begin{aligned}
t^{(\hat{e}_1)} &= t_1^{(\hat{e}_1)}\,\hat{e}_1 + t_2^{(\hat{e}_1)}\,\hat{e}_2 + t_3^{(\hat{e}_1)}\,\hat{e}_3 = t_j^{(\hat{e}_1)}\,\hat{e}_j \\
t^{(\hat{e}_2)} &= t_1^{(\hat{e}_2)}\,\hat{e}_1 + t_2^{(\hat{e}_2)}\,\hat{e}_2 + t_3^{(\hat{e}_2)}\,\hat{e}_3 = t_j^{(\hat{e}_2)}\,\hat{e}_j \\
t^{(\hat{e}_3)} &= t_1^{(\hat{e}_3)}\,\hat{e}_1 + t_2^{(\hat{e}_3)}\,\hat{e}_2 + t_3^{(\hat{e}_3)}\,\hat{e}_3 = t_j^{(\hat{e}_3)}\,\hat{e}_j
\end{aligned} \tag{4.3}$$

The nine components of the stress vector are:

$$t_j^{(\hat{e}_i)} = \sigma_{ij} \tag{4.4}$$

They define nine components of a tensor called the stress tensor, the explicit tensor denoted is denoted Σ such that the representation of its components and its matrix form are written explicitly as:

$$\Sigma = \begin{pmatrix} \sigma_{11} & \sigma_{12} & \sigma_{13} \\ \sigma_{21} & \sigma_{22} & \sigma_{23} \\ \sigma_{31} & \sigma_{32} & \sigma_{33} \end{pmatrix} \tag{4.5}$$

or

$$\left[\sigma_{ij}\right] = \begin{bmatrix} \sigma_{11} & \sigma_{12} & \sigma_{13} \\ \sigma_{21} & \sigma_{22} & \sigma_{23} \\ \sigma_{31} & \sigma_{32} & \sigma_{33} \end{bmatrix} \tag{4.6}$$

Regarding the three coordinate planes, the components of the stress tensor can be presented graphically, as indicated in Figure 4.1. The components perpendicular to the planes (σ_{11}, σ_{22}, σ_{33}) are called normal stresses. Those components that act tangent to the planes (σ_{12}, σ_{13}, σ_{21}, σ_{23}, σ_{31}, σ_{32}) are called shear or shear stresses. A stress is positive when it acts in the same direction as the coordinate axes and in a plane whose exterior normal points towards one of the positive directions of the coordinate axes.

The component σ_{ij} represents the component of the stress tensor that acts in a plane whose normal is parallel to the *i*-th coordinate axis and in the *j*-th direction, in other words, the *i* defines the plane of application and the *j* the direction in which it is applied, as shown in Figure 4.1.

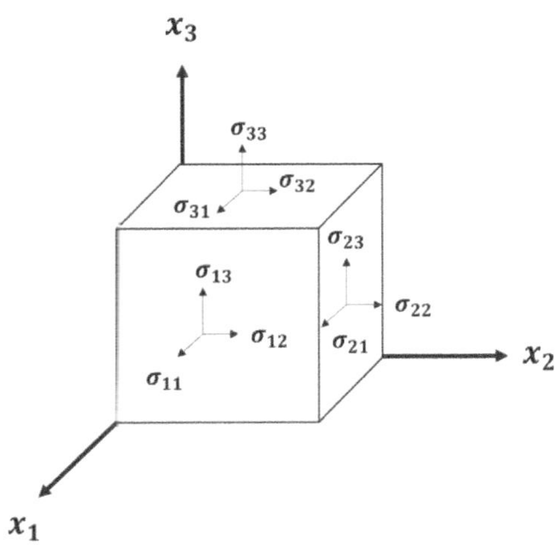

Figure 4.1. Components of the stress tensor.

In general, and compact form, the stress vector can be written as a function of the stress tensor as follows:

$$t_i^{(\hat{n})} = \sigma_{ij}\, n_j \tag{4.7}$$

where n_j represents the components of the normal vector in direction j, in vector-tensor notation it is written as:

$$t^{(\hat{n})} = \hat{n} \cdot \Sigma \tag{4.8}$$

or in matrix form:

$$\left[t_{1j}^{(\hat{n})}\right] = \left[n_{1k}\right]\left[\sigma_{kj}\right] \tag{4.9}$$

which we can write explicitly as follows:

$$\left[t_1^{(\hat{n})}, t_2^{(\hat{n})}, t_3^{(\hat{n})}\right] = \left[n_1, n_2, n_3\right] \begin{bmatrix} \sigma_{11} & \sigma_{12} & \sigma_{13} \\ \sigma_{21} & \sigma_{22} & \sigma_{23} \\ \sigma_{31} & \sigma_{32} & \sigma_{33} \end{bmatrix} \tag{4.10}$$

This system is equivalent to the system of equations given in (4.11):

$$\begin{aligned} t_1^{(\hat{n})} &= n_1\, \sigma_{11} + n_2\, \sigma_{21} + n_3\, \sigma_{31} \\ t_2^{(\hat{n})} &= n_1\, \sigma_{12} + n_2\, \sigma_{22} + n_3\, \sigma_{31} \\ t_3^{(\hat{n})} &= n_1\, \sigma_{13} + n_2\, \sigma_{23} + n_3\, \sigma_{33} \end{aligned} \tag{4.11}$$

4.1.2 Force and moment balance equations

The equilibrium conditions for an arbitrary volume V of a continuous medium, subjected to a system of surface forces $t_i^{(\hat{n})}$ and mass forces B_i (including inertial forces), require that the force and the resulting moments acting on the volume are zero, therefore, the sum of the mass and surface forces gives rise to the integral equation (4.12):

$$\int_S t_i^{(\hat{n})} dS + \int_V \rho B_i dV = 0 \tag{4.12}$$

Note that stress has units of force per unit area, so in the first integral the stress multiplied by a differential of area generates units of force, and in the second integral the density multiplied by the differential of volume provides units of mass, and B is force per unit of mass, in this way the second integral also represents a force as it should be, thinking in this way that this type of integrals are constructed.

Substituting $t_i^{(\hat{n})}$ for $\sigma_{ij}\, n_j$ in equation (4.12) and converting the resulting surface integral to a volume integral using the Gauss divergence theorem, we obtain equation (4.13):

$$\int_V (\sigma_{ji,i} + \rho B_i) dV = 0 \tag{4.13}$$

Since the volume V is arbitrary, the integrand of equation (4.13) is identically zero, this implies that:

$$\sigma_{ji,i} + \rho B_i = 0 \tag{4.14}$$

Equations (4.14) are called the equilibrium equations.

In the absence of distributed moments, the equilibrium of the moments with respect to the origin requires that the given condition be met in equation (4.15),

$$\int_S \epsilon_{ijk} \, x_j t_k^{(\hat{n})} dS + \int_V \epsilon_{ijk} \, x_j \rho B_k dV = 0 \tag{4.15}$$

In which x_i are the components of the position vector of the surface and volume elements. By again substituting $t_i^{(\hat{n})}$ for σ_{ij} n_j, and applying the Gauss divergence theorem in equation (4.15) to convert the surface integral to a volume integral, and then combining the two volume integrals, the condition given in equation (4.16) results.

$$\int_V \epsilon_{ijk} \, \sigma_{jk} dV = 0 \tag{4.16}$$

Since the integration volume V in the integral (4.16) is arbitrary, this implies that the integrand of (4.16) is zero, as shown by equation (4.17)

$$\epsilon_{ijk} \, \sigma_{jk} = 0 \tag{4.17}$$

The components of equation (4.17) are represented in the equations given in expression (4.18)

$$\sigma_{12} = \sigma_{21}, \sigma_{23} = \sigma_{32}, \sigma_{13} = \sigma_{31} \tag{4.18}$$

or in compact form

$$\sigma_{ij} = \sigma_{ji} \tag{4.19}$$

Equation (4.19) proves that the stress tensor as defined is symmetric. Because of this symmetry, the equilibrium equations (4.14) can be written as equations (4.20):

$$\sigma_{ij,j} + \rho b_i = 0 \tag{4.20}$$

which can be developed as the equations given in (4.21):

$$\begin{aligned}
&\frac{\partial \sigma_{11}}{\partial x_1} + \frac{\partial \sigma_{12}}{\partial x_2} + \frac{\partial \sigma_{13}}{\partial x_3} + \rho B_1 = 0 \\
&\frac{\partial \sigma_{21}}{\partial x_1} + \frac{\partial \sigma_{22}}{\partial x_2} + \frac{\partial \sigma_{23}}{\partial x_3} + \rho B_2 = 0 \\
&\frac{\partial \sigma_{31}}{\partial x_1} + \frac{\partial \sigma_{32}}{\partial x_2} + \frac{\partial \sigma_{33}}{\partial x_3} + \rho B_3 = 0
\end{aligned} \tag{4.21}$$

4.1.3 Laws of stress transformation

According to the law of transformation of Cartesian tensors of order one and order two, the components of the stress vector, $t_i^{(\hat{n})}$, and stress tensor, σ_{pq}, referring to the unrotated system are related to the components of the another rotated system $\overline{t}_i^{(\hat{n})}$ and $\overline{\sigma}_{ij}$ by the relations given in Table 4.1.

Table 4.1. Vector and stress tensor transformations.

	Transformation	Symbolic Form	Matrix Form
Stress vector	$\overline{t}_i^{(\hat{n})} = a_{ij} t_i^{(\hat{n})}$	$\overline{t}^{(\hat{n})} = A \cdot t^{(\hat{n})}$	$[\overline{t}_{i1}^{(\hat{n})}] = [a_{ij}][t_{j1}^{(\hat{n})}]$
Stress tensor	$\overline{\sigma}_{ij} = a_{ip}\, a_{jq}\, \sigma_{pq}$	$\overline{\Sigma} = A \cdot \Sigma \cdot A_c$	$[\sigma_{ij}] = [a_{ip}][\sigma_{pq}][a_{qj}]$

Explicitly the matrix forms of the stress vector and tensor can be written as:

$$\begin{bmatrix} \overline{t}_1^{(\hat{n})} \\ \overline{t}_2^{(\hat{n})} \\ \overline{t}_3^{(\hat{n})} \end{bmatrix} = \begin{bmatrix} a_{11} & a_{12} & a_{13} \\ a_{21} & a_{22} & a_{23} \\ a_{31} & a_{32} & a_{33} \end{bmatrix} \begin{bmatrix} t_1^{(\hat{n})} \\ t_2^{(\hat{n})} \\ t_3^{(\hat{n})} \end{bmatrix} \tag{4.22}$$

and

$$\begin{bmatrix} \overline{\sigma}_{11} & \overline{\sigma}_{12} & \overline{\sigma}_{13} \\ \overline{\sigma}_{21} & \overline{\sigma}_{22} & \overline{\sigma}_{23} \\ \overline{\sigma}_{31} & \overline{\sigma}_{32} & \overline{\sigma}_{33} \end{bmatrix} = \begin{bmatrix} a_{11} & a_{12} & a_{13} \\ a_{21} & a_{22} & a_{23} \\ a_{31} & a_{32} & a_{33} \end{bmatrix} \begin{bmatrix} \sigma_{11} & \sigma_{12} & \sigma_{13} \\ \sigma_{21} & \sigma_{22} & \sigma_{23} \\ \sigma_{31} & \sigma_{32} & \sigma_{33} \end{bmatrix} \begin{bmatrix} a_{11} & a_{21} & a_{31} \\ a_{12} & a_{22} & a_{32} \\ a_{13} & a_{23} & a_{33} \end{bmatrix} \tag{4.23}$$

4.1.4 Principal stresses and stress invariants

The equation $t_i^{(\hat{n})} = \sigma_{ij}\, n_j$ for which the components of the stress tensor are σ_{ij} at a point P associates to each direction n_i a stress vector $t_i^{(\hat{n})}$ .. The directions for which $t_i^{(\hat{n})}$ and n_i are collinear are called the direction of principal stresses. For a main address,

$$t_i^{(\hat{n})} = \sigma n_i \tag{4.24}$$

where σ is the magnitude of the stress vector and is called the principal stress value.

Substituting equation (4.24) into the equation $t_i^{(\hat{n})} = \sigma_{ij}\, n_j$ and using the identities $n_i = \delta_{ij}\, n_j$ y $\sigma_{ij} = \sigma_{ji}$, we obtain:

$$(\sigma_{ij} - \delta_{ij}\, \sigma)\, n_j = 0 \tag{4.25}$$

In the three equations given in expression (4.25), there are four unknown quantities, which are the three direction cosines n_i and the value of the principal stress σ.

For the solutions of equations (4.25) to be non-trivial, such as $n_j = 0$, the determinant of the coefficients, $|\sigma_{ij} - \delta_{ij}\, \sigma|$, must be null, explicitly:

$$|\sigma_{ij} - \delta_{ij}\, \sigma| = 0 \tag{4.26}$$

or equivalently

$$\begin{bmatrix} \sigma_{11}-\sigma & \sigma_{12} & \sigma_{13} \\ \sigma_{21} & \sigma_{22}-\sigma & \sigma_{23} \\ \sigma_{31} & \sigma_{32} & \sigma_{33}-\sigma \end{bmatrix} = 0 \tag{4.27}$$

The expansion of this determining generates the cubic polynomial in σ, given by:

$$\sigma^3 - \mathrm{I}_\Sigma\, \sigma^2 + \mathrm{II}_\Sigma\, \sigma - \mathrm{III}_\Sigma = 0 \tag{4.28}$$

where

$$\mathrm{I}_\Sigma = \sigma_{ii} = tr\,\Sigma = 0 \tag{4.29}$$

$$\mathrm{II}_\Sigma = \frac{1}{2}(\sigma_{ii}\sigma_{jj} - \sigma_{ij}\sigma_{ij}) \tag{4.30}$$

$$\mathrm{III}_\Sigma = |\sigma_{ij}| = \det \Sigma \tag{4.31}$$

The expressions for I_Σ, II_Σ, and III_Σ, are known respectively as the first, second, and third stress invariants. The three roots of equation (4.28), denoted as $\sigma_{(1)}, \sigma_{(2)}, \sigma_{(3)}$ are values of the three principal stresses. For each $\sigma_{(k)}$ associated with each principal stress, there is a direction of principal stress for which the direction cosines are solutions of the equations.

$$(\sigma_{ij} - \sigma_{(k)}\,\delta_{ij})\, n_j^{(k)} = 0, \quad (k = 1,2,3) \tag{4.32}$$

In equation (4.32) the subscripts or superscripts enclosed in parentheses are only distinctive and therefore do not participate in any addition operation. Explicitly developing equations (4.32) for the first principal direction (k = 1), we obtain:

$$\begin{aligned} (\sigma_{11} - \sigma_{(1)})n_1^{(1)} + \sigma_{12}n_2^{(1)} + \sigma_{13}n_3^{(1)} &= 0 \\ \sigma_{21}n_1^{(1)} + (\sigma_{22} - \sigma_{(1)})n_2^{(1)} + \sigma_{23}n_3^{(1)} &= 0 \\ \sigma_{31}n_1^{(1)} + \sigma_{32}n_2^{(1)} + (\sigma_{33} - \sigma_{(1)})n_3^{(1)} &= 0 \end{aligned}$$

Analogously for the second (k = 2) and third (k = 3) main direction, we write:

$$\begin{aligned} (\sigma_{11} - \sigma_{(2)})n_1^{(2)} + \sigma_{12}n_2^{(2)} + \sigma_{13}n_3^{(2)} &= 0 \\ \sigma_{21}n_1^{(2)} + (\sigma_{22} - \sigma_{(2)})n_2^{(2)} + \sigma_{23}n_3^{(2)} &= 0 \\ \sigma_{31}n_1^{(2)} + \sigma_{32}n_2^{(2)} + (\sigma_{33} - \sigma_{(2)})n_3^{(2)} &= 0 \end{aligned}$$

and

$$\begin{aligned} (\sigma_{11} - \sigma_{(3)})n_1^{(3)} + \sigma_{12}n_2^{(3)} + \sigma_{13}n_3^{(3)} &= 0 \\ \sigma_{21}n_1^{(3)} + (\sigma_{22} - \sigma_{(3)})n_2^{(3)} + \sigma_{23}n_3^{(3)} &= 0 \\ \sigma_{31}n_1^{(3)} + \sigma_{32}n_2^{(3)} + (\sigma_{33} - \sigma_{(3)})n_3^{(3)} &= 0 \end{aligned}$$

Because the stress tensor is real and symmetric, the values of the principal stresses are also real and constitute the eigenvalues of the coefficient matrix, therefore, they are the elements of the diagonal matrix

$$\left[\sigma_{ij}\right]=\begin{bmatrix}\sigma_{(1)} & 0 & 0\\ 0 & \sigma_{(2)} & 0\\ 0 & 0 & \sigma_{(3)}\end{bmatrix} \tag{4.33}$$

or symbolically

$$\left[\sigma_{ij}\right]=\begin{bmatrix}\sigma_{\mathrm{I}} & 0 & 0\\ 0 & \sigma_{\mathrm{II}} & 0\\ 0 & 0 & \sigma_{\mathrm{III}}\end{bmatrix} \tag{4.34}$$

In this second symbolic form, roman numerals are used as subscripts to indicate that the main stress are ordered, $\sigma_{\mathrm{I}} > \sigma_{\mathrm{II}} > \sigma_{\mathrm{III}}$.

4.1.5 Cauchy Stress Quadric

Let σ_{ij} be the values of the stress tensor at a point P of the continuous medium, referred to directions parallel to the local Cartesian axes, let us call these axes $P\xi_1\ \xi_2\ \xi_3$, as represented in Figure 4.2, then the equation

$$\sigma_{ij}\ \xi_i\ \xi_j = \pm k^2,\ k\ cte \tag{4.35}$$

The equation given in (4.35) geometrically represents similar quadric surfaces that have a common center at P.

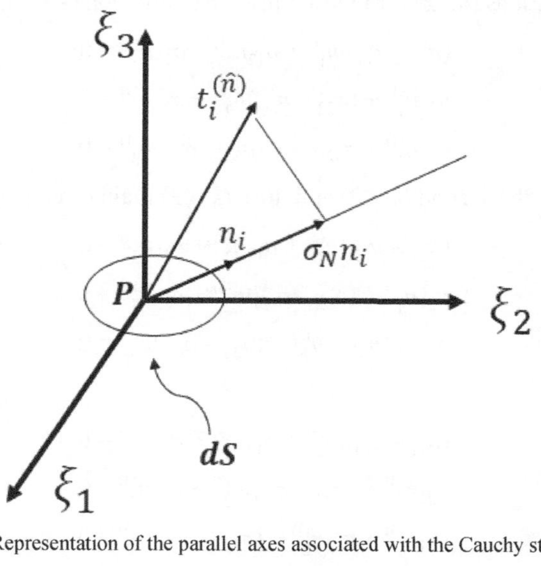

Figure 4.2. Representation of the parallel axes associated with the Cauchy stress quadric.

Choosing the plus or minus sign ensures that the surfaces are real. The position vector $\vec{r}$ of an arbitrary point lying on the surface of the quadric has components $\xi_i = rn_i$, where n_i is the unit normal vector in the direction of $\vec{r}$. The normal component $\sigma_N\, n_i$ of the stress vector $t_i^{(\hat{n})}$ at point P has a magnitude given by

$$\sigma_N = t_i^{(\hat{n})}\, n_i = t^{(\hat{n})} \cdot \hat{n} = \sigma_{ij}\, n_i\, n_j \tag{4.36}$$

If the constant k^2 in equation (4.35) is made equal to $\sigma_N r^2$, the resulting quadric is given by

$$\sigma_{ij}\, \xi_i\, \xi_j = \pm\sigma_N r^2 \tag{4.37}$$

Equation (4.37) is called the Cauchy stress quadric. This definition implies that the magnitude of σ_N of the component of the normal stress in the element dS perpendicular to the position vector $\vec{r}$ of a point located in the Cauchy stress quadric, is inversely proportional to r^2, that is, $\sigma_N = k^2/r^2$.

Furthermore, since the principal stress directions coincide with the principal axes of the Cauchy stress quadric, the principal stress values contain the maximum and minimum normal components at the point. In a space of principal stresses, that is, a space whose axes are in the directions of principal stresses and whose unit coordinates of measurement are the stresses ($t_1^{(\hat{n})}$, $t_2^{(\hat{n})}$, $t_3^{(\hat{n})}$) as represented in Figure 4.3, the arbitrary stress vector $t_i^{(\hat{n})}$ has the following component

$$t_1^{(\hat{n})} = \sigma_{(1)} n_1;\ t_2^{(\hat{n})} = \sigma_{(2)} n_2;\ \ t_3^{(\hat{n})} = \sigma_{(3)} n_3 \tag{4.38}$$

According to equation (4.7) $t_i^{(\hat{n})} = \sigma_{ij}\, n_j$.

But since $(n_1)^2 + (n_2)^2 + (n_3)^2 = 1$, for the unit vector $\hat{n}$, equation (4.38) implies in the stress space that the following equation is satisfied:

$$\frac{\left(t_1^{(\hat{n})}\right)^2}{\left(\sigma_{(1)}\right)^2} + \frac{\left(t_2^{(\hat{n})}\right)^2}{\left(\sigma_{(2)}\right)^2} + \frac{\left(t_3^{(\hat{n})}\right)^2}{\left(\sigma_{(3)}\right)^2} = 1 \tag{4.39}$$

Equation (4.39) is an ellipsoid and is known as the Lamé stress ellipsoid.

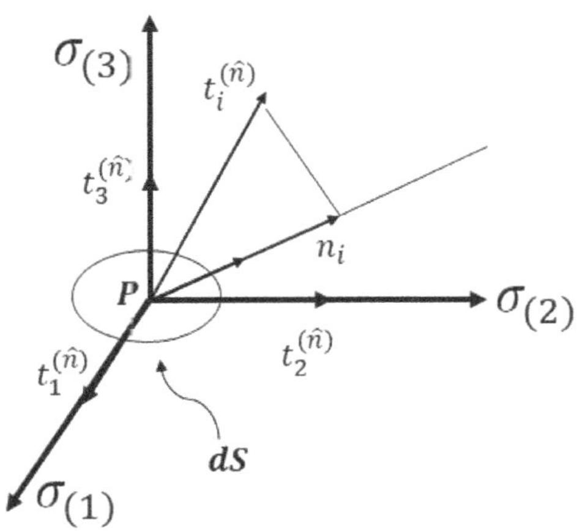

Figure 4.3. Principal stress space.

4.1.6 Maximum and minimum shear stress values

If the stress vector $t_i^{(\hat{n})}$ is decomposed into its orthogonal components. Normal and tangential, to the surface element dS on which it acts, the magnitude of a normal component can be determined from equation (4.36) and the magnitude of a shear or tangential component is given by

$$\sigma_S^2 = t_i^{(\hat{n})} \cdot t_i^{(\hat{n})} - \sigma_N^2 \tag{4.40}$$

This decomposition is presented in Figure 4.4, where the axes have been chosen in the principal stress directions and it is assumed that the principal stresses are ordered according to $\sigma_{\mathrm{I}} > \sigma_{\mathrm{II}} > \sigma_{\mathrm{III}}$. Then in equation (4.9) the components of $t_i^{(\hat{n})}$ are:

$$t_1^{(\hat{n})} = \sigma_I n_1;\ t_2^{(\hat{n})} = \sigma_{II} n_2;\ \ t_3^{(\hat{n})} = \sigma_{III} n_3 \tag{4.41}$$

and from (4.36), the magnitude of the normal component is:

$$\sigma_N = \sigma_I\, n_1^2 + \sigma_{II}\, n_2^2 + \sigma_{III}\, n_3^2 \tag{4.42}$$

Substituting equations (4.41) and (4.42) into (4.40), the squared magnitude of the shear as a function of the direction cosines n_i is given by equation (4.43).

$$\sigma_S^2 = \sigma_I^2\, n_1^2 + \sigma_{II}^2\, n_2^2 + \sigma_{III}^2 n_3^2 - (\sigma_I\, n_1^2 + \sigma_{II}\, n_2^2 + \sigma_{III}\, n_3^2)^2 \tag{4.43}$$

The maximum and minimum values of σ_S can be obtained from equation (4.43) by the Lagrange multiplier method. This method consists of constructing the Lagrange function, as shown in (4.44).

$$F = \sigma_S^2 - \lambda n_i\, n_i \tag{4.44}$$

in which the scalar λ is called the Lagrange multiplier. The condition for the stationary values (maximum and minimum) of F is that the derivative with respect to is zero,

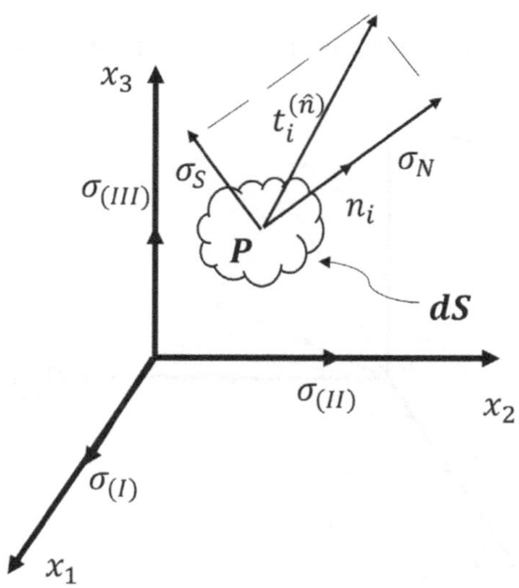

Figure 4.4. Decomposition into its normal and tangential components of the stress vector.

$\partial F/\partial n_i = 0$. In this way, performing the partial derivatives and equating them to zero, the following system of equations is obtained:

$$n_1\{\sigma_I^2 - 2\sigma_I(\sigma_I n_1^2 + \sigma_{II} n_2^2 + \sigma_{III} n_3^2) + \lambda\} = 0 \quad (4.45)$$

$$n_2\{\sigma_{II}^2 - 2\sigma_{II}(\sigma_I n_1^2 + \sigma_{II} n_2^2 + \sigma_{III} n_3^2) + \lambda\} = 0 \quad (4.46)$$

$$n_3\{\sigma_{III}^2 - 2\sigma_{III}(\sigma_I n_1^2 + \sigma_{II} n_2^2 + \sigma_{III} n_3^2) + \lambda\} = 0 \quad (4.47)$$

These equations, together with the condition $n_i n_i = 1$, can be solved for λ and the direction cosines associated with n_1, n_2, n_3, which in turn are associated with the extreme values of the shear stress.

It can be proven that a set of solutions for equations (4.45), (4.46) and (4.47) and the associated stresses (4.43) are:

$$n_1 = \pm 1,\ n_2 = 0,\ n_3 = 0;\ \text{for which } \sigma_S = 0 \quad (4.48)$$

$$n_1 = 0,\ n_2 = \pm 1,\ n_3 = 0;\ \text{for which } \sigma_S = 0 \quad (4.49)$$

$$n_1 = 0,\ n_2 = 0,\ n_3 = \pm 1;\ \text{for which } \sigma_S = 0 \quad (4.50)$$

The values of the shear stresses in (4.48), (4.49) and (4.50) are clearly minimum values. Furthermore, since (4.24) indicates that the shear components vanish in the principal planes, the directions given by (4.48), (4.49) and (4.50) are the principal stress directions.

It can be proven that a second set of solutions to equations (4.45), (4.46), and (4.47) is:

$$n_1 = 0, n_2 = \pm 1/\sqrt{2}, n_3 = \pm 1/\sqrt{2};\text{for which } \sigma_S = (\sigma_{II} - \sigma_{III})/2 \quad (4.51)$$

$$n_1 = \pm 1/\sqrt{2}, n_2 = 0, n_3 = \pm 1/\sqrt{2};\text{for which } \sigma_S = (\sigma_{III} - \sigma_I)/2 \quad (4.52)$$

$$n_1 = \pm 1/\sqrt{2}, n_2 = \pm 1/\sqrt{2}, n_3 = 0;\text{for which } \sigma_S = (\sigma_I - \sigma_{II})/2 \quad (4.53)$$

Equations (4.51), (4.52) and (4.53) give the value of the maximum cision, which is equal to the semi-difference of the largest and smallest principal stresses.

4.1.7 Spherical and deviator stress tensors

It is often useful to decompose the stress tensor σ_{ij} into two component tensors, one which is the hydrostatic or spherical stress tensor and the other which is the deviatoric stress tensor.

The stress or spherical tensor is defined as:

$$\Sigma_M = \sigma_M I = \begin{pmatrix} \sigma_M & & \\ & \sigma_M & \\ & & \sigma_M \end{pmatrix} \quad (4.54)$$

where $\sigma_M = -p = \sigma_{kk}/3$ is the average normal stress.

The deviatoric stress tensor is defined as:

$$\Sigma_D = \begin{pmatrix} \sigma_{11}-\sigma_M & \sigma_{12} & \sigma_{13} \\ \sigma_{21} & \sigma_{22}-\sigma_M & \sigma_{23} \\ \sigma_{31} & \sigma_{32} & \sigma_{33}-\sigma_M \end{pmatrix} \equiv \begin{pmatrix} S_{11} & S_{12} & S_{13} \\ S_{21} & S_{22} & S_{23} \\ S_{31} & S_{32} & S_{33} \end{pmatrix} \quad (4.55)$$

This decomposition is expressed by the equations

$$\sigma_{ij} = \frac{\delta_{ij}\sigma_{kk}}{3} + S_{ij}, \text{or} \sum = \sigma_M I + \Sigma_D \quad (4.56)$$

The principal directions of the deviatoric stress tensor S_{ij} are the same as those of the stress tensor σ_{ij}. The values of the main deviatoric forces are:

$$S_{(k)} = \sigma_{(k)} - \sigma_M \quad (4.57)$$

The characteristic equation of the deviatoric stress tensor, comparable to that of the stress tensor (4.28), is the cubic equation

$$S^3 + II_{\Sigma_D} S - III_{\Sigma_D} = 0 \quad (4.58)$$

or equivalently

$$S^3 + (S_I S_{II} + S_{II} S_{III} + S_{III} S_I)S - S_I S_{II} S_{III} = 0 \quad (4.59)$$

It can be easily proven that the first invariant of the deviatoric stress tensor I_{Σ_D} is identically null, which is why the quadratic term is missing in equation (4.59).

4.2 Strain tensor

Introduction

In the study of the kinematics of continuous media, the term "point" will be used exclusively to designate a position in a fixed space. The word "particle" will denote a small volume element or "material point", of a continuous medium, that is, a point is a place in space and a particle is a small part of a continuous material medium. A continuous medium having a volume V bounded by a surface *S*, at a given moment of time *t*, will occupy a region *R* of physical space.

The identification of the particles of the continuous medium with the points of space they occupy at instant t with respect to an appropriate set of coordinate axes is said to specify the configuration of the continuous medium at that instant.

The term strain refers to a change in the shape of the continuous medium, between an initial configuration (undeformed) and a subsequent configuration (deformed). In strain studies, special significance is placed on the initial and final configurations. No attention is paid to intermediate configurations or to a particular sequence of configurations through which strain takes place.

In contrast, the word flow is used to designate the continuous state of motion of a continuous medium. Of course, a prior configuration is inherent in flow investigations for which a time-dependent velocity field is specified. This section studies the main characteristics and properties of the strain tensor.

4.2.1 Position vector. Displacement vector

Figure 4.5 represents the undeformed configuration of a material continuous medium at time $t = 0$ together with the deformed configuration of the same continuous medium later $t = t$. For the present development it is convenient to refer the initial and final configurations to separate coordinate axes as can be seen in Figure 4.5.

$\vec{X}$ is representing the position vector in the initial configuration of a representative particle of the continuous medium that occupies a point P_0 in space and is defined as:

$$\vec{X} = X_1\,\hat{I}_1 + X_2\,\hat{I}_2 + X_3\,\hat{I}_3 \tag{4.60}$$

Regarding the rectangular Cartesian coordinate axes $OX_1X_2X_3$. In the expression (4.60) Capital letters are used as indices and as such will appear in several equations later, but their use as addition indices is only used in this section. In the rest of the book the subscripts and superscripts in capital letters only serve as labels. The coordinates (X_1,X_2,X_3) are called material coordinates. In the deformed configuration the particle originally located at P_0 now appears at point P and has the position vector.

$$\vec{x} = x_1\,\hat{e}_1 + x_2\,\hat{e}_2 + x_3\,\hat{e}_3 \tag{4.61}$$

Referring to the rectangular Cartesian coordinate axes $Ox_1x_2x_3$. Here lowercase letters are used as subscripts to indicate their connection with the coordinates (x_1,x_2,x_3) which give the current position of the particle and are called spatial coordinates.

The relative orientation of the material axes $OX_1X_2X_3$ and spatial axes $ox_1x_2x_3$ is specified by the direction cosines α_{kK} and y α_{Kk}, which are defined by the scalar products of the unit vectors as follows:

$$\hat{e}_k \cdot \hat{I}_K = \hat{I}_K \cdot \hat{e}_k = \alpha_{kK} = \alpha_{Kk} \tag{4.62}$$

In Figure 4.5, the vector $\vec{u}$ that joins the points P_0 and P (initial and final position of the particle, respectively), is known as the displacement vector. This vector can be expressed as

$$\vec{u} = u_k\,\hat{e}_k \tag{4.63}$$

or also like

$$\vec{U} = U_K\,\hat{I}_K \tag{4.64}$$

In which the components u_k and U_K are related through the direction cosines α_{kK}. According to expressions (4.62), the unit vector $\hat{e}_k$ is expressed in terms of the base material vectors $\hat{I}_K$ as

$$\hat{e}_k = \alpha_{kK}\,\hat{I}_K \tag{4.65}$$

Therefore, substituting (4.65) into (4.63), we obtain

$$\vec{u} = u_k(\alpha_{kK}\,\hat{I}_K) = U_K\,\hat{I}_K = \vec{U} \tag{4.66}$$

where

$$U_K = \alpha_{kK}\,u_k \tag{4.67}$$

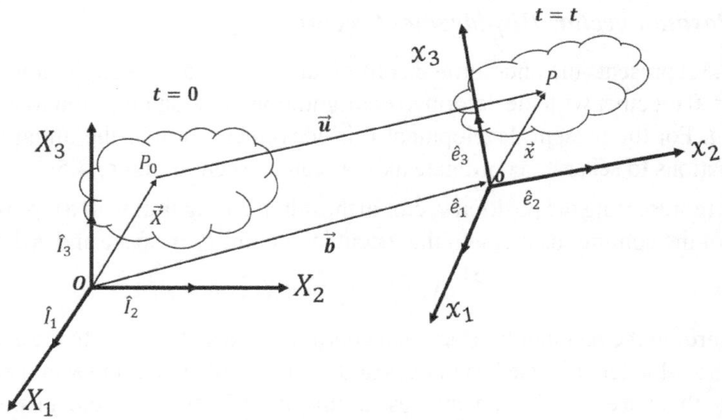

Figure 4.5. Deformed and undeformed configuration space of a continuous material medium.

Since the direction cosines are constant, expression (4.67) denotes that the components of the displacement vector transform as a Cartesian vector or tensor of order one. The vector $\vec{b}$ given in Figure 4.5 serves to locate the origin o with respect to the origin O. From the geometry of the figure, it can be established that:

$$\vec{u} = \vec{b} + \vec{x} - \vec{X} \tag{4.68}$$

Very often in the study of the mechanics of continuous media it is useful to consider the two coordinate systems $OX_1X_2X_3$ and $ox_1x_2x_3$ superimposed, so that $\vec{b} = 0$, in this way the expression (4.69);

$$\vec{u} = \vec{x} - \vec{X} \tag{4.69}$$

In the Cartesian component form the expression (4.69) is given by the general expression general

$$u_k = x_k - \alpha_{kK} X_K \tag{4.70}$$

Since the axes are superimposed, the unit triads of basis vectors for the two systems are identical, because the direction cosines $\alpha_{kK} = \delta_{kK}$, therefore, expression (4.70) becomes

$$u_k = x_k - X_K \tag{4.71}$$

an expression in which only lowercase subscripts appear.

In the rest of the book, unless otherwise specified, the spatial and material axes are assumed to be superimposed and then only lowercase indices will be used.

4.2.2 Lagrangian and Eulerian description

When a continuous medium undergoes strain (or flow), the particles of the continuous medium move along various paths in space. This movement can be expressed by equations of the form

$$x_i = x_i\,(X_1, X_2, X_3, t) = x_i\,(\vec{X}, t) \tag{4.72}$$

or symbolically

$$\vec{x} = \vec{x}\,(\vec{X}, t)$$

Expression (4.72) provides the current position of the particle x_i that occupied the point (X_1, X_2, X_3, t), at time $t = 0$.

Also (4.72) can be interpreted as the detailed form that a distribution takes from an initial configuration to a final configuration. Such a detailed distribution is assumed to be one-to-one and continuous, with continuous partial derivatives up to any desired order. The description of motion or strain expressed by (4.72) is known as the Lagrangian formulation.

Note that this formulation focuses on the particle, on the material as such, it follows the trajectory of a particle in the strain of the material.
If otherwise, the strain motion is described by equations of the form:

$$X_i = X_i\,(x_1, x_2, x_3, t) = X_i\,(\vec{x}, t) \tag{4.73}$$

or symbolically

$$\vec{X} = \vec{X}\,(\vec{x}, t)$$

In which the independent variables are the coordinates x_i and t, the description is known as the Eulerian formulation. Note that this formulation focuses on the points in space, not on the material, it describes the values that a fixed point in space takes with respect to the particles that occupy it in the strain process. Similarly, if the distribution (4.73) is also assumed continuous, one-to-one with continuous derivatives, then the two distributions are unique inverses, one with respect to the other. The necessary and sufficient condition for an inverse function to exist is that the Jacobian of the transformation does not cancel out, that is:

$$J = \left| \frac{\partial x_i}{\partial X_j} \right| \neq 0 \tag{4.74}$$

4.2.3 Strain gradients. Displacement gradients

Partial differentiation of equation (4.72) with respect to X_j produces a tensor $\partial x_i / \partial X_j$ which is called the material strain gradient. In symbolic, $\partial x_i / \partial X_j$, is represented by the dyadic symbolic notation

$$F = \vec{x}\nabla_{\vec{x}} \equiv \frac{\partial \vec{x}}{\partial X_1}\hat{e}_1 + \frac{\partial \vec{x}}{\partial X_2}\hat{e}_2 + \frac{\partial \vec{x}}{\partial X_3}\hat{e}_3 \tag{4.75}$$

In which the operator $\nabla_{\vec{x}} \equiv \frac{\partial}{\partial X_i}\hat{e}_i$ is applied to the right (as explicitly stated in the equation). The matrix form of F is written:

$$\mathcal{F} = \begin{bmatrix} x_1 \\ x_2 \\ x_3 \end{bmatrix}\left[\frac{\partial}{\partial X_1}, \frac{\partial}{\partial X_2}, \frac{\partial}{\partial X_3}\right] \equiv \begin{bmatrix} \frac{\partial x_1}{\partial X_1} & \frac{\partial x_1}{\partial X_2} & \frac{\partial x_1}{\partial X_3} \\ \frac{\partial x_2}{\partial X_1} & \frac{\partial x_2}{\partial X_2} & \frac{\partial x_2}{\partial X_3} \\ \frac{\partial x_3}{\partial X_1} & \frac{\partial x_3}{\partial X_2} & \frac{\partial x_3}{\partial X_3} \end{bmatrix} = \left[\frac{\partial x_i}{\partial X_j}\right] \tag{4.76}$$

Similarly, the partial differentiation of expression (4.73) with respect to x_j gives rise to the tensor $\partial X_i / \partial x_j$ which is called the spatial strain gradient. This tensor is represented by the dyadic

$$H = \vec{X}\nabla_{\vec{x}} \equiv \frac{\partial \vec{X}}{\partial x_1}\hat{e}_1 + \frac{\partial \vec{X}}{\partial x_2}\hat{e}_2 + \frac{\partial \vec{X}}{\partial x_3}\hat{e}_3 \tag{4.77}$$

The matrix form H is represented in the expression (4.78):

$$\mathcal{H} = \begin{bmatrix} X_1 \\ X_2 \\ X_3 \end{bmatrix}\left[\frac{\partial}{\partial x_1}, \frac{\partial}{\partial x_2}, \frac{\partial}{\partial x_3}\right] \equiv \begin{bmatrix} \frac{\partial X_1}{\partial x_1} & \frac{\partial X_1}{\partial x_2} & \frac{\partial X_1}{\partial x_3} \\ \frac{\partial X_2}{\partial x_1} & \frac{\partial X_2}{\partial x_2} & \frac{\partial X_2}{\partial x_3} \\ \frac{\partial X_3}{\partial x_1} & \frac{\partial X_3}{\partial x_2} & \frac{\partial X_3}{\partial x_3} \end{bmatrix} = \left[\frac{\partial X_i}{\partial x_j}\right] \tag{4.78}$$

The spatial and material strain tensors are related by means of the partial difference chain rule, such that:

$$\frac{\partial x_i}{\partial X_j}\frac{\partial X_j}{\partial x_k} = \frac{\partial X_i}{\partial x_j}\frac{\partial x_j}{\partial X_k} = \delta_{ik} \tag{4.79}$$

The partial differentiation of the displacement vector u_i with respect to both coordinates originates either the material displacement gradient $\partial u_i / \partial X_j$ or the spatial displacement gradient $\partial u_i / \partial x_j$. Deriving expression (4.71) as a coordinate difference, these tensors can be written as a function of material strain gradients, as:

$$\frac{\partial u_i}{\partial X_j} = \frac{\partial x_i}{\partial X_j} - \delta_{ij} \equiv J \equiv F - I \tag{4.80}$$

Or the spatial gradient like:

$$\frac{\partial u_i}{\partial x_j} = \delta_{ij} - \frac{\partial X_i}{\partial x_j} \equiv K \equiv H - I \tag{4.81}$$

Usually, the matrix forms of J and K are respectively

$$J = \begin{bmatrix} u_1 \\ u_2 \\ u_3 \end{bmatrix} \left[\frac{\partial}{\partial X_1}, \frac{\partial}{\partial X_2}, \frac{\partial}{\partial X_3} \right] \equiv \begin{bmatrix} \frac{\partial u_1}{\partial X_1} & \frac{\partial u_1}{\partial X_2} & \frac{\partial u_1}{\partial X_3} \\ \frac{\partial u_2}{\partial X_1} & \frac{\partial u_2}{\partial X_2} & \frac{\partial u_2}{\partial X_3} \\ \frac{\partial u_3}{\partial X_1} & \frac{\partial u_3}{\partial X_2} & \frac{\partial u_3}{\partial X_3} \end{bmatrix} = \left[\frac{\partial u_i}{\partial X_j} \right] \tag{4.82}$$

and

$$K = \begin{bmatrix} u_1 \\ u_2 \\ u_3 \end{bmatrix} \left[\frac{\partial}{\partial x_1}, \frac{\partial}{\partial x_2}, \frac{\partial}{\partial x_3} \right] \equiv \begin{bmatrix} \frac{\partial u_1}{\partial x_1} & \frac{\partial u_1}{\partial x_2} & \frac{\partial u_1}{\partial x_3} \\ \frac{\partial u_2}{\partial x_1} & \frac{\partial u_2}{\partial x_2} & \frac{\partial u_2}{\partial x_3} \\ \frac{\partial u_3}{\partial x_1} & \frac{\partial u_3}{\partial x_2} & \frac{\partial u_3}{\partial x_3} \end{bmatrix} = \left[\frac{\partial u_i}{\partial x_j} \right] \tag{4.83}$$

4.2.4 Strain Tensor. Finite Strain Tensors

Figure 4.6 represents the initial (undeformed) and final (deformed) configuration of a continuous medium referred to two superimposed rectangular Cartesian coordinate systems $OX_1X_2X_3$, for the undeformed system and $ox_1x_2x_3$ for the deformed system.

In the strain process, two closely spaced particles that occupy points P_0 and Q_0 before strain move to points P and Q, respectively, in the deformed configuration.

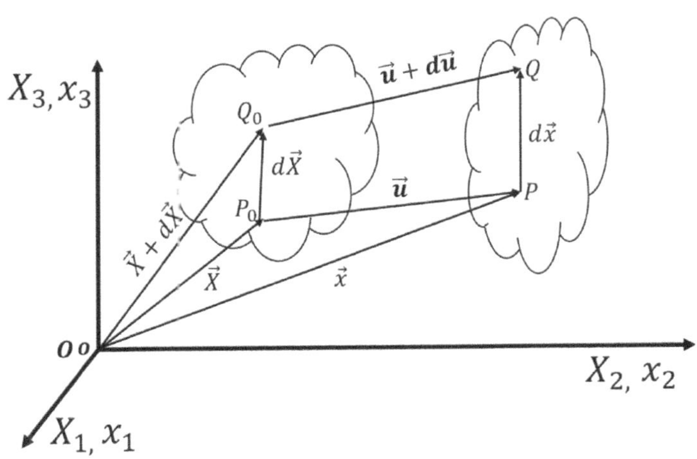

Figure 4.6. Strain configuration space of a material continuous medium.

The square of the differential element of length between P_0 and Q_0 is given by the following expression as:

$$(d\vec{X})^2 = d\vec{X} \cdot d\vec{X} = dX_i\, dX_i = \delta_{ij}\, dX_i\, dX_j \quad (4.84)$$

From expression (4.73), the differential distance dX_i is written as:

$$dX_i = \frac{\partial X_i}{\partial x_j} dx_j \quad (4.85)$$

Combining the expressions (4.84) and (4.85) we obtain

$$(d\vec{X})^2 = \frac{\partial X_k}{\partial x_i}\frac{\partial X_k}{\partial x_j} dx_i dx_j = C_{ij} dx_i dx_j \quad (4.86)$$

where

$$C_{ij} = \frac{\partial X_k}{\partial x_i}\frac{\partial X_k}{\partial x_j} \quad (4.87)$$

C_{ij} thus, defined is a tensor or order two and is known as the Cauchy strain tensor.

In the deformed configuration, the square of the differential element length between P and Q, is given by:

$$(d\vec{x})^2 = d\vec{x} \cdot d\vec{x} = dx_i\, dx_i = \delta_{ij}\, dx_i\, dx_j \quad (4.88)$$

From (4.72) the differential distance dx_i is here

$$x_i = \frac{\partial x_i}{\partial X_j} dX_j \quad (4.89)$$

In such a way that combining the expressions (4.88) and (4.89) we obtain

$$(d\vec{x})^2 = \frac{\partial x_k}{\partial X_i}\frac{\partial x_k}{\partial X_j} dX_i dX_j = G_{ij} dX_i dX_j \quad (4.90)$$

where G_{ij} is defined as:

$$G_{ij} = \frac{\partial x_k}{\partial X_i}\frac{\partial x_k}{\partial X_j} \quad (4.91)$$

G_{ij} thus, defined is a second-order tensor and is known as the Green strain tensor.

The difference between two very close particles of a continuous medium, $(d\vec{x})^2 - (d\vec{X})^2$, is used as a measure of the strain that takes place in the surroundings of the particles, when passing from one configuration initial to another final. If this difference is identically zero for all neighboring particles of a continuous medium, it is said that a rigid displacement has taken place. Using the expressions (4.90) for $(dx)^2$ and (4.84) for $(dX)^2$, this difference can be expressed as:

$$(d\vec{x})^2 - (d\vec{X})^2 = \left(\frac{\partial x_k}{\partial X_i}\frac{\partial x_k}{\partial X_j} - \delta_{ij}\right) dX_i dX_j = 2L_{ij} dX_i dX_j \quad (4.92)$$

where L_{ij} is defined as:

$$L_{ij} = \frac{1}{2}\left(\frac{\partial x_k}{\partial X_i}\frac{\partial x_k}{\partial X_j} - \delta_{ij} \right) \tag{4.93}$$

L_{ij} defined in this way is a tensor of order two and is called the Lagrangian (or Green) finite strain tensor.

Using the expressions (4.88) and (4.86) the same difference $(dx)^2 - (dX)^2$ can be expressed in the form:

$$(d\vec{x})^2 - (d\vec{X})^2 = \left(\delta_{ij} - \frac{\partial X_k}{\partial x_i}\frac{\partial X_k}{\partial x_j} \right) dx_i dx_j = 2E_{ij} dx_i dx \tag{4.94}$$

where E_{ij} is defined as:

$$E_{ij} = \frac{1}{2}\left(\delta_{ij} - \frac{\partial X_k}{\partial x_i}\frac{\partial X_k}{\partial x_j} \right) \tag{4.95}$$

E_{ij} defined in this way is a second-order tensor and is called the finite strain or Almansi tensor.

It is useful to express the Lagrangian and Eulerian finite strain tensors as functions of the displacement gradients. Then, if $\partial x_i / \partial X_j$ from equation (4.80) is substituted into (4.93), we obtain the Lagrangian finite strain tensor in the form

$$L_{ij} = \frac{1}{2}\left(\frac{\partial u_i}{\partial X_j} + \frac{\partial u_j}{\partial X_i} + \frac{\partial u_k}{\partial X_i}\frac{\partial u_k}{\partial X_j} \right) \tag{4.96}$$

In the same way, if $\partial X_i / \partial x_j$ from (4.81) is substituted into (4.95) the Eulerian finite strain tensor is obtained in the form:

$$E_{ij} = \frac{1}{2}\left(\frac{\partial u_i}{\partial x_j} + \frac{\partial u_j}{\partial x_i} - \frac{\partial u_k}{\partial x_i}\frac{\partial u_k}{\partial x_j} \right) \tag{4.97}$$

The matrix representations of (4.96) and (4.97) can be written directly from the expressions (4.82) and (4.83).

4.2.5 Infinitesimal strain tensor

Infinitesimal strains have as a basic condition the requirement that the displacement gradients are small compared to unity. If the displacement gradients are small compared to unity, their products are even smaller and these products can then be neglected in expressions (4.96) and (4.97) and thus the infinitesimal Lagrangian strain tensor is obtained, directly from the expression (4.96) which is denoted by l_{ij} and is given by the following expression:

$$l_{ij} = \frac{1}{2}\left(\frac{\partial u_i}{\partial X_j} + \frac{\partial u_j}{\partial X_i} \right) \tag{4.98}$$

Similarly, for expression (4.97) the products of the displacement gradients are neglected, and the Eulerian infinitesimal strain tensor is obtained, which is denoted by ϵ_{ij} and is given by the following expression:

$$\epsilon_{ij} = \frac{1}{2}\left(\frac{\partial u_i}{\partial x_j} + \frac{\partial u_j}{\partial x_i}\right) = \frac{1}{2}(u_{i,j} + u_{j,i}) \quad (4.99)$$

In this approximation in which the displacement gradients and displacements are very small, there is a very small difference between the spatial and material coordinates of a particle of a continuous medium, therefore the components of the material gradient $\partial u_i / \partial X_j$ and the spatial displacement gradient $\partial u_i / \partial x_j$ are almost equal, in such a way that the Lagrangian and Eulerian infinitesimal strain tensors can be considered equal, therefore we obtain that:

$$l_{ij} = \epsilon_{ij} \quad (4.100)$$

4.2.6 Linear rotation tensor

In Figure 4.7 the displacements of two neighboring particles are represented by the vectors $\vec{u}^{(P_0)}$ y $\vec{u}^{(Q_0)}$.

The vector

$$du_i = u_i^{(Q_0)} - u_i^{(P_0)} \quad (4.101)$$

It is called the relative displacement vector of the particle that is originally in Q_0 with respect to the one that is originally in P_0. Assuming adequate continuity conditions in the displacement field, a Taylor series expansion can be made for $u_i^{(P_0)}$ in the vicinity of P_0. Neglecting the higher order terms in this development, the relative displacement vector can be expressed as:

$$du_i = \left(\frac{\partial u_i}{\partial X_j}\right)_{P_0} dX_j \quad (4.102)$$

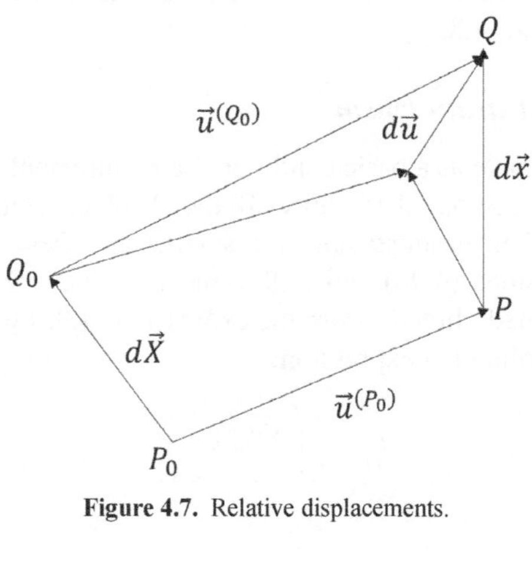

Figure 4.7. Relative displacements.

It should be considered that the derivatives are estimated at the point P_0. These derivatives are in fact the components of the material displacement gradient. Equation (4.102) is the Lagrangian form of the relative displacement vector.

It is also useful to define the unit relative displacement vector $\partial u_i / \partial X$ in which dX is the magnitude of the vector dX_i, therefore, if v_i is a unit vector in the direction of dX_i such that $dX_i = v_i dX$, so

$$\frac{du_i}{dX} = \frac{\partial u_i}{\partial X_j}\frac{dX_j}{dX} = \frac{\partial u_i}{\partial X_j} v_j \tag{4.103}$$

Since the material displacement gradient $\partial u_i / \partial X_j$ can be decomposed only into a symmetric and an antisymmetric part, the relative displacement vector du_i can be expressed as

$$du_i = \left[\frac{1}{2}\left(\frac{\partial u_i}{\partial X_j} + \frac{\partial u_j}{\partial X_i}\right) + \frac{1}{2}\left(\frac{\partial u_i}{\partial X_j} - \frac{\partial u_j}{\partial X_i}\right)\right] dX_j \tag{4.104}$$

The first term of expression (4.104) is known as the linear Lagrangian strain tensor l_{ij} and the second term is known as the linear Lagrangian rotation tensor, it is denoted by W_{ij} and is given by the following expression:

$$W_{ij} = \frac{1}{2}\left(\frac{\partial u_i}{\partial X_j} - \frac{\partial u_j}{\partial X_i}\right) \tag{4.105}$$

At a displacement for which the strain tensor l_{ij} is identically zero in the vicinity of the point P_0, the relative displacement at that point will be an infinitesimal rotation of the rigid body. This infinitesimal rotation is expressed by the rotation vector, given in the following expression:

$$\omega_i = \frac{1}{2} \epsilon_{ijk} W_{kj} \tag{4.106}$$

In this way, based on the relative displacement it can be established that:

$$du_i = \epsilon_{ijk}\, \omega_j\, dX_k \tag{4.107}$$

The development of the Lagrangian description of the relative displacement vector, the linear rotation tensor and the linear rotation vector is completely parallel to the development for the corresponding Eulerian quantities. According to this, the Eulerian description of the relative displacement vector is given by:

$$du_i = \frac{\partial u_i}{\partial x_j} dx_j \tag{4.108}$$

and of the unitary relative displacement vector

$$du_i = \frac{\partial u_i}{\partial x_j}\frac{dx_j}{dx} = \frac{\partial u_i}{\partial x_j}\mu_j \tag{4.109}$$

The decomposition of the Eulerian displacement gradient $\partial u_i / \partial x_j$ gives rise to the expression

$$\frac{du_i}{dx} = \left[\frac{1}{2}\left(\frac{\partial u_i}{\partial x_j} + \frac{\partial u_j}{\partial x_i} \right) + \frac{1}{2}\left(\frac{\partial u_i}{\partial x_j} - \frac{\partial u_j}{\partial x_i} \right) \right] dx_j \tag{4.110}$$

The first term in the bracket of expression (4.110) is the Eulerian linear strain tensor ϵ_{ij}. The second term is the linear Eulerian rotation tensor, it is denoted by ω_{ij} and is given by the expression (4.111):

$$\omega_{ij} = \frac{1}{2}\left(\frac{\partial u_i}{\partial x_j} - \frac{\partial u_j}{\partial x_i} \right) \tag{4.111}$$

From expression (4.111), the linear Eulerian rotation vector is defined as:

$$\omega_i = \frac{1}{2} \epsilon_{ijk}\, \omega_{kj} \tag{4.112}$$

Depending on the relative displacement it is given by the expression:

$$du_i = \frac{1}{2} \epsilon_{ijk}\, \omega_j\, dx_k \tag{4.113}$$

4.2.7 Interpretation of linear strain tensors

In the theory of infinitesimal strains, the Lagrangian finite strain tensor L_{ij} in expression (4.92) can be replaced by the linear Lagrangian strain tensor l_{ij}, and the following equation is then obtained

$$(dx)^2 - (dX)^2 = 2l_{ij}\, dX_i\, dX_j \tag{4.114}$$

Since for small strains $dx \approx dX$, equation (4.114) can be written as follows

$$\frac{dx - dX}{dX} = l_{ij} \frac{dX_i}{dX} \frac{dX_j}{dX} l_{ij} \nu_i \nu_j \tag{4.115}$$

The first member of equation (4.115) is interpreted as the change in length per unit of original length of the differential element and is called the normal strain for the line element that originally has the direction cosines dX_i / dX.

When equation (4.115) is applied to a differential element that is initially of length dX and is on the axis, say X_2, the normal strain of that element is obtained. In this case

$$\frac{dX_1}{dX} = \frac{dX_3}{dX} = 0, \quad \frac{dX_2}{dX} = 1 \tag{4.116}$$

and therefore equation (4.115) becomes:

$$\frac{dx - dX}{dX} = l_{22} = \frac{\partial u_2}{\partial X_2} \tag{4.117}$$

In this way, the normal strain of an element that is originally along the X_2 axis constitutes the l_{22} component. In an analogous manner, the values l_{11} and l_{33} are obtained for the axes X_1 and X_3. Therefore, the diagonal terms of the linear strain tensor represent normal strains in the coordinate directions.

The physical interpretation of the terms that do not occupy the diagonal of l_{ij} can be obtained by considering the line elements originally located along two coordinate axes. In Figure 4.8, the line elements P_0Q_0 and P_0M_0 that originally are along the X_2 and X_3 axes respectively, they become after strain into the line elements PQ and PM with respect to the parallel set of local axes with origin at P. The original right angle between the line elements becomes the angle θ.

From the definition of differential, we have that for u_i it is written as:

$$du_i = \left(\frac{\partial u_i}{\partial X_j}\right)_{P_0} dX_j \tag{4.118}$$

From the definition (4.118) and the small strain hypothesis, a first-order approximation for the unit vector in P and in the direction of Q would be

$$\hat{n}_2 = \frac{\partial u_1}{\partial X_2}\hat{e}_1 + \hat{e}_2 + \frac{\partial u_3}{\partial X_2}\hat{e}_3 \tag{4.119}$$

and for the unit vector at P in the direction of M:

$$\hat{n}_3 = \frac{\partial u_1}{\partial X_3}\hat{e}_1 + \frac{\partial u_2}{\partial X_3}\hat{e}_2 + \hat{e}_3 \tag{4.120}$$

Therefore,

$$\cos\theta = \hat{n}_2 \cdot \hat{n}_3 = \frac{\partial u_1}{\partial X_2}\frac{\partial u_1}{\partial X_3} + \frac{\partial u_2}{\partial X_3} + \frac{\partial u_3}{\partial X_2} \tag{4.121}$$

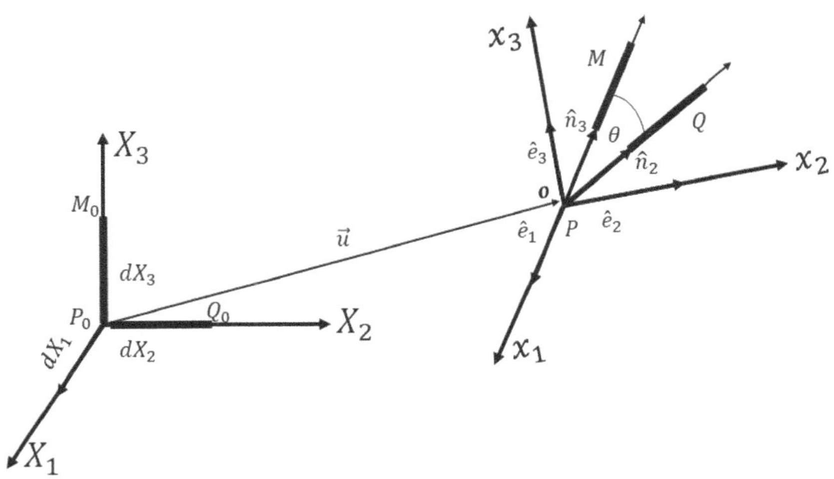

Figure 4.8. Strains of differential elements not parallel to the axes.

In the linear strain approximation, the gradients are small so the product of gradients can be neglected, therefore, expression (4.121) reduces to:

$$\cos\theta = \frac{\partial u_2}{\partial X_3} + \frac{\partial u_3}{\partial X_2} = 2l_{23} \tag{4.122}$$

On the other hand, taking the change from the original right angle, due to strain, in the angle between the elements as $\gamma_{23} = \pi/2 - \theta$ and remembering that, in the linear approximation, γ_{23}, is very small, it follows that:

$$\gamma_{23} \approx \sin\left(\pi/2 - \theta\right) = \cos\theta = 2l_{23} \tag{4.123}$$

Therefore, the non-diagonal components of the linear strain tensor represent half of the original right-angle change of one element with another. These strain components are called shear strains or distortions, and because of the factor 2 in expression (4.123), these components of the tensor are equal to half of the shear strains familiar from engineering.

A parallel and identical development to that presented for the interpretation of the components of l_{ij} can be made with the linear Eulerian strain tensor ϵ_{ij}.

The essential difference in the deductions lies in the choice of the line elements that in the Eulerian description must lie along the coordinate axes in the deformed state. The diagonal terms of ϵ_{ij} are the normal strains and the others are the shear strains.

In the strains for which $l_{ij} = \epsilon_{ij}$ is accepted as valid, no distinction is made between the Eulerian and Lagrangian interpretations.

4.2.8 Transformation properties of strain tensors

The various strain tensors defined previously and given by expressions (4.96), (4.97), (4.98) and (4.99), are all second order tensors, this implies that they obey transformation laws of the form:

$$\overline{L}_{ij} = b_{ip}\, b_{jq}\, L_{pq} \tag{4.124}$$

$$\overline{l}_{ij} = b_{ip}\, b_{jq}\, l_{pq} \tag{4.125}$$

$$\overline{E}_{ij} = a_{ip}\, a_{jq}\, E_{pq} \tag{4.126}$$

$$\overline{\epsilon}_{ij} = a_{ip}\, a_{jq}\, \epsilon_{pq} \tag{4.127}$$

By analogy with Cauchy quadrics, the Lagrangian and Eulerian linear strain quadrics with respect to the local Cartesian coordinates η_i and ξ_i at points P_0 and P respectively can be given as indicated in Figure 4.9.

Thus, the equation of the Lagrangian strain quadric is given by

$$l_{ij}\, \eta_i\, \eta_j = \pm h^2 \tag{4.128}$$

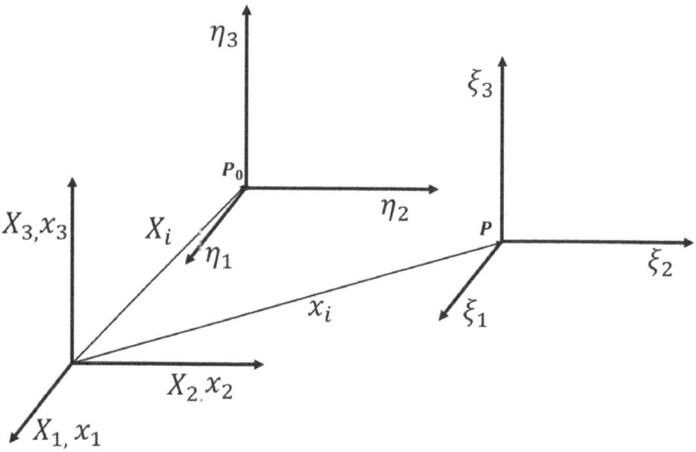

Figure 4.9. Linear transformation quadrics.

and the equation of the Eulerian strain quadric is

$$\epsilon_{ij}\, \xi_i\, \xi_j = \pm g^2 \tag{4.129}$$

The local strains in the vicinity of P_0 are calculated from the definition of the strain ellipsoid at that point, therefore, for an undeformed continuous medium, the equation of the boundary surface of an infinitesimal sphere of radius R is given as a function of the local material coordinates given according to (4.84), i.e.:

$$(d\vec{X})^2 = \delta_{ij}\, dX_i\, dX_j = R^2 \tag{4.130}$$

After strain, the surface equation of the same material particles is given according to (4.86) by:

$$(d\vec{X})^2 = C_{ij}\, dx_i\, dx_j = R^2 \tag{4.131}$$

which represents an ellipsoid known as the material strain ellipsoid. Thus, a spherical volume of a continuum in a continuum medium in an undeformed state is transformed by strain into an ellipsoid at P_0. Similarly, an infinitesimal spherical volume at P in the deformed medium originated because of an ellipsoidal volume element in the undeformed state. For a sphere of radius r located at P. The equations of these surfaces in terms of local coordinates are given by expressions (4.88) and (4.90) respectively, i.e.:

For the sphere

$$(d\vec{x})^2 = \delta_{ij}\, dx_i\, dx_j = r^2 \tag{4.132}$$

For the ellipsoid

$$(d\vec{x})^2 = G_{ij}\, dx_i\, dx_j = r^2 \tag{4.133}$$

The ellipsoid (4.133) is called a spatial strain ellipsoid. Strain ellipsoids are often called Cauchy strain ellipsoids.

4.2.9 Main strains. Strain invariants

The linear strain tensors, Lagrangian and Eulerian, are symmetric second-order Cartesian tensors and therefore, to determine their strains and principal directions, a procedure analogous to that used for stresses is followed, which is a standard procedure of linear algebra to find the eigenvalues and associated eigenvectors of the matrix in question.

Physically, a principal direction of a strain tensor is a direction for which the orientation of an element located at a given point is not altered by a pure strain, therefore, the value of a principal strain is simply the unit relative displacement (normal strain) that takes place in a principal direction.

For the Lagrangian strain tensor l_{ij}, the unit relative displacement vector is given by (4.103), therefore, the following expression can be written:

$$\frac{du_i}{dX} = (l_{ij} + W_{ij})v_j \tag{4.134}$$

Let us denote the normal strain in the direction of the unit vector n_i as $l_i^{(\hat{n})}$, in this way for a pure strain ($W_{ij} = 0$), (134) establishes that:

$$l_i^{(\hat{n})} = l_{ij}n_j \tag{4.135}$$

If direction n_i is a principal direction with principal strain l, then

$$l_i^{(\hat{n})} = l\delta_{ij}\, n_j \tag{4.136}$$

From relations (4.135) and (4.136) it can then be established that:

$$(l_{ij} - l\delta_{ij})\, n_j = 0 \tag{4.137}$$

which together with the condition of the unit vectors n_i, $n_i n_i = 1$, provide the necessary equations for the eigenvalues l and its eigenvectors or direction cosines n_i.

There exist non-trivial solutions of (4.137) if and only if the determinant of the coefficients vanishes, that is:

$$|l_{ij} - l\delta_{ij}| = 0 \tag{4.138}$$

whose development provides the cubic characteristic equation of l_{ij}, such that:

$$\mathrm{l}^3 - \mathrm{I_L}\, \mathrm{l}^2 + \mathrm{II_L}\, \mathrm{l} - \mathrm{III_L} = 0 \tag{4.139}$$

where

$$\mathrm{I_L} = \mathrm{l_{ii}} = \mathrm{tr\,L}, \quad \mathrm{II_L} = \frac{1}{2}(\mathrm{l_{ii}\, l_{ji}} - \mathrm{l_{ij}\, l_{ij}}), \quad \mathrm{III_L} = |\mathrm{l_{ii}}| = \det \mathrm{L} \tag{4.140}$$

They are respectively the first, second, and third Lagrangian strain invariants.

The roots of equation (4.139) establish the principal strains denoted $l_{(1)}$, $l_{(2)}$, and $l_{(3)}$.

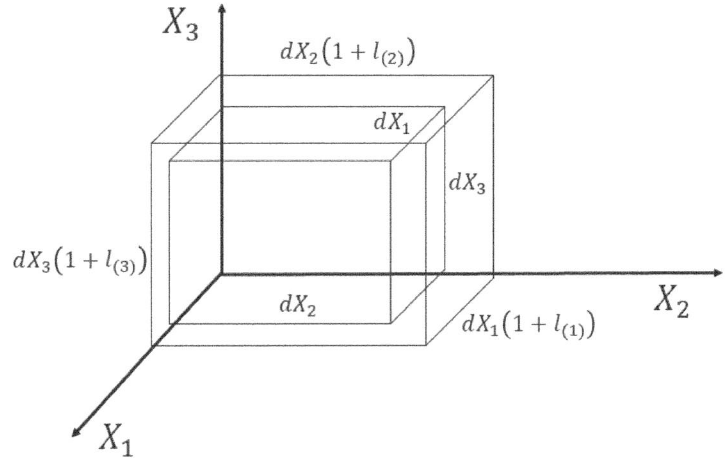

Figure 4.10. Strain invariants.

The first invariant of the Lagrangian strain tensor can be expressed in terms of the principal strains as:

$$I_L = l_{ii} = l_{(1)} + l_{(2)} + l_{(3)} \tag{4.141}$$

To understand the physical interpretation of (4.141), consider a differential rectangular parallelepiped whose edges are parallel to the principal strain directions, as shown in Figure 4.10.

The change in volume per unit of original volume of this parallelepiped is called cubic expansion and is calculated using the following expression:

$$D_0 = \frac{\Delta V_0}{V_0} = \frac{dX_1(1+l_{(1)})dX_2(1+l_{(2)})dX_3(1+l_{(3)}) - dX_1 dX_2 dX_3}{dX_1 dX_2 dX_3} \tag{4.142}$$

In the small strain approximation, the products can be ignored in expression (4.142) and only first order terms taken, therefore, expression (4.142), approximates the sum of the principal strains, such that:

$$D_0 = l_{(1)} + l_{(2)} + l_{(3)} = I_L \tag{4.143}$$

For this reason, it is called invariant, because regardless of the value of the strain, if it is small, the change in volume per unit volume is constant.

When the Eulerian strain tensor ϵ_{ij} and its associated unit relative displacement vector $\epsilon_i^{(\hat{n})}$ are considered, the principal directions and strains denoted $\epsilon_{(1)}$, $\epsilon_{(2)}$ and $\epsilon_{(3)}$. are determined in the same way as their Lagrangian counterparts. The Eulerian strain invariants can be expressed in terms of the principal strains as:

$$\begin{aligned} \mathrm{I_E} &= \epsilon_{(1)} + \epsilon_{(2)} + \epsilon_{(3)} \\ \mathrm{II_E} &= \epsilon_{(1)}\epsilon_{(2)} + \epsilon_{(2)}\epsilon_{(3)} + \epsilon_{(3)}\epsilon_{(1)} \\ \mathrm{III_E} &= \epsilon_{(1)}\epsilon_{(2)}\epsilon_{(3)} \end{aligned} \tag{4.144}$$

and the cubic dilation corresponding to the Eulerian description is:

$$\frac{\Delta V}{V} = D = \epsilon_{(1)} + \epsilon_{(2)} + \epsilon_{(3)} = I_E \quad (4.145)$$

4.2.10 Spherical and deviator strain tensors

Each of the linear Lagrangian and Eulerian strain tensors can be split into a spherical and a deviatoric tensor in the same way as the stress tensor was split. In this way, if the components of the Lagrangian and Eulerian deviatoric tensors are denoted by d_{ij} and e_{ij} respectively, then the strain tensors can be written as:

$$l_{ij} = d_{ij} + \delta_{ij}\frac{l_{kk}}{3} \quad (4.146)$$

and

$$\epsilon_{ij} = \epsilon_{ij} + \delta_{ij}\frac{\epsilon_{kk}}{3} \quad (4.147)$$

The deviatoric tensors are associated with shear strain, so the cubic dilation is zero. Therefore, it is not surprising that the first invariants d_{ij} and ϵ_{ij} of the deviatoric strain tensors are identically zero.

4.2.11 Compatibility equations for linear strains

If in the strain tensor given in expression (4.99), i.e.,

$$\epsilon_{ij} = \frac{1}{2}\left(\frac{\partial u_i}{\partial x_j} + \frac{\partial u_j}{\partial x_i}\right)$$

Its components are expressed as functions of the coordinates, the six equations generated from the expression (4.99), mathematically can be considered as a system of six partial differential equations to determine the three displacement components u_i, but said system is indeterminate and, in general, does not have a solution for any choice of the strain components of the tensor ϵ_{ij}.

Therefore, if the components of the displacement vector u_i have single and continuous values, some conditions must be imposed on the strain components. The necessary and sufficient conditions for such a displacement field are expressed by the following equations.

$$\frac{\partial^2 \epsilon_{ij}}{\partial x_k \partial x_m} + \frac{\partial^2 \epsilon_{km}}{\partial x_i \partial x_j} - \frac{\partial^2 \epsilon_{ik}}{\partial x_j \partial x_m} - \frac{\partial^2 \epsilon_{jm}}{\partial x_i \partial x_k} = 0 \quad (4.148)$$

In total there are 81 equations given by the expression (4.148), but only six are independent, which can be written as:

$$\frac{\partial^2 \epsilon_{11}}{\partial x_2^2} + \frac{\partial^2 \epsilon_{22}}{\partial x_1^2} = 2\frac{\partial^2 \epsilon_{12}}{\partial x_1 \partial x_2}$$

$$\frac{\partial^2 \epsilon_{22}}{\partial x_3^2} + \frac{\partial^2 \epsilon_{33}}{\partial x_2^2} = 2\frac{\partial^2 \epsilon_{23}}{\partial x_2 \partial x_3} \tag{4.149}$$

$$\frac{\partial^2 \epsilon_{33}}{\partial x_1^2} + \frac{\partial^2 \epsilon_{11}}{\partial x_3^2} = 2\frac{\partial^2 \epsilon_{31}}{\partial x_3 \partial x_1}$$

$$\frac{\partial}{\partial x_1}\left(-\frac{\partial \epsilon_{23}}{\partial x_1} + \frac{\partial \epsilon_{31}}{\partial x_2} + \frac{\partial \epsilon_{12}}{\partial x_3}\right) = \frac{\partial^2 \epsilon_{11}}{\partial x_2 \partial x_3}$$

$$\frac{\partial}{\partial x_2}\left(+\frac{\partial \epsilon_{23}}{\partial x_1} - \frac{\partial \epsilon_{31}}{\partial x_2} + \frac{\partial \epsilon_{12}}{\partial x_3}\right) = \frac{\partial^2 \epsilon_{22}}{\partial x_3 \partial x_1}$$

$$\frac{\partial}{\partial x_3}\left(+\frac{\partial \epsilon_{23}}{\partial x_1} + \frac{\partial \epsilon_{31}}{\partial x_2} - \frac{\partial \epsilon_{12}}{\partial x_3}\right) = \frac{\partial^2 \epsilon_{33}}{\partial x_1 \partial x_2}$$

4.3 Fundamental laws and equations

Introduction

The fundamental laws of continuum mechanics are essentially conservation laws of some physical quantity. In this book we consider only the conservation laws dealing with mass, momentum, and energy. With respect to energy, we first develop a purely mechanical energy balance and then develop an energy balance that includes both mechanical and thermal energies, making use of the first law of thermodynamics. Conservation laws are often called laws of equilibrium and are usually stated in the context of global (integral) relations derived from consideration of the conservation of some property of the body as a whole. The global equations can then be used to develop associated field equations that are valid at all points within the body and at its boundaries; for example, the local equations of motion are derived from a global statement of the conservation of linear momentum.

Constitutive equations, which reflect the internal constitution of a material, define specific types of material behavior. They are fundamental in that they serve as a starting point for studies in the disciplines of elasticity, plasticity, and various idealized fluids. Constitutive equations establish the relationship between kinematics and stresses in a material. These equations are the subject of this section. Before beginning a discussion of global conservation laws, it is useful to develop expressions for motion and flow in a continuous strain process and then for the material derivatives of certain integrals.

4.3.1 Movement and flow

Motion and flow are terms used to describe the continuous or instantaneous change in the configuration of a continuous medium. Flow sometimes implies the idea of a motion leading to permanent strain, as, for example, in plasticity studies. However, in fluid flow the word denotes continuous motion. As indicated in (4.72) and (4.73), the motion of a continuous medium can be expressed either in terms of material coordinates (Lagrangian description) according to

$$x_i = x_i\,(X_1, X_2, X_3, t) = x_i\,(\vec{X}, t) \tag{4.150}$$

or by the inverse of these equations in terms of spatial coordinates (Eulerian description) according to the expression:

$$X_i = X_i\,(x_1, x_2, x_3, t) = X_i\,(\vec{x}, t) \tag{4.151}$$

The necessary and sufficient condition for the existence of the inverse functions (4.151) is that the Jacobian determinant does not vanish

$$J = \left|\frac{\partial x_i}{\partial X_j}\right| \neq 0 \tag{4.152}$$

Since (4.150) and (4.151) are inverses of each other, any physical property of the continuum that is expressed with respect to a specific particle (material or Lagrangian description) can also be expressed with respect to the particular position in space occupied by the particle (spatial or Eulerian description). For example, if the material description of the density ρ is given by

$$\rho = \rho(X_i, t) = \rho(\vec{X}, t) \tag{4.153}$$

The spatial description is obtained by replacing X_i in this equation by the function given in (4.151). In this way the spatial description of the density is

$$\rho = \rho(X_i\,(x_i, t), t) = \rho^*\,(x_i, t) \equiv \rho(\vec{X}(\vec{x}, t), t) = \rho^*\,(\vec{x}, t) \tag{4.154}$$

where the symbol ρ^* is used here to highlight that the functional form of the Eulerian description is not necessarily the same as the Lagrangian form.

The time-related relationship of the variation of any property of a continuum relative to the specific particles of the moving continuum is called the material derivative of that property. The material derivative (also known as the substantial or convective derivative) can be thought of as a change with respect to time that would be measured by an observer traveling with the specific particles being studied.

The instantaneous position x_i of a particle is itself a property of the particle. The material derivative of the position of a particle is the instantaneous velocity of the particle. Therefore, adopting the symbol d/dt or the superimposed dot ($\cdot$) to represent the material derivative operation (some books use D/Dt), the velocity vector is defined as

$$v_i = \frac{dx_i}{dt} = \dot{x}_i \equiv \vec{v} = \frac{d\vec{v}}{dt} = \dot{\vec{x}} \tag{4.155}$$

In general, if $P_{ij\,...}$ is a scalar, vector or tensor property of a continuous medium that can be expressed as a position function of the coordinates of the point and if the Lagrangian description is given by

$$P_{ij\,...} = P_{ij\,...}\,(\vec{X}, t) \tag{4.156}$$

The material derivative of the property is expressed by

$$\frac{dP_{ij...}}{dt} = \frac{\partial P_{ij...}(\vec{X},t)}{\partial t} \tag{4.157}$$

The second member of equation (4.157) is sometimes written as

$$\left[\frac{\partial P_{ij...}(\vec{X},t)}{\partial t}\right]_{\vec{X}} \tag{4.158}$$

To emphasize that the coordinates $\vec{X}$ remain constant, that is, when taking the derivative the same particles are involved. When the property $P_{ij...}$ is expressed by the spatial description in the form

$$P_{ij...} = P_{ij...}(\vec{x}, t) \tag{4.159}$$

……..The material derivative is given by

$$\frac{dP_{ij...}(\vec{x},t)}{dt} = \frac{\partial P_{ij...}(\vec{x},t)}{\partial t} + \frac{\partial P_{ij...}(\vec{x},t)}{\partial x_k}\frac{\partial x_k}{\partial t} \tag{4.160}$$

where the last term on the right side appears because the specific particles are changing in space. The first term on the right side of (4.160) gives the rate of change at a particular position and is therefore conveniently called the local variation. This term is sometimes written as

$$\left[\frac{\partial P_{ij...}(\vec{x},t)}{\partial t}\right]_{\vec{x}} \tag{4.161}$$

to emphasize that $\vec{x}$ remains constant in this derivation. The second term on the right of (4.160) is called the convective variation since it expresses the contribution due to the motion of the particles in the variable field of the property. From (4.155), the material derivative (4.160) can be expressed as:

$$\frac{dP_{ij...}(\vec{x},t)}{dt} = \frac{\partial P_{ij...}(\vec{x},t)}{\partial t} + v_k\frac{\partial P_{ij...}(\vec{x},t)}{\partial x_k} \tag{4.162}$$

which immediately suggests the introduction of the material derivative operator

$$\frac{d}{dt} = \frac{\partial}{\partial t} + v_k\frac{\partial}{\partial x_k} \equiv \frac{d}{dt} = \frac{\partial}{\partial t} + v\cdot\nabla_{\vec{x}} \tag{4.163}$$

It is used to take the material derivatives of quantities expressed in spatial coordinates.

4.3.1.1 Speed and acceleration

A definition of the velocity vector is given in expression (4.155), according to $v_i = {dx_i}/{dt}$. Another way of defining the same vector is obtained from expression (4.69) which gives $x_i = u_i + X_i$. In this way the velocity can be defined by

$$v_i = \frac{dx_i}{dt} = \frac{d(u_i + X_i)}{dt} = \frac{du_i}{dt} \qquad (4.164)$$

Since X_i is independent of time. In expression (4.164), if the displacement is expressed in the Lagrangian form $u_i = u_i(\vec{X}, t)$, then

$$v_i = \dot{u}_i = \frac{du_i(\vec{X},t)}{dt} = \frac{\partial u_i(\vec{X},t)}{\partial t} \qquad (4.165)$$

.......If, on the other hand, the displacement is in Eulerian form $u_i = u_i(\vec{x}, t)$, then

$$v_i(\vec{x},t) = \dot{u}_i(\vec{x},t) = \frac{du_i(\vec{x},t)}{dt} = \frac{\partial u_i(\vec{x},t)}{\partial t} + v_k(\vec{x},t)\frac{\partial u_i(\vec{x},t)}{\partial x_k} \qquad (4.166)$$

.......In (4.166) the velocity is given implicitly since it appears as a factor of the second term on the right. The function

$$v_i = v_i(\vec{x}, t) \qquad (4.167)$$

It is said to specify the instantaneous velocity field.

.......The material derivative of the velocity is the acceleration. If the velocity is given in the Lagrangian form (4.165), then

$$a_i = \dot{v}_i = \frac{dv_i(\vec{X},t)}{dt} = \frac{\partial v_i(\vec{X},t)}{\partial t} \qquad (4.168)$$

.......If the velocity is given in the Eulerian form (4.166), then

$$a_i(\vec{x},t) = \frac{dv_i(\vec{x},t)}{dt} = \frac{\partial v_i(\vec{x},t)}{\partial t} + v_k(\vec{x},t)\frac{\partial v_i(\vec{x},t)}{\partial x_k} \qquad (4.169)$$

4.3.1.2 Strain rate and vorticity and physical interpretation

In this section it is important to highlight the difference between the concepts of trajectory and streamline, before going on to review the description of the strain rate.

.......A trajectory is the curve or path traveled by a particle in a flow or motion. A streamline is a tangent curve that at any point is in the direction of the velocity at that point. The motion of a continuous medium is called stationary motion if the velocity field is independent of time, such that ${\partial v_i}/{\partial t} = 0$. For stationary motion, trajectory lines coincide with streamlines.

…….The spatial gradient of the instantaneous velocity field defines the velocity gradient tensor $\partial v_i / \partial x_j$ which will be denoted by Y_{ij}. This tensor can be decomposed into its symmetric and antisymmetric parts as follows:

$$Y_{ij} = \frac{\partial v_i}{\partial x_j} = \frac{1}{2}\left(\frac{\partial v_i}{\partial x_j} + \frac{\partial v_j}{\partial x_i}\right) + \frac{1}{2}\left(\frac{\partial v_i}{\partial x_j} - \frac{\partial v_j}{\partial x_i}\right) = D_{ij} + V_{ij} \tag{4.170}$$

This decomposition is valid even if v_i and $\partial v_i / \partial x_j$ are finite quantities. The symmetric tensor

$$D_{ij} = D_{ji} = \frac{1}{2}\left(\frac{\partial v_i}{\partial x_j} + \frac{\partial v_j}{\partial x_i}\right) \tag{4.171}$$

It is called the strain rate tensor. Other texts call it the strain rate tensor, the extension rate tensor, and the specific strain rate tensor. The antisymmetric tensor

$$V_{ij} = -V_{ji} = \frac{1}{2}\left(\frac{\partial v_i}{\partial x_j} - \frac{\partial v_j}{\partial x_i}\right) \tag{4.172}$$

It is called the spin tensor, vorticity tensor or rotational velocity tensor. It can be easily proven that the strain rate tensor is the material derivative of the Eulerian linear strain tensor (4.99). Thus, if in the equation

$$\frac{d\epsilon_{ij}}{dt} = \frac{1}{2}\frac{d}{dt}\left(\frac{\partial u_i}{\partial x_j} + \frac{\partial u_j}{\partial x_i}\right) \tag{4.173}$$

The derivatives with respect to time and coordinates can be interchanged, then $d/dt\left(\partial u_i / \partial x_j\right)$ is replaced by $d/dx_j\left(\partial u_i / \partial t\right)$ and equation (4.173) takes the form

$$\frac{d\epsilon_{ij}}{dt} = \frac{1}{2}\frac{d}{dt}\left(\frac{\partial v_i}{\partial x_j} + \frac{\partial v_j}{\partial x_i}\right) = D_{ij} \tag{4.174}$$

Using a similar procedure, it can be proven that the vorticity tensor is equal to the material derivative of the Eulerian linear rotation tensor. The result is expressed by the equation

$$\frac{d\omega_{ij}}{dt} = \frac{1}{2}\frac{d}{dt}\left(\frac{\partial v_i}{\partial x_j} - \frac{\partial v_j}{\partial x_i}\right) = V_{ij} \tag{4.175}$$

An interesting interpretation can be given to equation (4.174) when it is rewritten in the form

$$d\epsilon_{ij} = D_{ij}\, dt \tag{4.176}$$

The first member of equation (4.176) represents the components known as natural strain increments which are widely used in flow problems and in plasticity theory.

4.3.1.3 Physical interpretation of strain rate tensors

The velocities of two nearby particles at two points P and Q of a moving continuum are given by v_i and $v_i + dv_i$ respectively. The relative velocity of the particle at point Q with respect to that at point P is therefore:

$$dv_i = \frac{\partial v_i}{\partial x_j} dx_j \tag{4.177}$$

in which the partial derivatives are to be evaluated at point P. As a function of D_{ij} and V_{ij}, equation (4.177) becomes:

$$dv_i = (D_{ij} + V_{ij})dx_j \tag{4.178}$$

If the strain rate tensor is identically zero ($D_{ij} \equiv 0$), then,

$$dv_i = V_{ij}dx_j \tag{4.179}$$

And the motion in the vicinity of P is rigid body rotation. For this reason, it is said that a field is irrotational, and the vorticity field vanishes at any point in the field. The vector associated with the vorticity tensor is defined by:

$$q_i = \epsilon_{ijk}\, v_{k,j} \tag{4.180}$$

.......It is known as the vorticity vector or vortex, which can be symbolically written as $\vec{q} = \nabla_{\vec{X}} \times \vec{V} = rot\ \vec{V}$, indicating that the vorticity vector is the curl of the velocity field. The vector is defined as half of the vorticity vector,

$$\Omega_i = \frac{1}{2} q_i = \epsilon_{ijk}\, v_{k,j} \tag{4.181}$$

.......It is called the rotation velocity vector. For the rotation of a rigid body, such as that described by equation (4.179), the relative velocity of a neighboring particle separated from P by dx_i, is given by:

$$dv_i = \epsilon_{ijk}\, \Omega_i\, dx_i \tag{4.182}$$

The diagonal elements of the tensor D_{ij} are known as the velocity components of extension or elongation. For a pure strain, equation (4.178) gives

$$dv_i = D_{ij}\, dx_j \tag{4.183}$$

And since the rate at which the length of a line element dx_i changes per unit instantaneous length is given by:

$$d_i^{(v)} = \frac{dv_i}{dx} = D_{ij} \frac{dx_j}{dx} = D_{ij} v_j \tag{4.184}$$

The strain rate in the direction of the unit vector v_i is:

$$d = d_i^{(v)}\, v_j = D_{ij}\, v_j\, v_i \tag{4.185}$$

And if v_i is in the direction of a coordinate axis for example $\hat{e}_1$, then

$$d = d_{11} \tag{4.186}$$

.......Since D_{ij} is a symmetric second-order tensor, the concepts of principal axes, principal values, invariants, a strain rate quadric, and a strain rate deviator tensor are associated with it. Compatibility equations for the strain rate tensor components, analogous to those already presented, can also be developed.

4.3.1.4 Material derivatives of volume, surface and line elements

During the motion from an initial configuration at time $t = 0$ to a configuration present at time $t = t$, the particles of the continuum that occupied the differential volume element dV_0 in the initial state now occupy the differential volume element dV. If the rectangular parallelepiped indicated in Figure 4.11 is taken as the initial volume element, then

$$dV = dX_1\,\hat{e}_1 \times dX_2\,\hat{e}_2 \cdot dX_3\,\hat{e}_3 = dX_1\,dX_2\,dX_3 \tag{4.187}$$

Due to the motion, this parallelepiped is translated and distorted, but because the motion is assumed to be continuous the volume element is not subdivided. Indeed, due to the relation (4.89) between the spatial and material line elements, the "Particle Line" formed by dX_1 now forms the differential line segment

$$dx_i^{(1)} = \left(\frac{\partial x_i}{\partial X_1}\right) dX_1 \tag{4.188}$$

Similary dX_2 becomes

$$dx_i^{(2)} = \left(\frac{\partial x_i}{\partial X_2}\right) dX_2 \tag{4.189}$$

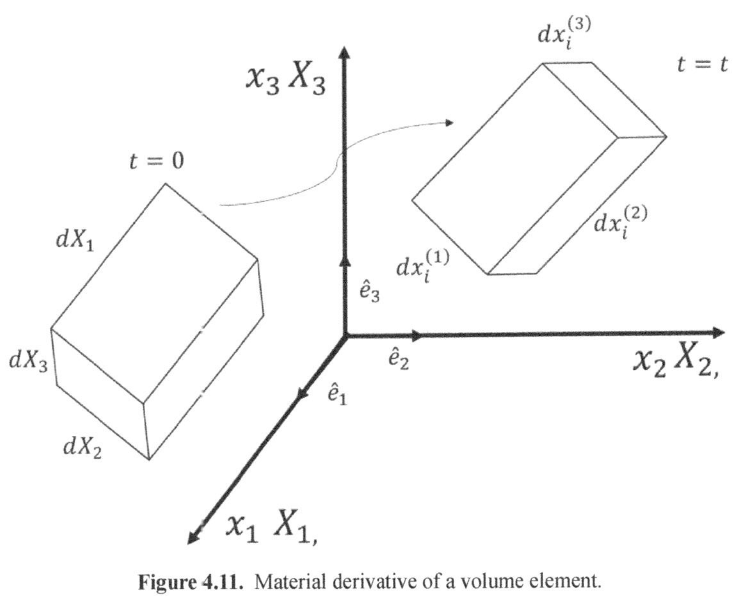

Figure 4.11. Material derivative of a volume element.

and dX_3 in

$$dx_i^{(3)} = \left(\frac{\partial x_i}{\partial X_3}\right) dX_3 \tag{4.190}$$

Therefore, the differential volume element dV is a rotated parallelepiped with edges $x_i^{(1)}$, $dx_i^{(2)}$, $dx_i^{(3)}$ and a volume given by the scalar triple product

$$dV = d\vec{x}^{(1)} \times d\vec{x}^{(2)} \cdot d\vec{x}^{(3)} = \epsilon_{ijk}\, dx_i^{(1)}\, dx_i^{(2)}\, dx_i^{(3)} \tag{4.191}$$

From the expression (4.191), we can write

$$dV = \epsilon_{ijk} \frac{\partial x_i}{\partial X_1} \frac{\partial x_j}{\partial X_2} \frac{\partial x_k}{\partial X_3} dX_1 dX_2 dX_3 = J dV_0 \tag{4.192}$$

Where $J = \left| \partial x_i / \partial X_j \right|$ is the Jacobian. Using expression (4.192), the material derivative of dV as:

$$\frac{d}{dt}(dV) = \frac{d}{dt}(J dV_0) = \frac{dJ}{dt}(dV_0) \tag{4.193}$$

Since dV_0 is independent of time, then $\frac{d}{dt}(dV_0) = 0$.

On the other hand, let us denote as $x_{i,P}$ the derivative $\partial x_i / \partial X_p$ (see Table 3.3, Chapter 3) so that we can write J as:

$$J = \epsilon_{PQR}\, x_{1,P}\, x_{2,Q}\, x_{3,R} \tag{4.194}$$

And the derivative, $\dot{J}$, becomes the sum of the three determinants,

$$\dot{J} = \epsilon_{PQR}(\dot{x}_{1,P}\, x_{2,Q}\, x_{3,R} + x_{1,P}\, \dot{x}_{2,Q}\, x_{3,R} + x_{1,P}\, x_{2,Q}\, \dot{x}_{3,R}) \tag{4.195}$$

besides

$$\dot{x}_{1,P} = v_{1,s}\, x_{s,P};\ \dot{x}_{2,Q} = v_{2,s}\, x_{s,Q};\ \dot{x}_{1,P} = v_{3,s}\, x_{s,R} \tag{4.196}$$

In this way the expression (4.195) can be written as:

$$\dot{J} = \epsilon_{PQR}\, (v_{1,s}\, x_{s,P}\, x_{2,Q}\, x_{3,R} + x_{1,P}\, v_{2,s}\, x_{s,Q}\, x_{3,R} + x_{1,P}\, x_{2,Q}\, v_{3,s}\, x_{s,R}) \tag{4.197}$$

The sum in s of the expression (4.197) generates nine 3×3 determinants, the three non-zero determinants give

$$\dot{J} = v_{1,1}\, J + v_{2,2}\, J + v_{3,3}\, J = v_{s,s}\, J \tag{4.198}$$

therefore

$$\dot{J} = J \nabla_{\vec{x}} \cdot \vec{v} \equiv \frac{dJ}{dt} = J \frac{\partial v_i}{dx_i} \tag{4.199}$$

and so

$$\frac{d(\ln J)}{dt} = div\,\vec{v} \tag{4.200}$$

from (4.193) and (4.199) we obtain

$$\frac{d}{dt}(dV) = \frac{\partial v_i}{\partial x_i} dV \tag{4.201}$$

For the initial configuration of a continuum, a differential element of area having magnitude dS^0 can be represented in terms of its unit normal vector n_i by the expression $dS^0 n_i$. For the present configuration of the moving continuum, the particles that initially form the area $dS^0 n_i$ now occupy an area element represented by the vector $dS\, n_i$ or dS_i. It can be shown that:

$$dS_i = J\frac{\partial X_j}{\partial x_i} dX_j \tag{4.202}$$

Therefore, the material derivative of the area element is

$$\frac{d}{dt}(dS_i) = \frac{d}{dt}\left(J\frac{\partial X_j}{\partial x_i}\right) dX_j = \frac{\partial v_j}{\partial x_j} dS_i - \frac{\partial v_j}{\partial x_i} dS_j \tag{4.203}$$

The material derivative of the squared length of the differential line element dx_i, can be calculated as follows:

$$\frac{d}{dt}(dx^2) = \frac{d}{dt}(dx_i dx_i) = 2\frac{d(x_i)}{dt} dx_i \tag{4.204}$$

However, since $dx_i = \left(\partial x_i \big/ \partial X_j\right)$,

$$\frac{d(x_i)}{dt} = \frac{d}{dt}\left(\frac{\partial x_i}{\partial X_j}\right) dX_j = \frac{\partial v_i}{\partial X_j} dX_j = \frac{\partial v_i}{\partial x_k}\frac{\partial x_k}{\partial X_j} dX_j = \frac{\partial v_i}{\partial x_k} dx_k \tag{4.205}$$

with this result, the expression (4.204) becomes:

$$\frac{d}{dt}(dx^2) = 2\frac{v_i}{x_k} dx_k dx_i \tag{4.206}$$

The expression on the right side of (4.206) in the indicial form is symmetric in i and k, and therefore can be written as:

$$\frac{\partial v_i}{\partial x_k} dx_k dx_i + \frac{\partial v_k}{\partial x_i} dx_i dx_k = \left(\frac{\partial v_i}{\partial x_k} + \frac{\partial v_k}{\partial x_i}\right) dx_i dx_k \tag{4.207}$$

or of (4.171)

$$\frac{d}{dt}(dx^2) = 2D_{ij} dx_k dx_i \tag{4.208}$$

4.3.1.5 Material derivatives of volume, surface and line integrals

Not all properties of a continuum can be defined for a specific particle as functions of the coordinates such as those given by (4.156) and (4.159). Some properties are defined as integrals over a finite region of a continuum. Let any scalar, vector or tensor property be represented by the volume integral

$$P_{ij\ldots}(t) = \int_V P^*_{ij}\ldots(\vec{x},t)dV \qquad (4.209)$$

where V is the volume occupied by the considered part of the continuous medium at the instant t: The material derivative of $P_{ij\ldots}(t)$ and since the derivation is with respect to a defined part of the continuous medium, the operations of derivation and integration can be interchanged, therefore, we have

$$\frac{d\left[P_{ij\ldots}(t)\right]}{dt} = \frac{d}{dt}\int_V P^*_{ij}\ldots(\vec{x},t)dV = \int_V \frac{d}{dt}\left[P^*_{ij}\ldots(\vec{x},t)\right]dV \qquad (4.210)$$

In expression (4.210) after performing the derivation and using (4.198), we obtain

$$\frac{d}{dt}\int_V P^*_{ij}\ldots(\vec{x},t)dV = \int_V \left[\frac{dP^*_{ij}\ldots(\vec{x},t)}{dt} + P^*_{ij}\ldots(\vec{x},t)\frac{\partial v_p}{\partial x_p}\right]dV \qquad (4.211)$$

Using the material derivation operator expression given in (4.163), the expression (4.211) can be written as:

$$\frac{d}{dt}\int_V P^*_{ij}\ldots(\vec{x},t)dV = \int_V \left[\frac{\partial P^*_{ij}\ldots(\vec{x},t)}{\partial t} + \frac{\partial}{\partial x_p}\left(v_p P^*_{ij}\ldots(\vec{x},t)\right)\right]dV \qquad (4.212)$$

.......Using the Gauss divergence theorem, the second term of the integral of the second member of (4.212) can be expressed as a surface integral, in this way the following expression is obtained:

$$\frac{d}{dt}\int_V P^*_{ij}\ldots(\vec{x},t)dV = \int_V \left[\frac{\partial P^*_{ij}\ldots(\vec{x},t)}{\partial t}\right]dV + \int_S v_p\left[P^*_{ij}\ldots(\vec{x},t)\right]dS_p \qquad (4.213)$$

The physical interpretation of equation (4.213) states that the rate of growth of the property $P_{ij\ldots}(t)$ in a portion of the material that instantaneously occupies the volume V, is equal to the sum of the amount of the property created within V plus the flux $v_p[P^*_{ij\ldots}(\vec{x}, t)]$ through the boundary S of V.

.......In the same way as the material derivative of volume integrals has just been found, the material derivative of surface and line integrals is found, then, for any tensor property of a continuous medium represented by the surface integral

$$Q_{ij}\ldots(t) = \int_S Q^*_{ij}\ldots(\vec{x},t)dS_p \qquad (4.214)$$

where S is the surface occupied by the considered part of the continuous medium at the instant t, then as already done:

$$\frac{d}{dt}\int_S Q^*_{ij}\ldots(\vec{x},t)dS_p = \int_S \frac{d}{dt}\left[Q^*_{ij}\ldots(\vec{x},t)\right]dS_p \tag{4.215}$$

.......Using (4.203), differentiation of (4.215) gives the following result:

$$\begin{aligned}\frac{dQ_{ij}\ldots(t)}{dt} &= \int_S\left[\frac{dQ^*_{ij}\ldots(\vec{x},t)}{dt} + \frac{\partial v_p}{\partial x_q}Q^*_{ij}\ldots(\vec{x},t)\right]dS_p \\ &-\int_S Q^*_{ij}\ldots\frac{\partial v_p}{\partial x_q}dS_p\end{aligned} \tag{4.216}$$

.......For properties expressed in line integrals such as:

$$R_{ij}\ldots(t) = \int_C R^*_{ij}\ldots(\vec{x},t)dx_p \tag{4.217}$$

.......The material derivative is given by

$$\frac{d}{dt}\int_C R^*_{ij}\ldots(\vec{x},t)dx_p = \int_C \frac{d}{dt}\left[R^*_{ij}\ldots(\vec{x},t)\right]dx_p \tag{4.218}$$

.......Deriving the integral of the second member as established in (4.218) and using the expression (4.205), the partial derivative is obtained as:

$$\frac{d}{dt}\left[R_{ij}\ldots(t)\right] = \int_C \frac{d}{dt}\left[R^*_{ij}\ldots(\vec{x},t)\right]dx_p + \int_C \frac{\partial v_p}{\partial x_q}\left[R^*_{ij}\ldots(\vec{x},t)\right]dx_p \tag{4.219}$$

4.3.2 Conservation laws

4.3.2.1 Conservation of the mass and the continuity equation

The mass of a continuous material occupying a volume V at time t is defined by the integral:

$$m = \int_V \rho(\vec{x},t)dV \tag{4.220}$$

where $\rho(\vec{x}, t)$ is the mass density, a continuous function in space and time. The law of conservation of mass states that the mass of a specific part of the continuous medium remains constant, and therefore the material derivative of expression (4.220) is zero. Then, from (4.208) with $P^*_{ij}\ldots(\vec{x}, t) = \rho(\vec{x}, t)$, the derivative of the mass is:

$$\frac{dm}{dt} = \frac{d}{dt}\int_V \rho(\vec{x},t)dV = \int_V\left[\frac{d\rho(\vec{x},t)}{dt} + \rho(\vec{x},t)\frac{\partial v_p}{\partial x_p}\right]dV = 0 \tag{4.221}$$

Since (4.221) is valid for an arbitrary volume V, the integrand must necessarily be zero, that is:

$$\frac{d\rho(\vec{x},t)}{dt} + \rho(\vec{x},t)\frac{\partial v_p}{\partial x_p} = 0 \tag{4.222}$$

It can also be written as:

$$\frac{d\rho}{dt} + \rho v_{k,k} = 0 \equiv \frac{d\rho}{dt} + \rho(\nabla \cdot \vec{v}) = 0 \qquad (4.223)$$

Equation (4.223) is called the continuity equation; using the material derivation operator it can also be written as:

$$\frac{d\rho}{dt} + (\rho v_k)_k = 0 \equiv \frac{d\rho}{dt} + \nabla \cdot (\rho \vec{v}) = 0 \qquad (4.224)$$

For an incompressible continuous medium, the mass density of each particle is independent of time, such that $d\rho/dt = 0$ and therefore, expression (4.223) gives:

$$v_{k,k} = 0 \equiv \nabla \cdot \vec{v} = 0 \equiv div\, \vec{v} = 0 \qquad (4.225)$$

The velocity field $\vec{v}(\vec{x}, t)$ of an incompressible continuum can therefore be expressed by the equation

$$v_i = \epsilon_{ijk}\, s_{k,j} \equiv \vec{v} = \nabla \times \vec{s} \equiv \vec{v} = rot\, \vec{s} \qquad (4.226)$$

where $\vec{s}(\vec{x}, t)$ is a continuous vector field in space and time and is called the vector potential of $\vec{v}$.

.......The continuity equation can also be expressed in its material or Lagrangian form. Conservation of mass requires that:

$$\int_{V_0} \rho_0(\vec{X},0)dV_0 = \int_{V_0} \rho(\vec{x}(\vec{X},t),t)JdV_0 = \int_{V_0} \rho(\vec{X},t)JdV_0 \qquad (4.227)$$

Since this relationship is valid for any arbitrary volume V_0, we have:

$$\rho_0 = \rho J \qquad (4.228)$$

which implies that the product ρJ is independent of time since V is arbitrary, therefore, its derivative with respect to time is zero, that is:

$$\frac{d}{dt}(\rho J) = 0 \qquad (4.229)$$

Equation (4.229) is the Lagrangian differential form of the continuity equation.

4.3.2.2 Conservation of the amount of motion. Equations of motion

Consider a moving continuous medium occupying a volume V at time t. Suppose that the mass forces b_i per unit volume are known and that the stress vector $t_i^{(\hat{n})}$ acts on a differential element dS of the boundary surface, and that the velocity field $v_i = du_i/dt$ t, is prescribed throughout the region occupied by the continuous medium. In this case, the total linear momentum of the mass system contained in V is given by:

$$P_i(t) = \int_V \rho v_i dV \qquad (4.230)$$

The principle of linear momentum, based on Newton's second law, states that the variation per unit of time of an arbitrary part of a continuous medium is equal to the resultant force acting on that part considered, therefore, if the internal forces between the particles of the continuous medium obey Newton's third law of action and reaction, the principle of momentum for this mass system is expressed by:

$$\int_S t_i^{(\hat{n})} dS + \int_V \rho B_i dV = \frac{d}{dt}\int_V \rho v_i dV \tag{4.231}$$

Substituting $t_i^{(\hat{n})} = \sigma_{ji}\, n_i$ into the first integral of (4.231) and transforming this surface integral into a volume integral using the Gauss divergence theorem, the expression (4.231) becomes:

$$\int_V (\sigma_{ji,j} + \rho B_i) dV = \frac{d}{dt}\int_V \rho v_i dV \tag{4.232}$$

By calculating the material derivative of (4.232), the continuity equation can be used in the form given by expression (4.229), therefore:

$$\begin{aligned}\frac{d}{dt}\int_V \rho v_i dV &= \frac{d}{dt}\int_{V_0} \rho v_i J dV_0 \\ &= \int_{V_0}\left[v_i \frac{d(\rho J)}{dt} + \rho J \frac{dv_i}{dt}\right] dV_0 = \int_V \frac{dv_i}{dt}\rho dV\end{aligned} \tag{4.233}$$

Replacing the second term of equation (4.232) by the corresponding term of equation (4.233), regrouping terms we obtain the principle of linear momentum in its integral form:

$$\int_V (\sigma_{ji,j} + \rho B_i - \rho \dot{v}_i) dV = 0 \tag{4.234}$$

Since the integration volume is arbitrary, the integrand of equation (4.234) must be zero. Therefore, we obtain:

$$\sigma_{ji,j} + \rho B_i = \rho \dot{v}_i \tag{4.235}$$

which are known as the equations of motion. In the case of static equilibrium, the acceleration components cancel out and equations (4.235) take the form:

$$\sigma_{ji,j} + \rho B_i = 0 \tag{4.236}$$

These are called the mechanical equilibrium equations.

4.3.2.3 Energy conservation

Considering only mechanical quantities, the principle of conservation of energy of the continuous medium can be deduced directly from the equation of motion (4.235), to obtain the expression of conservation of energy the scalar product between equation (4.235), the velocity vector v_i is calculated and the result is integrated in the volume V. Then

$$\int_V \rho v_i \dot{v}_i dV = \int_V v_i \sigma_{ji,j} dV + \int_V \rho v_i B_i dV \tag{4.237}$$

But how

$$\int_V \rho v_i \dot{v}_i dV = \frac{d}{dt}\int_V \rho \frac{v_i v_i}{2} dV = \frac{d}{dt}\int_V \rho \frac{v^2}{2} dV = \frac{dK}{dt} \tag{4.238}$$

which represents the variation per unit of time of the kinetic energy K in the continuous medium. In addition $v_i\, \sigma_{ji,j} = (v_i\, \sigma_{ji})_{,i} - v_{i,j}\sigma_{ji}$ and from equation (4.170) $v_{i,j} = D_{ij} + V_{ij}$ which equation (4.237) gives:

$$\frac{dK}{dt} + \int_V D_{ij}\sigma_{ji} dV = \int_V (v_i \sigma_{ji})_{,j} dV + \int_V \rho v_i B_i dV \tag{4.239}$$

since $v_{i,j}\sigma_{ji} = 0$.

.......Transforming the first integral of the second term of equation (4.239) into a surface integral by means of the Gauss divergence theorem and making use of the identity $t_i^{(\hat{n})} = \sigma_{ji}n_j$, the energy equation appears in the form

$$\frac{dK}{dt} + \int_V D_{ij}\sigma_{ji} dV = \int_S v_i t_i^{(\hat{n})} dS + \int_V \rho v_i B_i dV \tag{4.240}$$

The first member of equation (4.240) relates the variation of the total energy per unit of time of the continuous medium, with the variation of the amount of work done by the mass and surface forces that appear in the second member of the equation. The integral of the first member is known as the variation of the internal mechanical energy per unit of time and is denoted dU/dt. Therefore, equation (4.240) can be written as:

$$\frac{dK}{dt} + \frac{dU}{dt} = \frac{\bar{d}W}{dt} \tag{4.241}$$

where $\frac{dW}{dt}$ represents the change in the amount of work, and the symbol $\bar{d}$ indicates that this quantity is not an exact differential. If both mechanical and non-mechanical energies are to be considered, the principle of conservation of energy must be used in its most general form. In this form the principle of conservation states that the change in kinetic energy plus internal energy per unit time is equal to the sum of the change in work plus any other energy supplied or extracted per unit time. Such supplied energies may include thermal, chemical or electromagnetic energy. In the following, only mechanical and thermal energies will be considered, and the principle of energy takes the form of the well-known first law of thermodynamics.

For a continuous thermo-mechanical medium, it is customary to express the variation of the internal energy per unit of time as the integral expression:

$$\frac{dU}{dt} = \frac{d}{dt}\int_V \rho u dV = \int_V \rho \dot{u} dV \tag{4.242}$$

where u is called the specific internal energy. We define the vector c_i as the heat flux per unit area and time in heat conduction, and p is taken as the heat radiation constant per unit mass and time, the rate of growth of the amount of heat in the continuous medium is given by:

$$\frac{dQ}{dt} = -\int_S c_i n_i dS + \int_V \rho p dV \tag{4.243}$$

Therefore, the energy principle for a continuous thermo-mechanical medium is:

$$\frac{dK}{dt} + \frac{dU}{dt} = \frac{\bar{d}W}{dt} + \frac{\bar{d}Q}{dt} \tag{4.244}$$

In integral form we can write:

$$\begin{aligned} &\frac{d}{dt}\int_V \rho \frac{v_i v_i}{2} dV + \int_V \rho \dot{u} dV \\ &= \int_S v_i t_i^{(\hat{n})} dS + \int_V \rho v_i B_i dV - \int_S c_i n_i dS + \int_V \rho p dV \end{aligned} \tag{4.245}$$

Transforming the surface integrals of (4.245) into volume integrals by Gauss's divergence theorem, and again using the fact that V is arbitrary, one arrives at the local form of the energy equation.

$$\frac{d}{dt}\left(\frac{v^2}{2} + u\right) = \frac{1}{\rho}(\sigma_{ij} v_i)_{,j} + B_i v_i - \frac{1}{\rho} c_{i,i} + p \tag{4.246}$$

Within the arbitrarily small volume element for which the local energy equation is valid, the momentum balance given by (4.246) must also be satisfied. Therefore, taking the scalar product of (4.235) and the velocity $\dot{v}_i\, v_i = v_i\, \sigma_{ji,i} + \rho v_i\, B_i$ and, by performing algebraic operations, subtracting this product from (4.246), the reduced form of the local energy equation results:

$$\frac{du}{dt} = \frac{1}{\rho} \sigma_{ij} D_{ij} - \frac{1}{\rho} c_{i,i} + p \tag{4.247}$$

This equation expresses the variation of internal energy per unit of time as the sum of the work per unit of time due to the stresses or power of stress plus the heat added to the continuous medium.

4.4 Linear elasticity

In the linear elasticity approximation, the displacement gradients and displacements are assumed to be very small, according to equations (4.98) and (4.100), this implies that

$$l_{ij} = \epsilon_{ij} = \frac{1}{2}\left(\frac{\partial u_i}{\partial x_j} + \frac{\partial u_j}{\partial x_i}\right) = \frac{1}{2}(u_{i,j} + u_{j,i}) \tag{4.248}$$

In the following, strain processes are assumed to be adiabatic (no heat loss or gain) and isothermal (at constant temperature) unless specifically stated otherwise. This section studies Hooke's Law, its symmetries, constants and applications to fields such as elastostatics, elastodynamics, hypoelasticity, hyperelasticity and linear thermoelasticity.

4.4.1 Hooke's law

The constitutive equations for a linear elastic solid relate the stress and strain tensors through the expression

$$\sigma_{ij} = C_{ijkm}\,\epsilon_{km} \tag{4.249}$$

which is known as the generalized Hooke's law. In equations (4.249), the tensor of elastic constants C_{ijkm} has 81 components, but due to the symmetry of the stress and strain tensors, there are at most 36 distinct elastic constants. To write Hooke's law using these 36 constants, the system of assigning double indices to the stress and strain components is often replaced by a simple notation system with a rank 6 index, as follows

$$\begin{aligned} &\sigma_{11} = \sigma_1 \quad \sigma_{23} = \sigma_{32} = \sigma_4 \\ &\sigma_{22} = \sigma_2 \quad \sigma_{13} = \sigma_{31} = \sigma_5 \\ &\sigma_{33} = \sigma_3 \quad \sigma_{12} = \sigma_{21} = \sigma_6 \end{aligned} \tag{4.250}$$

$$\begin{aligned} &\epsilon_{11} = \epsilon_1 \quad 2\epsilon_{23} = 2\epsilon_{32} = \epsilon_4 \\ &\epsilon_{22} = \epsilon_2 \quad 2\epsilon_{13} = 2\epsilon_{31} = \epsilon_5 \\ &\epsilon_{33} = \epsilon_3 \quad 2\epsilon_{12} = 2\epsilon_{21} = \epsilon_6 \end{aligned} \tag{4.251}$$

Hooke's law can be written

$$\sigma_k = C_{KM}\,\epsilon_M \quad (K,M = 1,2,3,4,5,6) \tag{4.252}$$

where C_{KM} represents the 36 elastic constants, where the subscripts in Latin capitals are used to highlight that the range of these indices is 6. If thermal effects are not considered, the energy equations (4.247) can be written as:

$$\frac{du}{dt} = \frac{1}{\rho}\sigma_{ij}D_{ij} = \frac{1}{\rho}\sigma_{ij}\,\dot{\epsilon}_{ij} \tag{4.253}$$

In this case, the internal energy is purely mechanical and is called strain energy (per unit mass). From equation (4.253) we can write:

$$du = \frac{1}{\rho}\sigma_{ij}\,d\,\epsilon_{ij} \tag{4.254}$$

and if u is considered as a function of the nine strain components, $u = (\mu_{ij})$, its differential will be:

$$du = \frac{\partial u}{\partial \epsilon_{ij}}\,d\epsilon_{ij} \tag{4.255}$$

Comparing this equation with the previous one, we observe that:

$$\frac{1}{\rho}\sigma_{ij} = \frac{\partial u}{\partial \epsilon_{ij}} \tag{4.256}$$

The strain energy density u^* (per unit volume) is defined as:

$$u^* = \rho u \tag{4.257}$$

and since ρ can be considered constant in the theory of small strains, u^* has the property

$$\sigma_{ij} = \rho\frac{\partial u}{\partial \epsilon_{ij}} = \frac{\partial u^*}{\partial \epsilon_{ij}} \tag{4.258}$$

Furthermore, a zero-strain energy state can be arbitrarily chosen; and since the stress must vanish with strain, the simplest form of the strain energy function that leads to a linear stress-strain relationship is the quadratic form

$$u^* = \frac{1}{2} C_{ijkm}\, \epsilon_{ij}\, \epsilon_{km} \tag{4.259}$$

Using the relation (4.249), this equation can be written

$$u^* = \frac{1}{2} \sigma_{ij}\, \epsilon_{ij} \tag{4.260}$$

In the simple index assignment system, this equation can be written as

$$u^* = \frac{1}{2} C_{KM}\, \epsilon_K\, \epsilon_M \tag{4.261}$$

where $C_{KM} = C_{MK}$. Due to this symmetry of C_{KM}, the number of independent elastic constants is at most 21 if there exists a function for the strain energy.

4.4.2 Elastic symmetry

If the elastic properties are independent of the reference frame, such a material is said to be elastically isotropic. A material that is not isotropic is called anisotropic. Since the elastic properties of a Hookean solid are expressed through the coefficients C_{KM}, a general anisotropic body will have a matrix of elastic constants of the form

$$C_{KM} = \begin{bmatrix} C_{11} & C_{12} & C_{13} & C_{14} & C_{15} & C_{16} \\ C_{21} & C_{22} & C_{23} & C_{24} & C_{25} & C_{26} \\ C_{31} & C_{32} & C_{33} & C_{34} & C_{35} & C_{36} \\ C_{41} & C_{42} & C_{43} & C_{44} & C_{45} & C_{46} \\ C_{51} & C_{52} & C_{53} & C_{54} & C_{55} & C_{56} \\ C_{61} & C_{62} & C_{63} & C_{64} & C_{65} & C_{66} \end{bmatrix} \tag{4.262}$$

When there is a strain energy function for a body, $C_{KM} = C_{MK}$, and the 36 constants of (4.262) reduce to 21.

.......Let us consider two coordinate systems that are related under the following transformation.

$$\bar{x}_1 = x_1, \bar{x}_2 = x_2, \bar{x}_3 = -x_3 \tag{4.263}$$

In matrix form this coordinate transformation can be written as:

$$\begin{bmatrix} \bar{x}_1 \\ \bar{x}_2 \\ \bar{x}_3 \end{bmatrix} = \begin{bmatrix} 1 & 0 & 0 \\ 0 & 1 & 0 \\ 0 & 0 & -1 \end{bmatrix} \begin{bmatrix} x_1 \\ x_2 \\ x_3 \end{bmatrix} \tag{4.264}$$

The transformation matrix would be:

$$a_{ij} = \begin{bmatrix} 1 & 0 & 0 \\ 0 & 1 & 0 \\ 0 & 0 & -1 \end{bmatrix} \tag{4.265}$$

At a point there exists a plane of elastic symmetry when the elastic constants have the same values for each pair of coordinate systems that are mirror images of each other with respect to the plane. The axes of these coordinate systems are called equivalent elastic directions. If the plane x_1x_2 is one of elastic symmetry, the constants C_{KM} are invariant under the transformation (4.264). By inserting the values of the matrix (4.265) into the transformation of the linear stress tensors given in Table 4.1 and strain tensors (4.127), the elastic matrix of a material having the plane x_1x_2 as a plane of symmetry is obtained. It is written:

$$C_{KM} = \begin{bmatrix} C_{11} & C_{12} & C_{13} & 0 & 0 & C_{16} \\ C_{21} & C_{22} & C_{23} & 0 & 0 & C_{26} \\ C_{31} & C_{32} & C_{33} & 0 & 0 & C_{36} \\ 0 & 0 & 0 & C_{44} & C_{45} & 0 \\ 0 & 0 & 0 & C_{54} & C_{55} & 0 \\ C_{61} & C_{62} & C_{63} & 0 & 0 & C_{66} \end{bmatrix} \tag{4.266}$$

The 20 matrix constants (4.266) are reduced to 13 when there is a strain energy function: If a material has three mutually perpendicular elastic symmetry planes, the material is called orthotropic and its elastic matrix is of the form.

$$C_{KM} = \begin{bmatrix} C_{11} & C_{12} & C_{13} & & & \\ C_{21} & C_{22} & C_{23} & & 0 & \\ C_{31} & C_{32} & C_{33} & & & \\ & & & C_{44} & 0 & 0 \\ & 0 & & 0 & C_{55} & 0 \\ & & & 0 & 0 & C_{66} \end{bmatrix} \tag{4.267}$$

which has 12 independent constants, or 9 if $C_{KM} = C_{MK}$. An axis of elastic symmetry of order N is said to exist at a point when there is a set of equivalent elastic directions that can be superimposed by a rotation of an angle $2\pi/N$ about the axis.

4.4.3 Elastic constants

Bodies that are elastically equivalent in all directions possess complete symmetry and are called isotropic. In this case every plane and every axis have elastic symmetry. In case of isotropy, the number of independent elastic constants is reduced to 2, and the elastic matrix is symmetric regardless of the existence of a strain energy function. Choosing the two independent constants as the well-known Lamé constants λ and μ, the matrix (4.267) reduces to the elastic and isotropic form

$$C_{KM} = \begin{bmatrix} \lambda+2\mu & \lambda & \lambda & & & \\ \lambda & \lambda+2\mu & \lambda & & 0 & \\ \lambda & \lambda & \lambda+2\mu & & & \\ & & & \mu & 0 & 0 \\ & 0 & & 0 & \mu & 0 \\ & & & 0 & 0 & \mu \end{bmatrix} \qquad (4.268)$$

In this way, in terms of λ and μ, taking into account the matrix (4.268) Hooke's law (4.249) can be written for an isotropic body as:

$$\sigma_{ij} = \lambda\delta_{ij}\,\epsilon_{kk} + 2\mu\epsilon_{ij} \qquad (4.269)$$

This equation can be inverted to express the strains as a function of the stresses as follows:

$$\epsilon_{ij} = -\frac{\lambda}{2\mu(3\lambda+2\mu)}\delta_{ij}\sigma_{kk} + \frac{1}{2\mu}\delta_{ij} \qquad (4.270)$$

For a uniaxial stress state in the x_1 direction, the engineering constants E and v can be introduced through the relations $\sigma_{11} = E\epsilon_{11}$ and $\epsilon_{22} = \epsilon_{33} = -v\epsilon_{11}$. The constant E is known as the Young's modulus, and v is called the Poisson's ratio. In terms of these constants, Hooke's Law for isotropic bodies becomes

$$\sigma_{ij} = \frac{E}{1+v}\left(\epsilon_{ij} + \frac{v}{1-2v}\delta_{ij}\epsilon_{kk}\right) \qquad (4.271)$$

Or inverted

$$\epsilon_{ij} = \frac{1+v}{E}\sigma_{ij} + \frac{v}{E}\delta_{ij}\,\sigma_{kk} \qquad (4.272)$$

If the state of stress originated by a uniform hydrostatic pressure is considered, it is possible to define the volumetric modulus as:

$$K = \frac{E}{3(1-2v)} \qquad (4.273)$$

Or equivalently

$$K = \frac{3\lambda+2\mu}{3} \qquad (4.274)$$

Which relates the pressure to the cubic expansion of a body so loaded. For the pure shear state, the modulus of rigidity G relates the shear stress and strain components G is in fact equal to μ and it can be proven that:

$$\mu = K = \frac{E}{3(1+2v)} \tag{4.275}$$

4.4.4 Elastostatic-Elastodynamic

The mathematical model of the elastostatic problem of a homogeneous and isotropic body consists of the equilibrium equations, Hooke's law and the displacement-strain relations, i.e., respectively of equations (4.14), (4.99) and (4.269), explicitly:

$$\sigma_{ji,i} + \rho B_i = 0$$

$$\epsilon_{ij} = \frac{1}{2}\left(\frac{\partial u_i}{\partial x_j} + \frac{\partial u_j}{\partial x_i}\right) = \frac{1}{2}(u_{i,j} + u_{j,i})$$

$$\sigma_{ij} = \lambda\delta_{ij}\,\epsilon_{kk} + 2\mu\epsilon_{ij}$$

These equations must be satisfied at all points of the body, and they must also satisfy certain conditions of stress and/or displacement at the boundary surface of the body. In the theory of elasticity, boundary value problems are classified according to the boundary conditions of the problem for which the following are given:

1. Displacements at all parts of the boundary.
2. Stresses (surface tractions) at all parts of the boundary.
3. Displacements at one part of the boundary and stresses at the remaining parts.

For all these cases, the mass forces at all parts of the continuous medium are assumed to be known. For those problems in which the components of the displacements at all parts of the boundary are given by an equation of the form

$$\mu = g_i(\vec{X}) \tag{4.276}$$

The displacement-strain relationships (4.99) can be substituted into Hooke's law (4.269) and the result is in turn substituted into (4.14), thus obtaining the mathematical model that solves this type of problems.

$$\mu u_{i,jj} + (\lambda + \mu)\, u_{j,ji} + \rho B_i = 0 \tag{4.277}$$

Or in vector form

$$\mu\nabla^2 \vec{u} + (\lambda + \mu)\nabla\nabla \cdot \vec{u} + \rho\vec{B} = 0 \tag{4.278}$$

Equations (4.278) are called the Navier-Cauchy equations. The solution to this type of problem is therefore based on the displacement vector u_i, which satisfies equation (4.278) in all parts of the continuum and fulfills the boundary conditions (4.276). The mathematical model for solving problems involving surface tractions in all parts of the boundary, given by equations (4.7), is obtained by combining these stresses with the compatibility equations (4.149), Hooke's law (4.272) and the equilibrium equation (4.14). In this way, we obtain:

$$\sigma_{ij,kk} = \frac{E}{1+\nu}\sigma_{kk,ij} + \rho(B_{i,j} + B_{j,i}) + \frac{\nu}{1-\nu}\delta_{ij}\rho B_{k,k} \tag{4.279}$$

Equations (4.279) are called the Beltrami-Michell compatibility equations. The solution to this type of model is obtained by specifying the stress tensor that satisfies equations (4.279) throughout the boundary of the continuum and satisfies conditions (4.7). But problems where the boundary conditions are mixed must be solved by the mathematical model consisting of equations (4.14), (4.99) and (4.269). The solution to this model gives the field of stresses and displacements in all parts of the continuum. The stress components have to satisfy conditions (4.7) somewhere on the boundary, while the displacements satisfy conditions (4.276).

For mathematical modeling of elastodynamic problems, the equilibrium equations (4.14) must be replaced by the equations of motion (4.235) and the boundary and initial conditions must be specified. The mathematical model that solves the elastodynamic problem is analogous to the mathematical model for solving elastostatic problems, and is:

$$\mu u_{i,jj} + (\lambda + \mu)\, u_{j,ji} + \rho B_i = \rho \ddot{u}_i \tag{4.280}$$

The solutions to the mathematical model (4.280) appear in the form $u_i = u_i(\vec{x}, t)$ and have to satisfy not only the initial conditions of the motion, frequently expressed by equations such as:

$$u_i = u_i(\vec{x}, 0);\ \dot{u}_i = \dot{u}_i\,(\vec{x}, 0) \tag{4.281}$$

but also, the boundary concitions, or displacements:

$$\mu = g_i\,(\vec{x}, 0) \tag{4.282}$$

Or the superficial tractions

$$t_i^{(\hat{n})} = t_i^{(\hat{n})}(\vec{x}, 0) \tag{4.283}$$

4.4.5 Hyperelasticity - Hypolasticity

The continuum constitutive equations for elastic materials define these materials in a special sense as hyperelastic and hypoelastic materials. Materials are hyperelastic if they possess a strain function U such that the material derivative of this function is equal to the work due to stresses per unit time and volume. In this case the equation is of the form:

$$\frac{dU}{dt} = \frac{1}{\rho}\sigma_{ij}D_{ij} = \frac{1}{\rho}\sigma_{ij}\dot{\epsilon}_{ij} \tag{4.284}$$

Where D_{ij} is the strain rate tensor. A material is hypoelastic if the variation in stress per unit time is a linear and homogeneous function of the strain rate. Then equation is written:

$$\Delta\sigma_{ij} = K_{ijkm} D_{km} \tag{4.285}$$

where the variation of the stress tensor per unit time $\Delta\sigma_{ij}$ here is defined as:

$$\Delta\sigma_{ij} = \frac{d\sigma_{ij}}{dt} - \sigma_{iq}V_{qj} - \sigma_{jq}V_{qi} \tag{4.286}$$

where V_{qi} is the vortex tensor (4.180).

4.4.6 Linear thermoelasticity

If, in addition to considering the effects of the stress tensor, the thermal effects are considered, the linear strain tensor ϵ_{ij} can be written as the sum of both effects.

$$\epsilon_{ij} = \epsilon_{ij}^{(S)} + \epsilon_{ij}^{(T)} \tag{4.287}$$

where $\epsilon_{ij}^{(S)}$ represents the contribution of the stress field and $\epsilon_{ij}^{(T)}$ the contribution of the temperature field. It is established experimentally that a change in temperature from a reference temperature T_0 to a temperature T, the components of the strain tensor of an elementary volume in an unforced isotropic body can be written as:

$$\epsilon_{ij}^{(T)} = \alpha(T - T_0)\,\delta_{ij} \tag{4.288}$$

where α is the coefficient of linear thermal expansion. Replacing equations (4.288), together with Hooke's law (4.270) in (4.287), we have

$$\epsilon_{ij} = \frac{1}{2\mu}\left(\sigma_{ij} - \frac{\lambda}{(3\lambda + 2\mu)}\delta_{ij}\sigma_{kk}\right) + \alpha(T - T_0)\delta_{ij} \tag{4.289}$$

Equations (4.289) are known as the Duhamel-Neumann relations and can be inverted to obtain the thermoelastic constitutive equations:

$$\sigma_{ij} = \lambda\delta_{ij}\,\epsilon_{kk} + 2\mu\epsilon_{ij} - (3\lambda + 2\mu)\alpha\delta_{ij}\,(T - T_c) \tag{4.290}$$

Heat conduction in an isotropic elastic solid is described by Fourier's law of heat conduction.

$$c_i = -\kappa T_{,i} = -\kappa\frac{\partial T}{\partial x_i} \tag{4.291}$$

where κ is the thermal conductivity of the body, which must be positive to ensure a positive variation in entropy production. If the constant specific heat of strain $c_{(v)}$ is introduced by means of the equation

$$-c_{i,i} = -\frac{\partial c_i}{\partial x_i} = \rho c_{(v)}T \tag{4.292}$$

and the internal energy is assumed to be a function of the strain components ϵ_{ij} and the temperature T, the energy equation (4.247) can be written as:

$$\kappa T_{,ii} = \rho c_{(v)}\,\dot{T} + (3\lambda + 2\mu)\alpha T_0\,\dot{\epsilon}_{ii} \tag{4.293}$$

This equation is known as the coupled heat equation. In vector form it can be written as:

$$\kappa \nabla^2 T = \rho c_{(v)}\,\dot{T} + (3\lambda + 2\mu)\alpha T_0\,\dot{\epsilon} \tag{4.294}$$

In summary, the system of equations that structure the mathematical model to describe and solve thermoelastic problems in general for an isotropic body are the following:

- Equations of motion

$$\sigma_{ji,i} + \rho B_i = \ddot{u}_i$$

- Thermoelastic constitutive equations

$$\sigma_{ij} = \lambda\delta_{ij}\,\epsilon_{kk} + 2\mu\epsilon_{ij} - (3\lambda + 2\mu)\alpha\delta_{ij}\,(T - T_c)$$

- Displacement – strain relationships

$$\epsilon_{ij} = \frac{1}{2}\left(\frac{\partial u_i}{\partial x_j} + \frac{\partial u_j}{\partial x_i}\right) = \frac{1}{2}(u_{i,j} + u_{j,i})$$

- Couple heat equation

$$\kappa T_{,ii} = \rho c_{(v)}\,\dot{T} + (3\lambda + 2\mu)\alpha T_0\,\dot{\epsilon}_{ii}$$

This system of equations must be solved for the stress, displacement and temperature fields subjected to the appropriate initial and boundary conditions. There is a large collection of problems in which coupling, and inertia effects can be neglected. In these cases, the general thermoelastic problem is decomposed into two separate problems which must be solved successively and independently. Then for the uncoupled, quasi-static thermoelastic problem the basic equations are

- Heat conduction equations

$$\kappa T_{,ii} = \rho c_{(v)}\,\dot{T} \quad \text{or} \quad \kappa\nabla^2 T = \rho c_{(v)}\,\dot{T}$$

- Equilibrium equations

$$\sigma_{ji,i} + \rho B_i = 0$$

- Thermoelastic stress-strain equations

$$\sigma_{ij} = \lambda\delta_{ij}\,\epsilon_{kk} + 2\mu\epsilon_{ij} - (3\lambda + 2\mu)\alpha\delta_{ij}\,(T - T_c)$$

- Displacement-strain relationships

$$\epsilon_{ij} = \frac{1}{2}\left(\frac{\partial u_i}{\partial x_j} + \frac{\partial u_j}{\partial x_i}\right) = \frac{1}{2}(u_{i,j} + u_{j,i})$$

Table 4.2. Annex: symbolic expressions of the main mathematical equations.

Number	Tensor form	Symbolic form
(4.1)	$\rho B_i = p_i$	$\rho \vec{B} = \vec{p}$
(4.2)	$-t_i^{(\hat{n})} = t_i^{(-\hat{n})}$	$-t^{(\hat{n})} = t^{(-\hat{n})}$
(4.7)	$t_i^{(\hat{n})} = \sigma_{ij}\, n_j$	$t^{(\hat{n})} = \hat{n} \cdot \Sigma$
(4.12)	$\int_S t_i^{(\hat{n})} dS + \int_V \rho B_i dV = 0$	$\int_S t^{(\hat{n})} dS + \int_V \rho B dV = 0$
(4.13)	$\int_V (\sigma_{ji,i} + \rho B_i) dV = 0$	$\int_V (\nabla \cdot \Sigma + \rho B) dV = 0$
(4.14)	$\sigma_{ji,i} + \rho B_i = 0$	$\nabla \cdot \Sigma + \rho B = 0$
(4.15)	$\int_S \epsilon_{ijk} x_j t_k^{(\hat{n})} dS + \int_V \epsilon_{ijk} x_j \rho B_k dV = 0$	$\int_S x \times t^{(\hat{n})} dS + \int_V x \times \rho B dV = 0$
(4.16)	$\int_V \epsilon_{ijk} \sigma_{jk} dV = 0$	$\rho \int_V \Sigma_v\, dV = 0$
(4.17)	$\epsilon_{ijk}\, \sigma_{jk} = 0$	$\Sigma_v = 0$
(4.24)	$t_i^{(\hat{n})} = \sigma n_i$	$t^{(\hat{n})} = \sigma \hat{n}$
(4.25)	$(\sigma_{ij} - \delta_{ij}\, \sigma) n_j = 0$	$(\Sigma - I_\sigma) \cdot \hat{n} = 0$
(4.32)	$(\sigma_{ij} - \sigma_{(k)}\, \delta_{ij})\, n_j^{(k)} = 0$	$(\Sigma - \sigma_{(k)}\, I) \cdot \hat{n}^{(k)}$
(4.80)	$\frac{\partial u_i}{\partial X_j} = \frac{\partial x_i}{\partial X_j} - \delta_{ij}$	$J \equiv \vec{u} \nabla_{\vec{X}} = F - I$
(4.81)	$\frac{\partial u_i}{\partial x_j} = \delta_{ij} - \frac{\partial X_i}{\partial x_j}$	$K \equiv \vec{u} \nabla_{\vec{x}} = H - I$
(4.86)	$(dX)^2 = C_{ij}\, dx_i\, dx_j$	$(dX)^2 = dx \cdot C \cdot dx$
(4.87)	$C_{ij} = \frac{\partial X_k}{\partial x_i} \frac{\partial X_k}{\partial x_j}$	$C = H_c \cdot H$
(4.89)	$x_i = \frac{\partial x_i}{\partial X_j} dX_j$	$dx = F \cdot dX$
(4.90)	$(dx)^2 = G_{ij}\, dX_i\, dX_j$	$(dx)^2 = dX \cdot G \cdot dX$
(4.91)	$G_{ij} = \frac{\partial x_k}{\partial X_i} \frac{\partial x_k}{\partial X_j}$	$G = F_c \cdot F$
(4.92)	$(dx)^2 - (dX)^2 = 2L_{ij}\, dX_i\, dX_j$	$(dx)^2 - (dX)^2 = dX \cdot 2L_G \cdot dX$
(4.94)	$(dx)^2 - (dX)^2 = 2E_{ij}\, dx_i\, dx$	$(dx)^2 - (dX)^2 = dx \cdot 2E_A \cdot dx$
(4.95)	$E_{ij} = \frac{1}{2}\left(\delta_{ij} - \frac{\partial X_k}{\partial x_i} \frac{\partial X_k}{\partial x_j} \right)$	$E_A = \frac{1}{2}(I - H_c \cdot H)$

Table 4.2 contd. ...

...Table 4.2 contd.

Number	Tensor form	Symbolic form
(4.96)	$L_{ij}=\frac{1}{2}\left(\frac{\partial u_i}{\partial X_j}+\frac{\partial u_j}{\partial X_i}+\frac{\partial u_k}{\partial X_i}\frac{\partial u_k}{\partial X_j}\right)$	$L_G=\frac{1}{2}(J+J_c+J_c\cdot J)$
(4.97)	$E_{ij}=\frac{1}{2}\left(\frac{\partial u_i}{\partial x_j}+\frac{\partial u_j}{\partial x_i}-\frac{\partial u_k}{\partial x_i}\frac{\partial u_k}{\partial x_j}\right)$	$E_A=\frac{1}{2}(K+K_c-K_c\cdot K)$
(4.98)	$l_{ij}=\frac{1}{2}\left(\frac{\partial u_i}{\partial X_j}+\frac{\partial u_j}{\partial X_i}\right)$	$L=\frac{1}{2}(J+J_c)$
(4.99)	$\epsilon_{ij}=\frac{1}{2}\left(\frac{\partial u_i}{\partial x_j}+\frac{\partial u_j}{\partial x_i}\right)$	$E=\frac{1}{2}(K+K_c)$
(4.101)	$du_i=u_i^{(Q_0)}-u_i^{(P_0)}$	$d\vec{u}=u^{(Q_0)}-u^{(P_0)}$
(4.102)	$du_i=\left(\frac{\partial u_i}{\partial X_j}\right)_{P_0}dX_j$	$d\vec{u}=(\vec{u}\nabla_{\vec{X}})_{P_0}\cdot d\vec{X}$
(4.103)	$\frac{du_i}{dX}=\frac{\partial u_i}{\partial X_j}\frac{dX_j}{dX}=\frac{\partial u_i}{\partial X_j}v_j$	$\frac{d\vec{u}}{dX}=\vec{u}\nabla_{\vec{X}}\cdot\hat{v}=J\cdot\hat{v}$
(4.104)	$du_i=\left[\frac{1}{2}\left(\frac{\partial u_i}{\partial X_j}+\frac{\partial u_j}{\partial X_i}\right)+\frac{1}{2}\left(\frac{\partial u_i}{\partial X_j}-\frac{\partial u_j}{\partial X_i}\right)\right]dX_j$	$d\vec{u}=\left[\frac{1}{2}(\vec{u}\nabla_{\vec{x}}+\nabla_{\vec{x}}\vec{u})+\frac{1}{2}(\vec{u}\nabla_{\vec{x}}-\nabla_{\vec{x}}\vec{u})\right]\cdot dX$
(4.105)	$W_{ij}=\frac{1}{2}\left(\frac{\partial u_i}{\partial X_j}-\frac{\partial u_j}{\partial X_i}\right)$	$W=\frac{1}{2}(\vec{u}\nabla_{\vec{X}}-\nabla_{\vec{X}}\vec{u})$
(4.106)	$\omega_i=\frac{1}{2}\epsilon_{ijk}W_{kj}$	$W=\frac{1}{2}\nabla_{\vec{X}}\times\vec{u}$
(4.107)	$du_i=\epsilon_{ijk}\,\omega_j\,dX_k$	$d\vec{u}=\vec{w}\times d\vec{X}$
(4.108)	$du_i=\frac{\partial u_i}{\partial x_j}dx_j$	$d\vec{u}=\vec{K}\cdot d\vec{x}$
(4.109)	$du_i=\frac{\partial u_i}{\partial x_j}\frac{dx_j}{dx}=\frac{\partial u_i}{\partial x_j}\mu_j$	$d\vec{u}=\vec{K}\cdot\hat{\mu}$
(4.110)	$\frac{du_i}{dx}=\left[\frac{1}{2}\left(\frac{\partial u_i}{\partial x_j}+\frac{\partial u_j}{\partial x_i}\right)+\frac{1}{2}\left(\frac{\partial u_i}{\partial x_j}-\frac{\partial u_j}{\partial x_i}\right)\right]dx_j$	$d\vec{u}=\left[\frac{1}{2}(\vec{u}\nabla_{\vec{x}}+\nabla_{\vec{x}}\vec{u})+\frac{1}{2}(\vec{u}\nabla_{\vec{x}}-\nabla_{\vec{x}}\vec{u})\right]\cdot d\vec{x}$
(4.111)	$\omega_{ij}=\frac{1}{2}\left(\frac{\partial u_i}{\partial x_j}-\frac{\partial u_j}{\partial x_i}\right)$	$\Omega=\frac{1}{2}(\vec{u}\nabla_{\vec{x}}-\nabla_{\vec{x}}\vec{u})$

Table 4.2 contd. ...

...Table 4.2 contd.

Number	Tensor form	Symbolic form
(4.112)	$\omega_i = \frac{1}{2} \epsilon_{ijk} \omega_{kj}$	$\omega = \frac{1}{2} \nabla_{\vec{X}} \times \vec{u}$
(4.113)	$du_i = \frac{1}{2} \epsilon_{ijk} \omega_j dx_k$	$d\vec{u} = \vec{\omega} \times d\vec{x}$
(4.154)	$l_{ij} \eta_i \eta_j = \pm h^2$	$\eta \cdot L \cdot \eta = \pm h^2$
(4.155)	$\epsilon_{ij} \xi_i \xi_j = \pm g^2$	$\xi \cdot E \cdot \xi = \pm g^2$
(4.164)	$v_i = \frac{dx_i}{dt} = \frac{d(u_i + X_i)}{dt} = \frac{du_i}{dt}$	$\vec{v} = \frac{d\vec{x}}{dt} = \frac{d(\vec{u} + \vec{X})}{dt} = \frac{d\vec{u}}{dt}$
(4.165)	$v_i = \dot{u}_i = \frac{du_i(\vec{X},t)}{dt} = \frac{\partial u_i(\vec{X},t)}{\partial t}$	$\vec{v} = \dot{\vec{u}} = \frac{du(\vec{X},t)}{dt} = \frac{\partial u(\vec{X},t)}{\partial t}$
(4.166)	$v_i(\vec{x},t) = \dot{u}_i(\vec{x},t) = \frac{du_i(\vec{x},t)}{dt} = \frac{\partial u_i(\vec{x},t)}{\partial t} + v_k(\vec{x},t)\frac{\partial u_i(\vec{x},t)}{\partial x_k}$	$v(\vec{x},t) = \dot{u}(\vec{x},t) = \frac{du(\vec{x},t)}{dt} = \frac{\partial u(\vec{x},t)}{\partial t} + v(\vec{x},t)\nabla_{\vec{x}} u(\vec{x},t)$
(4.168)	$a_i = \dot{v}_i = \frac{dv_i(\vec{X},t)}{dt}$	$\vec{a} = \dot{\vec{v}} = \frac{dv(\vec{X},t)}{dt} = \frac{\partial v(\vec{X},t)}{\partial t}$
(4.169)	$a_i(\vec{x},t) = \frac{dv_i(\vec{x},t)}{dt} = \frac{\partial v_i(\vec{x},t)}{\partial t} + v_k(\vec{x},t)\frac{\partial u_i(\vec{x},t)}{\partial x_k}$	$a(\vec{x},t) = \frac{dv(\vec{x},t)}{dt} = \frac{\partial v(\vec{x},t)}{\partial t} + v(\vec{x},t)\nabla_{\vec{x}} v(\vec{x},t)$
(4.170)	$Y_{ij} = \frac{1}{2}\left(\frac{\partial v_i}{\partial x_j} + \frac{\partial v_j}{\partial x_i}\right) + \frac{1}{2}\left(\frac{\partial v_i}{\partial x_j} - \frac{\partial v_j}{\partial x_i}\right) = D_{ij} + V_{ij}$	$Y = \frac{1}{2}(\vec{v}\nabla_{\vec{x}} + \nabla_{\vec{x}}\vec{v}) + \frac{1}{2}(\vec{v}\nabla_{\vec{x}} - \nabla_{\vec{x}}\vec{v}) = D + V$
(4.171)	$D_{ij} = \frac{1}{2}\left(\frac{\partial v_i}{\partial x_j} + \frac{\partial v_j}{\partial x_i}\right)$	$D = \frac{1}{2}(\vec{v}\nabla_{\vec{x}} + \nabla_{\vec{x}}\vec{v})$
(4.172)	$V_{ij} = -V_{ji} = \frac{1}{2}\left(\frac{\partial v_i}{\partial x_j} - \frac{\partial v_j}{\partial x_i}\right)$	$V = \frac{1}{2}(\vec{v}\nabla_{\vec{x}} - \nabla_{\vec{x}}\vec{v})$
(4.173)	$\frac{d\epsilon_{ij}}{dt} = \frac{1}{2}\frac{d}{dt}\left(\frac{\partial u_i}{\partial x_j} + \frac{\partial u_j}{\partial x_i}\right)$	$\frac{dE}{dt} = \frac{1}{2}\frac{d}{dt}(\vec{u}\nabla_{\vec{x}} - \nabla_{\vec{x}}\vec{u})$
(4.174)	$\frac{d\epsilon_{ij}}{dt} = \frac{1}{2}\frac{d}{dt}\left(\frac{\partial v_i}{\partial x_j} + \frac{\partial v_j}{\partial x_i}\right) = D_{ij}$	$\frac{dE}{dt} = \frac{1}{2}\frac{d}{dt}(v\nabla_{\vec{x}} + \nabla_{\vec{x}}v) = D$

Table 4.2 contd. ...

...Table 4.2 contd.

Number	Tensor form	Symbolic form
(4.208)	$\frac{d}{dt}(dx^2) = 2D_{ij}dx_k dx_i$	$\frac{d}{dt}(dx^2) = 2d\vec{x} \cdot D \cdot d\vec{x}$
(4.231)	$\int_S t_i^{(\vec{n})} dS + \int_V \rho B_i dV = \frac{d}{dt}\int_V \rho v_i dV$	$\int_S t^{(\vec{n})} dS + \int_V \rho \vec{B} dV = \frac{d}{dt}\int_V \rho \vec{v} dV$
(4.234)	$\int_V (\sigma_{ji,j} + \rho B_i - \rho v_i) dV = 0$	$\int_V (\nabla_X \cdot \Sigma + \rho \vec{B} - \rho \dot{\vec{v}}) dV = 0$

Bibliography

Clayton, J.D. (2019). Applications of tensor analysis in continuum mechanics.
Hetnarski, R.B. and Ignaczak, J. (2006). Mathematical theory of elasticity. J. Therm. Stress 29: 505–506.
Kassir, M. (2018). Applied Elasticity and Plasticity. CRC Press, Taylor & Francis Group Boca Raton.
Surana, K.S. (2022). Classical Continuum Mechanics. CRC Press.

CHAPTER 5

Material Properties from the Symmetry Group

In Chapter 2, we studied the structure of crystalline solids from the point of view of their symmetries. This study is important because the macroscopic behavior of solids is closely related to their structure at the atomic scale. We are then talking about a microstructure-properties correlation, which gives meaning to solid-state physics.

In this chapter we are going to connect three things, the symmetries of the solid, the properties of the solid that arise from these symmetries, and the tensor modeling of the properties of the solid that arise precisely from the symmetries, that is, the connection that also exists between the symmetries of the solid and its tensor properties.

5.1 Equilibrium tensor properties of solids

5.1.1 Tensor property

It is useful to describe a property as a coefficient that linearly relates an intensive parameter (or cause) to an extensive parameter (or effect). When the intensive and extensive parameters are tensor components, the property is also a tensor quantity, i.e.:

$$\begin{bmatrix} \text{Effect} \\ \text{tensor} \\ \text{parameter} \end{bmatrix} = \begin{bmatrix} \text{Tensor} \\ \text{property} \\ \text{component} \end{bmatrix} \begin{bmatrix} \text{Cause} \\ \text{tensor} \\ \text{parameter} \end{bmatrix} \tag{5.1}$$

The tensor property components can be physically interpreted as the rate of change of one parameter with respect to the other as implied by equation (5.1). One or both parameters defining the property components will usually be a field tensor such as the stress tensor or strain tensor, the electric or magnetic field strength, or the temperature (a scalar or rank zero tensor). The degree to which a property of a crystal exists with respect to the application of a given field and depends on the electronic and subatomic structure of the atoms and the way they are arranged in the lattice.

Four typical examples of crystal properties that are directly related by the cause-and-effect tensor parameter are shown in Table 5.1. Here, the pyroelectric

DOI: 10.1201/9781003344438-5

Table 5.1. Examples of properties of crystals related by cause-and-effect tensor parameters.

Effect tensor parameter	Cause tensor parameter	Tensor equation	Tensor rank of coefficient	Property name
δP_i	δT	$\delta P_i = p_i \delta T$	First (p_i)	Pyroelectricity
$\delta \epsilon_{ij}$	δT	$\delta \epsilon_{ij} = \alpha_{ij} \delta T$	Second (α_{ij})	Thermal expansion
P_i	σ_{ij}	$P_i = d_{ijk} \sigma_{jk}$	Third (d_{ijk})	Piezoelectricity
ϵ_{ij}	σ_{kl}	$\epsilon_{ij} = s_{ijkl}\, \sigma_{kl}$	Fourth (s_{ijkl})	Elasticity

coefficient p_i is a vector-scalar tensor property, hence a vector, that relates the change in polarization δP_i to a change in temperature δT. The thermal expansion coefficient α_{ij} is a rank-2 tensor-scalar tensor property that relates a change in strain $\delta \epsilon_{ij}$ to a change in temperature δT. The piezoelectric coefficient d_{ijk} is a rank-3 tensor-scalar vector-scalar tensor property that relates the polarization P_i to the stress tensor σ_{ij}. Finally, the elastic flexibility coefficient s_{ijkl} is a rank-4 tensor-tensor property that relates the stress σ_{kl} to the strain ϵ_{ij} (Hooke's law).

Thus, it is seen that the tensor rank of a property coefficient is the sum of the tensor ranks of the parameters it relates. We shall now study properties that can be represented by tensors and described with reference to equilibrium states and thermodynamically reversible changes in crystals. Suppose that in a crystalline material or any closed system, all measurable physical quantities are accurately recorded, if these values are observed to be macroscopically independent of time, it is said to be in thermodynamic equilibrium.

This requires, for example, the absence of heat (or entropy), charge, and mass transport and that all parts of the system are mechanically in static equilibrium. Thus, thermodynamic equilibrium prevails if the system under consideration is in thermal, electrical, magnetic, mechanical, and chemical equilibrium. Thermodynamic equilibrium allows microscopic fluctuations at the atomic level to occur if they are random and do not produce measurable time-dependent changes in the physical state of the system.

Properties of crystals that represent time-independent processes are called equilibrium properties. Such properties can be adequately described phenomenologically by using equilibrium thermodynamics, that is, thermostatics. Properties that represent time-dependent or spontaneous processes are called non-equilibrium or transport properties. Such properties can be described only by using the thermodynamics of irreversible processes; as we can see, the concept of thermodynamic reversibility is directly related to the concept of equilibrium.

The term reversible applies to a process within a system that has been displaced from equilibrium by only an infinitesimal amount. A reversible process may therefore be defined as the hypothetical passage of the system through a series of equilibrium states so close together as to be almost indistinguishable. Such a process is reversed by an appropriate infinitesimal change in the physical conditions imposed on the system. No ongoing process is reversible in the strict sense of the word. However, if the conditions imposed on the system are applied infinitesimally rapidly, then within experimental error the process may be considered reversible.

A reversible process is usually considered to be linear in an infinitesimal change of the cause-and-effect parameters. However, a reversible process can be nonlinear, provided that infinitesimal changes in the system are restored reversibly. In developing the equilibrium properties, we will assume that reversible processes are linear, and that thermodynamic equilibrium is established at each point of an infinitesimal change.

Note that all tensor property coefficients given in Table 5.1 are examples of equilibrium properties. A well-known example of a nonequilibrium or transport property is the resistivity ρ_{ij}, a second-rank tensor relating the electric field strength E_i to the current density J_j according to $E_i = \rho_{ij}J_j$, which, of course, is the tensor representation of Ohm's law.

5.1.2 Representation of first-order equilibrium properties

In this section, interaction diagrams and the nomenclature of first-order equilibrium properties associated with four conjugate pairs of parameters will be reviewed. For this purpose, let us first define what are called conjugate parameters, since general first-order interaction diagrams will be constructed from them. Four conjugate parameter pairs to be reviewed here are (T, S), (E, P), (B, M) and (σ, ϵ), which give rise to $\sum_{n=1}^{8}(n-1) = 28$ first-order properties including 4 main effects and 24 coupled effects. The conjugate parameter pairs are (temperature, entropy), (electric field strength, polarization), (magnetic flux density, magnetization), and (stress, strain), respectively.

Extensive parameters are those that depend on the amount of material in the solid phase system (entropy, polarization, magnetization, strain). Parameters that are independent of the amount of material in the solid phase are called intensive parameters (temperature, electric field, magnetic flux, mechanical stresses) and are associated only with the external conditions (fields) imposed on the system.

The 16 compliance equilibrium properties are presented in Table 5.2 in order of increasing tensor rank, first for the main effects and then for the interaction (coupled) effects. Note that each parameter pair is given in the form of a notation (extensive, intensive).

Some properties are of little or no interest for the developments in this text. Higher-order effects are usually of high tensor rank and are therefore more complicated. These include effects such as electrostriction, magnetostriction, and a variety of optical effects.

Next, we will consider an interaction diagram associated with the thermal and mechanical variables, that is, one representing the thermal, thermoelastic, and elastic effects. This interaction diagram is shown in Figure 5.1, where the arrows point to the dependent variables.

Note from Figure 5.1 that both the stress and temperature fields contribute to an elastic strain, in differential form we can write:

$$d\epsilon_{ij} = S^{T}_{ijkl}\, d\sigma_{kl} + \alpha^{\sigma}_{ij}\, dT \tag{5.2}$$

Table 5.2. The set of compliance effects and their associated parameter pairs.

Parameter pair [Tensor Rank] (extensive, intensive)	Associated property
Principal effects	
(S, T) [0]	Heat capacity
(P, E) [2]	Dielectric susceptibility
(M, B) [2]	Magnetic susceptibility
(ϵ, σ) [4]	Elasticity
Interaction (coupled) effects	
(S, E) [1]	Pyroelectric effect
(P, T) [1]	Converse pyroelectric effect
(S, B) [1]	Pyromagnetic effect (magnetocaloric effect)
(M, T) [1]	Converse pyromagnetic effect
(S, σ) [2]	Piezocaloric effect
(ϵ, T) [2]	Thermal expansion (Piezocaloric effect)
(P, B) [2]	Magnetoelectric susceptibility
(M, E) [2]	Direct piezoelectric effect
(ϵ, E) [3]	Converse piezoelectric effect (first-order electrostriction)
(M, σ) [3]	Direct piezomagnetic effect
(P, σ) [3]	Direct piezoelectric effect
(ϵ, B) [3]	Converse piezomagnetic effect (first-order magnetostriction)

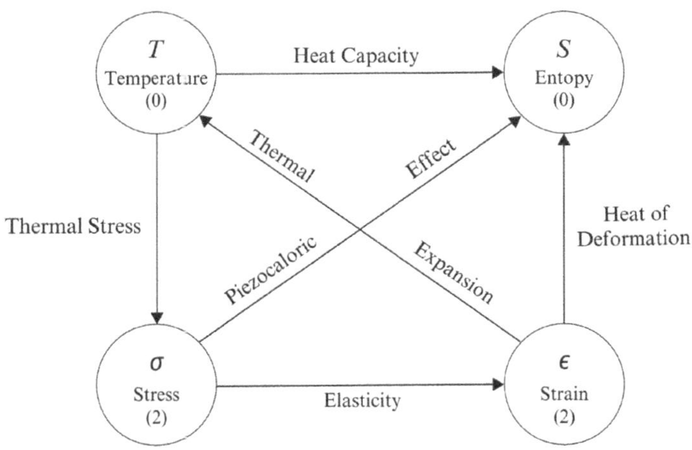

Figure 5.1. Interaction diagram representing the thermal, thermoelastic, and elastic effects.

or integrating this expression, we obtain:

$$\epsilon_{ij} = S^{T}_{ijkl}\,\sigma_{kl} + \alpha^{\sigma}_{ij}\,\Delta T \tag{5.3}$$

In these equations,

$$S^{T}_{i\,jkl} = \left(\frac{\partial \epsilon_{i\,j}}{\partial \sigma_{kl}}\right)_{T} \tag{5.4}$$

The tensor S^{T}_{ijkl} given in equation (5.4) is the isothermal (temperature independent) elastic compliance and α^{σ}_{ij} is the isopiestic (stress inariant) coefficient of thermal expansion and is given by:

$$\alpha^{\sigma}_{i\,j} = \left(\frac{\partial \epsilon_{i\,j}}{\partial T}\right)_{\sigma} \tag{5.5}$$

The tensor S^{T}_{ijkl} represents a fourth-rank tensor property that describes the main effect (elasticity) because it relates the conjugate parameters σ and ϵ. The tensor α^{σ}_{ij} represents a second-rank tensor property, on the other hand, that describes a coupled effect (thermal expansion) because it relates the cross parameters T and ϵ to a constant stress, σ. By a similar reasoning, it is observed in Figure 5.1 that a positive increase in stress, as well as an increase in temperature and strain, produces an increase in the entropy of a crystal. In the differential form, we can write:

$$dS = \alpha^{T}_{i\,j} d\sigma_{i\,j} + \left(\frac{C^{\sigma}}{T}\right) dT \tag{5.6}$$

or integrating this expression

$$\Delta S = \alpha^{T}_{i\,j}\sigma_{i\,j} + \left(\frac{C^{\sigma}}{T}\right)\Delta T \tag{5.7}$$

where

$$\alpha^{T}_{i\,j} = \left(\frac{\partial S}{\partial \sigma_{i\,j}}\right)_{T} \tag{5.8}$$

α^{T}_{ij} is the isothermal piezocaloric coefficient. Also:

$$C^{\sigma} = T\left(\frac{\partial S}{\partial T}\right)_{\sigma} \tag{5.9}$$

C^{σ} is the isopiestic (constant stress) heat capacity.

The properties just described have been defined in such a way that only the intensive parameters are held constant. Other coupled effects result from an association of the thermal and mechanical variables, which will keep both the intensive and the extensive parameters constant. Let us now relate the isothermal elastocaloric stress coefficient (heat of strain) to the isometric thermal stress coefficient, as represented by the vertical arrows in Figure 5.1.

5.1.3 Neumann's principle, Curie groups and its application to tensor properties

Neumann's principle is basic and the most important one for analyzing the effect of crystal symmetry on the tensor properties of crystals, and states that the symmetry elements of any physical property of a crystal must include all the symmetry elements of the point group of the crystal. This means that the component of a tensor representing a property must remain invariant under a coordinate transformation governed by a symmetry operation valid for the point group of the crystal. Mathematically, this is represented as

$$P_{ijk...} = P_{ijk...} \tag{5.10}$$

Among the symmetry groups that describe the tensor properties of the crystal, there are groups containing the axes of infinite order. Such symmetry groups are known as boundary groups or Curie groups. Using boundary groups, we can characterize the symmetry of the texture of polycrystalline aggregates, in which the arrangement of crystallites is ordered. The symmetry of artificial piezoelectrics: barium titanate ceramics and the texture of Seignette salt, are characterized by Curie groups denoted as ∞m and $\infty 2$, respectively. Some Curie groups are useful to describe the symmetry of physical agents, for example, the symmetry of the continuous homogeneous electric field can be characterized by the group ∞m; that of the continuous homogeneous magnetic field, by the group ∞/m; the mechanical stress of uniaxial tension or compression, by the symmetry group ∞/mmm; the compression in all directions by the group $\infty\infty m$. The phenomenon of the rotation of the polarization plane can be characterized with the limit group $\infty\infty$.

It also follows from Neumann's principle that a crystal can have certain physical properties only if its point symmetry group is a subset of the point symmetry group of the physical property under consideration.

For example, spontaneous electric polarization and pyroelectric properties can only be achieved by crystals whose symmetry is subordinated to the symmetry of the polar vector, that is, crystals belonging to the point groups: 1,2,3,4,6, *m*, *mm*, 3*mm*, 4*mm*, 6*mm*. This aspect is discussed in more detail in section 5.3.

Curie's principle: If a physical agent with a certain symmetry acts on a crystal, the symmetry of this crystal located in the field of action of the same varies and can be determined by means of the principle of superposition of symmetries, called the Curie principle. The crystal that is under the action of an external agent will possess those elements of symmetry that are common both to the crystal in the absence of the agent and to the agent in the absence of the crystal.

The description of the symmetry of the resulting phenomenon is important not only for the symmetry of the interacting phenomena, but also for the mutual arrangement of their symmetry elements. To illustrate Curie's principle, it is important to examine the question of restrictions imposed by the symmetry of crystals on the possibility of the existence of the pyroelectric effect in them. For the piezoelectric effect and the phenomenon of polarization in the external electric field, see Table 5.3.

Table 5.3. Crystal symmetry associated with the effect of the symmetry of the physical agent.

Effect	Equation	Symmetry groups before the agent's action	Symmetry of the agent	Crystal symmetry revealing the effect
Pyroelectric effect	$P_i = \gamma_i \Delta T$	∞m	$\infty\infty m$	Ten polar classes. The crystal itself ensures dissymmetry*, the existence of the vector P is necessary.
Piezoelectric effect	$P_i = d_{ijk} t_{jk}$	$\infty 2, \infty m, \bar{4}3m, \bar{6}m2$	mmm	Twenty non-concentric classes. The dissymmetry of the crystal must eliminate the symmetry center of the agent.
Polarization in the external field	$P_i = \chi_{ij} E_j$	$\infty\infty m$	∞m	Thirty-two-point symmetry groups. The necessary polar symmetry is ensured by the dissymmetry of the agent

* Dissymmetry is the absence of all elements of symmetry.

According to Table 5.3, by superimposing different physical fields that have different symmetry, the reaction of the crystal is the same.

The polarized crystal, according to Neumann's principle, must have the symmetry of one of the 10 polar classes, since only these classes have the unitary polar direction along which the vector P is located.

In crystals, polarization can be observed if, as a result of the superposition of the symmetry of the agent on the symmetry of the crystal, the symmetry of one of the 10 polar classes remains, according to the Curie principle; it does not matter whether this is achieved at the expense of the symmetry of the crystal, the highest being the symmetry of the agent (pyroelectric effect), or, on the contrary, at the expense of the polar symmetry of the agent, when polarization can be observed in the crystal of any symmetry and even in an isotropic field.

If, on the other hand, a particular component of a tensor property changes sign after a coordinate transformation, the component must vanish for that crystal according to the Neumann principle. Mathematically, this is expressed as

$$P_{ijk\ldots} = -P_{ijk\ldots} = 0 \tag{5.11}$$

The application of Neumann's principle to tensor properties requires the use of the transformation laws for tensors. The transformation laws for both polar and axial property tensors must hold for any symmetry operation allowed by a given point group. If s represents a symmetry operator, that is, the transformation matrix of a symmetry operation valid for a particular point group, then the generalized transformation laws become

$$P_{ijk\ldots} = s_{im}\, s_{jn}\, s_{ko} \ldots P_{mno} \tag{5.12}$$

for polar property tensors and

$$P_{ijk\ldots} = |s|\, s_{im}\, s_{jn}\, s_{ko} \ldots P_{mno} \tag{5.13}$$

Table 5.4. The 10 possible generating matrices are shown for both the s and σ forms.

Center of symmetry $s^{(1)} = \begin{bmatrix} -1 & 0 & 0 \\ 0 & -1 & 0 \\ 0 & 0 & -1 \end{bmatrix}$	$\sigma^{(1)} = \begin{bmatrix} 1 & 0 & 0 & 0 & 0 & 0 \\ 0 & 1 & 0 & 0 & 0 & 0 \\ 0 & 0 & 1 & 0 & 0 & 0 \\ 0 & 0 & 0 & 1 & 0 & 0 \\ 0 & 0 & 0 & 0 & 1 & 0 \\ 0 & 0 & 0 & 0 & 0 & 1 \end{bmatrix} = I$
Twofold CCW rotation about y $s^{(2)} = \begin{bmatrix} -1 & 0 & 0 \\ 0 & 1 & 0 \\ 0 & 0 & -1 \end{bmatrix}$	$\sigma^{(2)} = \begin{bmatrix} 1 & 0 & 0 & 0 & 0 & 0 \\ 0 & 1 & 0 & 0 & 0 & 0 \\ 0 & 0 & 1 & 0 & 0 & 0 \\ 0 & 0 & 0 & -1 & 0 & 0 \\ 0 & 0 & 0 & 0 & 1 & 0 \\ 0 & 0 & 0 & 0 & 0 & -1 \end{bmatrix}$
Twofold CCW rotation about z $s^{(3)} = \begin{bmatrix} -1 & 0 & 0 \\ 0 & -1 & 0 \\ 0 & 0 & 1 \end{bmatrix}$	$\sigma^{(3)} = \begin{bmatrix} 1 & 0 & 0 & 0 & 0 & 0 \\ 0 & 1 & 0 & 0 & 0 & 0 \\ 0 & 0 & 1 & 0 & 0 & 0 \\ 0 & 0 & 0 & -1 & 0 & 0 \\ 0 & 0 & 0 & 0 & -1 & 0 \\ 0 & 0 & 0 & 0 & 0 & 1 \end{bmatrix}$
Reflection plane $\perp$ to x $s^{(4)} = (m \perp x) = \begin{bmatrix} -1 & 0 & 0 \\ 0 & 1 & 0 \\ 0 & 0 & 1 \end{bmatrix}$	$\sigma^{(4)} = \begin{bmatrix} 1 & 0 & 0 & 0 & 0 & 0 \\ 0 & 1 & 0 & 0 & 0 & 0 \\ 0 & 0 & 1 & 0 & 0 & 0 \\ 0 & 0 & 0 & 1 & 0 & 0 \\ 0 & 0 & 0 & 0 & -1 & 0 \\ 0 & 0 & 0 & 0 & 0 & -1 \end{bmatrix}$
Reflection plane $\perp$ to y $s^{(5)} = (m \perp y) = \begin{bmatrix} 1 & 0 & 0 \\ 0 & -1 & 0 \\ 0 & 0 & 1 \end{bmatrix}$	$\sigma^{(5)} = \begin{bmatrix} 1 & 0 & 0 & 0 & 0 & 0 \\ 0 & 1 & 0 & 0 & 0 & 0 \\ 0 & 0 & 1 & 0 & 0 & 0 \\ 0 & 0 & 0 & -1 & 0 & 0 \\ 0 & 0 & 0 & 0 & 1 & 0 \\ 0 & 0 & 0 & 0 & 0 & -1 \end{bmatrix}$
Reflection plane $\perp$ to z $s^{(6)} = (m \perp z) = \begin{bmatrix} 1 & 0 & 0 \\ 0 & -1 & 0 \\ 0 & 0 & 1 \end{bmatrix}$	$\sigma^{(6)} = \begin{bmatrix} 1 & 0 & 0 & 0 & 0 & 0 \\ 0 & 1 & 0 & 0 & 0 & 0 \\ 0 & 0 & 1 & 0 & 0 & 0 \\ 0 & 0 & 0 & -1 & 0 & 0 \\ 0 & 0 & 0 & 0 & -1 & 0 \\ 0 & 0 & 0 & 0 & 0 & 1 \end{bmatrix}$

Table 5.4 contd. ...

...Table 5.4 contd.

Threefold CCW rotation about z $s^{(7)} = \begin{bmatrix} -1/2 & \sqrt{3}/2 & 0 \\ -\sqrt{3}/2 & -1 & 0 \\ 0 & 0 & 1 \end{bmatrix}$	$\sigma^{(7)} = \begin{bmatrix} 0.3 & 0.8 & 0 & 0 & 0 & -0.9 \\ 0.8 & 0.3 & 0 & 0 & 0 & 0.9 \\ 0 & 0 & 1 & 0 & 0 & 0 \\ 0 & 0 & 0 & -0.5 & -0.9 & 0 \\ 0 & 0 & 0 & 0.9 & -0.5 & 0 \\ 0.4 & -0.4 & 0 & 0 & 0 & -0.5 \end{bmatrix}$
Fourfold CCW rotation about x $s^{(8)} = \begin{bmatrix} 1 & 0 & 0 \\ 0 & 0 & 1 \\ 0 & -1 & 0 \end{bmatrix}$	$\sigma^{(8)} = \begin{bmatrix} 1 & 0 & 0 & 0 & 0 & 0 \\ 0 & 0 & 1 & 0 & 0 & 0 \\ 0 & 1 & 0 & 0 & 0 & 0 \\ 0 & 0 & 0 & -1 & 0 & 0 \\ 0 & 0 & 0 & 0 & 0 & -1 \\ 0 & 0 & 0 & 0 & 1 & 0 \end{bmatrix}$
Fourfold CCW rotation about z $s^{(9)} = \begin{bmatrix} 0 & 1 & 0 \\ -1 & 0 & 0 \\ 0 & 0 & 1 \end{bmatrix}$	$\sigma^{(9)} = \begin{bmatrix} 0 & 1 & 0 & 0 & 0 & 0 \\ 1 & 0 & 0 & 0 & 0 & 0 \\ 0 & 0 & 1 & 0 & 0 & 0 \\ 0 & 0 & 0 & 0 & -1 & 0 \\ 0 & 0 & 0 & 1 & 0 & -1 \\ 0 & 0 & 0 & 0 & 1 & 0 \end{bmatrix}$
Cyclic interchange of x_i $s^{(9)} = \begin{bmatrix} 0 & 1 & 0 \\ -1 & 0 & 0 \\ 0 & 0 & 1 \end{bmatrix}$	$\sigma^{(10)} = \begin{bmatrix} 0 & 1 & 0 & 0 & 0 & 0 \\ 0 & 0 & 0 & 0 & 0 & 0 \\ 1 & 0 & 1 & 0 & 0 & 0 \\ 0 & 0 & 0 & 0 & 1 & 0 \\ 0 & 0 & 0 & 0 & 0 & 1 \\ 0 & 0 & 0 & 1 & 0 & 0 \end{bmatrix}$

For axial tensors, where $|s|$ is the determinant of the symmetry operator.

For a given crystal class, there will be as many sets of equations for each tensor property as there are valid symmetry operations for the point group of that class. However, the number of operators derived from these symmetry operations can be greatly reduced to a minimum number required to generate the point group. These are called generating operators, which result from the 10 generating matrices (s) given in Table 5.4.

The generating matrices σ are the quadratic forms of the matrices (see chapter 3, equations (3.66) and (3.67)) s and are applicable only to symmetric tensor properties, here $s^{(0)} = (1) = (\delta_{ij})$, $\sigma\,(0) = I$, and $[\sigma^{(0)}]^{-1} = \tilde{\sigma}^{(0)} = I$ are identity (unit) matrices. CCW Counterclockwise (positive rotation).

The following numerical approximations have been made here $\sqrt{3}/2 = 0.9$, $\sqrt{3}/4 = 0.4$, $1/4 = 0.3$, $3/4 = 0.8$.

The minimum number of generating matrices (derived from the 10 generating matrices in Table 5.4) required for each of the 32 point groups are given in Table 5.5.

The generator matrices will be used to determine the non-zero components of a tensor property as required for a particular point set and according to the appropriate transformation law.

Table 5.5. Minimum generating matrices for the 32-point groups centrosymmetric crystal classes designated by $||X||$.

Crystal system	International symbol (short form)	Minimum generating matrices
Triclinic	1	$s^{(0)}$
	$\|\|\bar{1}\|\|$	$s^{(1)}$
Monoclinic	2	$s^{(2)}$
	m	$s^{(5)}$
	$\|\|2/m\|\|$	$s^{(2)}, s^{(5)}$
Orthorhombic	222	$s^{(2)}, s^{(3)}$
	$mm2$	$s^{(3)}, s^{(5)}$
	$\|\|mmm\|\|$	$s^{(2)}, s^{(5)}, s^{(6)}$
Tetragonal	4	$s^{(9)}$
	$\bar{4}$	$s^{(11)}$
	$\|\|4/m\|\|$	$s^{(6)}, s^{(9)}$
	422	$s^{(12)}, s^{(9)}$
	$4mm$	$s^{(4)}, s^{(9)}$
	$\bar{4}2m$	$s^{(12)}, s^{(11)}$
	$\|\|4/mmm\|\|$	$s^{(4)}, s^{(6)}, s^{(9)}$
$s^{(1)} \cdot s^{(9)} = s^{(11)}; s^{(1)} \cdot s^{(4)} = s^{(14)}$		
Trigonal	3	$s^{(7)}$
	$\|\|\bar{3}\|\|$	$s^{(13)}$
	32	$s^{(12)}, s^{(7)}$
	$3m$	$s^{(4)}, s^{(7)}$
	$\|\|\bar{3}m\|\|$	$s^{(4)}, s^{(13)}$
Hexagonal	6	$s^{(14)}$
	$\bar{6}$	$s^{(15)}$
	$\|\|6/m\|\|$	$s^{(6)}, s^{(14)}$
	622	$s^{(12)}, s^{(14)}$
	$6mm$	$s^{(6)}, s^{(14)}$
	$\bar{6}2m$	$s^{(4)}, s^{(15)}$
	$\|\|6/mmm\|\|$	$s^{(4)}, s^{(6)}, s^{(14)}$
Cubic	23	$s^{(12)}, s^{(10)}$
	$\|\|m3\|\|$	$s^{(4)}, s^{(10)}$
	432	$s^{(8)}, s^{(10)}$
	$\bar{4}3m$	$s^{(16)}, s^{(10)}$
	$\|\|m3m\|\|$	$s^{(4)}, s^{(10)}, s^{(17)}$
$s^{(1)} \cdot s^{(7)} = s^{(13)}; s^{(3)} \cdot s^{(7)} = s^{(14)}; s^{(1)} \cdot s^{(14)} = s^{(15)}; s^{(1)} \cdot s^{(8)} = s^{(16)}; s^{(4)} \cdot s^{(9)} = s^{(17)}$		

The work required to simplify the components of a given tensor can be greatly reduced by recognizing that a tensor of rank r transforms as the products of the r coordinates of a point. For example, the center of symmetry operator $s^{(1)} = (-\delta_{ij})$ requires that

$$\begin{Bmatrix} x_1 \rightarrow -x_1 \\ x_2 \rightarrow -x_2 \\ x_3 \rightarrow -x_3 \end{Bmatrix} \equiv \begin{Bmatrix} 1 \rightarrow -1 \\ 2 \rightarrow -2 \\ 3 \rightarrow -3 \end{Bmatrix} \tag{5.14}$$

This implies that, for a polar tensor property of odd rank, any combination of subscripts will result in $P' = -P = 0$ according to Neumann's principle.

Similarly, for an axial tensor property of even rank, any combination of subscripts will result in $P' = -P = 0$ also according to Neumann's principle. This simplification process is called the direct inspection method and is applicable to all symmetry point groups except those in the trigonometric and hexagonal crystal systems. For a particular crystal class in these two systems, it is necessary to operate a tensor property with each generating matrix in turn. The use of quadratic generation operators is then recommended, especially for tensor properties with a rank greater than two.

The macroscopic symmetry of a given crystal can dramatically affect the number of nonzero components of a tensor property that are allowed to exist in that crystal. This is easily demonstrated by applying Neumann's principle to a symmetric second-rank tensor property S_{ij}, which can represent any of several properties that the crystal may have.

Typical examples are the dielectric susceptibility and the coefficient of thermal expansion. We start with the triclinic crystal system and continue in the direction of increasing symmetry up to the cubic system. For this purpose, the generating matrices for the 11-point groups in Table 5.5, those possessing a symmetry center, and the generating matrices necessary for full symmetry, will be used. As we proceed from the triclinic to the cubic system, the direct inspection method will be used, except for the trigonometric and hexagonal systems as indicated above.

Triclinic system: Since all second-rank symmetric tensor properties are centrosymmetric, they cannot be made null by the center of symmetry operation. The symmetry transformation becomes

$$S' = s^{(1)} S s'^{(1)} = S = \begin{bmatrix} S_{11} & S_{12} & S_{31} \\ S_{12} & S_{22} & S_{23} \\ S_{31} & S_{23} & S_{33} \end{bmatrix} \tag{5.15}$$

thus, leaving the property S_{ij} unchanged, with all six independent components needed to represent it.

Monoclinic system: The monoclinic system requires the use of the generating operators $s^{(2)}$ and $s^{(5)}$ through the group ||2/m||. From $s^{(2)}$, S' is transformed as:

$$S' = s^{(2)} S s^{t(2)} = \begin{bmatrix} -1 & 0 & 0 \\ 0 & 1 & 0 \\ 0 & 0 & -1 \end{bmatrix} \begin{bmatrix} S_{11} & S_{12} & S_{31} \\ S_{12} & S_{22} & S_{23} \\ S_{31} & S_{23} & S_{33} \end{bmatrix} \begin{bmatrix} -1 & 0 & 0 \\ 0 & 1 & 0 \\ 0 & 0 & -1 \end{bmatrix}$$
$$= \begin{bmatrix} S_{11} & -S_{12} & S_{31} \\ -S_{12} & S_{22} & -S_{23} \\ S_{31} & -S_{23} & S_{33} \end{bmatrix} \tag{5.16}$$

Applying Neumann's principle requires that $S_{12} = S_{23} = 0$, giving the result

$$S = \begin{bmatrix} S_{11} & 0 & S_{31} \\ 0 & S_2 & 0 \\ S_{31} & 0 & S_{33} \end{bmatrix} \tag{5.17}$$

where $S_{22} = S_2$. Application of operator $s^{(5)}$ leaves the results of equation (5.18).

Similarly using the generating matrices given in Table 5.5 for the orthorhombic, tetragonal, trigonal and hexagonal systems, the corresponding matrix S is obtained as shown in Table 5.6, where in the tetragonal system $S_1 = S_{11} = S_{22}$ and in the cubic system $S = S_1 = S_2 = S_3$.

Table 5.6. Effect of crystal symmetry on symmetrical second-rank tensor properties.

System	Generating matrices	Independent components of the property matrix
Triclinic	$s^{(1)}$	$\begin{bmatrix} S_{11} & S_{12} & S_{31} \\ S_{12} & S_{22} & S_{23} \\ S_{31} & S_{23} & S_{33} \end{bmatrix}$
Monoclinic	$s^{(2)}, s^{(5)}$	$\begin{bmatrix} S_{11} & 0 & S_{31} \\ 0 & S_2 & 0 \\ S_{31} & 0 & S_{33} \end{bmatrix}$
Orthorhombic	$s^{(4)}, s^{(5)}, s^{(6)}$	$\begin{bmatrix} S_1 & 0 & 0 \\ 0 & S_2 & 0 \\ 0 & 0 & S_3 \end{bmatrix}$
Tetragonal, trigonal, and hexagonal	$s^{(1)}, s^{(3)}, s^{(4)}, s^{(6)}, s^{(7)}, s^{(9)}$	$\begin{bmatrix} S_1 & 0 & 0 \\ 0 & S_2 & 0 \\ 0 & 0 & S_3 \end{bmatrix}$
Cubic	$s^{(4)}, s^{(9)}, s^{(10)}$	$\begin{bmatrix} S & 0 & 0 \\ 0 & S & 0 \\ 0 & 0 & S \end{bmatrix}$

Referring to Table 5.4, it is clear that each of the 11 groups designated by $||X||$, contains the symmetry center operator $s^{(1)} = (-\delta_{ij})$, which requires that the symmetry transformations of (5.15) apply to any crystal system containing a symmetry center. As a result, Neumann's principle requires that for all classes of centrosymmetric crystals

$$T' = -T = 0 \Rightarrow \begin{cases} \text{for polar tensors of odd rank} \\ \text{for axial tensors of even rank} \end{cases} \quad (5.18)$$

Thus, all 11 classes of centrosymmetric crystals will lack odd-rank polar tensor properties and even-rank axial tensor properties. For example, the piezoelectric tensor d_{ikl}, is a third-rank polar tensor property whose property components vanish for any of the 11 centrosymmetric classes designated by $||X||$ in Table 5.5. Another example is the pyroelectric coefficient defined by $\Delta P_i = P_i\ \Delta T$, where all three components of p_i are null for any of the 11 centrosymmetric crystal classes.

All axial and axial-magnetic tensor properties of even rank will lack components for any of the centrosymmetric crystal classes. The 90 magnetic point groups come into play only for magnetic and axial-magnetic tensor properties of odd rank, further reducing the number of existing components beyond those strictly required by conventional symmetry arguments.

5.1.4 Examples of applications of Neumann's and Curie's principles

1. It is known that quartz crystals are piezoelectric, i.e., they become polarized under the action of mechanical stress. Which quartz sheets oriented perpendicular to the axis 3 or axis 2 should be selected as sensitive elements for piezoelectric uniaxial pressure sensors? Can quartz crystals be used as hydrostatic pressure sensors? Let us apply Curie's principle and Neumann's principle to answer these questions:

   Quartz crystals belong to class 32. As to the first question, polarization of crystals under the action of uniaxial compression arises only when a unitary direction appears in the crystal which is simultaneously a polar direction (a unitary or special direction is a direction which is impossible to repeat by the symmetry elements of the crystal. The polar direction is the direction in the crystal in which two ends cannot be made to coincide by any symmetry element of the given class). In class 32, the third order axis is a unitary direction, but it is not a polar direction due to the presence of the second order axes which are perpendicular to this axis.

Acting on the crystal a pressure that has the symmetry group ∞/mmm along the third order axis, we can write conventionally in the following form:

$$32 \cap \infty/mmm, \| 3 \quad (5.19)$$

where $\cap$ denotes the intersection or product of the groups; we obtain that the symmetry of the crystal in this case does not vary, that is:

$$32 \cap \infty/mmm, \| 3 = 32 \quad (5.20)$$

Therefore, by compressing a quartz sheet cut so that its active faces are perpendicular to the 3 axis, we do not obtain a polarization effect.

On the other hand, by compressing the quartz crystal along one of the 2 axis:

$$32 \cap \infty/mmm, \| 2 = 32 \tag{5.21}$$

In this case, one of the polar directions of the crystal lying in the plane perpendicular to the axis 3 will be singled out. Now it turns out that this direction is unitary and polar; therefore, the piezoelectric polarization vector will lie along it. It is obvious that to achieve the piezoelectric effect by uniaxial compression, it is necessary to cut the quartz plate in such a way that its active faces are perpendicular to one of the axes 2. As for the second question, only crystals that have a unitary polar direction in the absence of the agent can be used as hydrostatic pressure sensors, the symmetry of which is characterized by the group $\infty\infty m$. Therefore, quartz crystals are not suitable for these purposes.

2. A uniaxial tensile stress is applied to a cubic crystal of $m3m$ symmetry. Let us ask ourselves the following question: what symmetry will the crystal have, if the stress is applied along the directions: [001], [111], [110], [hk0]? According to the problem, the tensile stress, whose symmetry is ∞/mmm, is applied to the crystal of $m3m$ symmetry along the [001] direction which coincides with the fourth order axis. In this case, the uniaxial stress field of ∞/mmm symmetry and the crystal of $m3m$ symmetry interact with each other.

   If all symmetry elements of interacting phenomena intersect at one point, and applying the principle of superposition of symmetry, we discover that the symmetry group of the resulting phenomenon (in the given case the symmetry of the crystal) must contain a fourth-order axis, four planes passing through it, and a plane perpendicular to it: an inversion center and four second-order axes. The symmetry of the crystal must be $4/mmm$, that is, tetragonal.

When the stress is applied along the [111], direction, which coincides with the third-order axis, it leads to the phenomenon that the crystal retains the third-order axis along which the stress is exerted, three symmetry planes, an inversion center, and three second-order axes. In other words, a trigonal crystal of a rhombohedron symmetry: $\bar{3}m$. When the stress is applied in the [110] direction in a cubic crystal, of $m3m$ symmetry, along one of the second-order axes in the crystal only the set of mmm symmetry elements remain, which is typical of rhombic crystals.

Finally, applying the tensile stress in the [hk0] direction, that is, in an arbitrary way in one of the symmetry planes, in this case the crystal will have the symmetry $2/m$, that is, the crystal becomes monoclinic.

3. A small monocrystalline cube of rock salt is placed in a homogeneous continuous electric field such that the direction of the field intensity vector $\vec{E}$ coincides in turn with the crystallographic directions: [001], [111], [110], [hk0].

Rock salt crystals belong to the class $m3m$. The electric field symmetry is ∞m. If the direction of the electric field intensity vector coincides with the crystallographic direction [001], that is, the field is applied in the direction of the fourth-order axis, then, according to the principle of superposition of symmetry, the crystal symmetry drops to tetragonal, $4mm$. If the field is directed along [111], that is, in the direction of the third-order symmetry axis, the crystal symmetry drops to trigonal, $3m$. In the absence of an electric field, the optical indicatrix of the cubic crystal is spherical, and in this crystal, there is no double refraction. When an electric field is applied to the crystal, in the first and in the second case its optical indicatrix is deformed, transforming from a sphere into an ellipsoid of revolution. Double refraction occurs in the crystal: the crystal becomes optically uniaxial.

By superimposing the electric field along the [110] direction, i.e., along the second-order symmetry axis, the crystal symmetry drops to *mm* rhombic. If the field is superimposed along the [*hk*0] type directions, i.e., in an arbitrary direction, but lying in the symmetry plane, in this case the crystal becomes monoclinic, with *m* symmetry.

4. A thin section *Z* is made of sphalerite (see Figure 5.2), let us ask ourselves the following question: Will this sheet become polarized if a uniaxial compressive stress is applied to its active faces?

The compressive action of the *Z* section of stephalerite means that the uniaxial mechanical stress characterized with the symmetry group ∞/mmm acts on the crystal of symmetry: $\bar{4}3m$ in the [001] direction. This lowers the symmetry of the crystal to the $\bar{4}2m$ class. Since in this case the highlighted polar directions do not appear, the piezoelectric polarization cannot arise either.

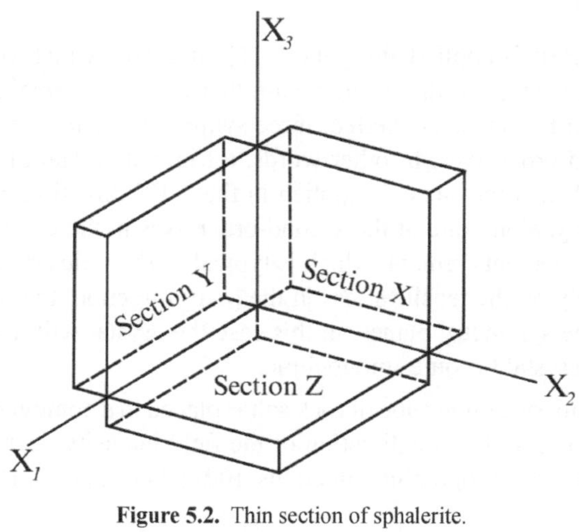

Figure 5.2. Thin section of sphalerite.

Let us answer the same question, but now the sheet is subjected to pure displacement stresses acting in the plane of active faces and characterized by the *mmm* symmetry group. In this case, the action of the pure displacement stress characterized by the *mmm* symmetry group lowers the crystal symmetry to class 2, in which the second order axis is a special polar direction. In this case, the thin Z section of sphalerite can reveal the piezoelectric effect, and this fact is experimentally validated.

5.2 Pyroelectric effect

Dielectric crystals which, in the absence of an external electric field, have a non-zero spontaneous polarization throughout the temperature range of their existence are called pyroelectric. Crystals in which the direction of spontaneous polarization cannot be changed by the superposition of an external electric field, even at stress values close to the magnitude of the disruptive stress, are called pyroelectric crystals proper. Examples of crystal materials include: tourmaline, lithium sulphate, potassium tartrate, etc.

Crystals in which the direction of spontaneous polarization can be changed by applying an external field belong to ferroelectrics. Ferroelectric materials constitute a subgroup of pyroelectrics and have several properties that are inherent only to this group of crystals. The most important ferroelectrics are Seignette salt, barium titania, potassium dihydrophosphate, triglycine sulfate, among others.

Due to the presence of spontaneous polarization, bound charges are found on the opposite faces of the cut sheet of a pyroelectric crystal. Their density σ depends on the magnitude and direction of the spontaneous polarization vector relative to the faces of the sheet.

$$\sigma = (\vec{P}^s \cdot \vec{n}) = P_i^s n_i = |P^s| \cos \alpha \tag{5.22}$$

where $\vec{n}$ is a unit vector normal to the face of the sheet, α is the angle between the vectors $\vec{P}^s$ and $\vec{n}$ as shown in Figure 5.3. The maximum charge density will be observed on the faces of a sheet cut perpendicularly to the direction of spontaneous polarization.

However, the potential difference arising between the opposite charges of the film cannot be revealed, since the bound charges turn out to be compensated by the air charges and due to the non-zero conductivity of the pyroelectric crystals.

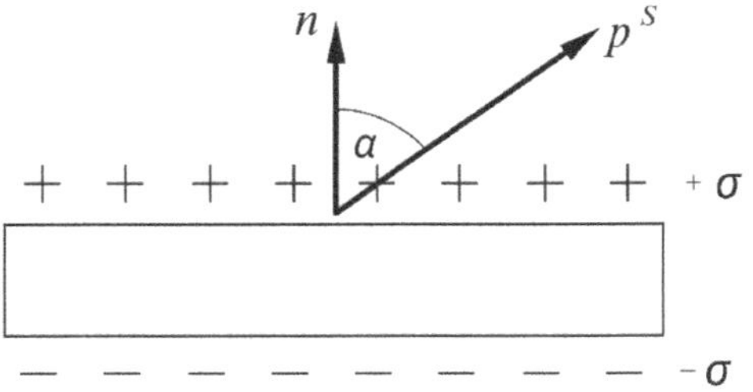

Figure 5.3. Relation of surface charge density σ to spontaneous polarization $\vec{P}^s$.

In real crystals, $\sigma \neq (\vec{P}^s \cdot \vec{n})$ due to the compensation of surface charges at the expense of electroconductivity of crystals. By uniformly changing the temperature of pyroelectrics, the magnitude of their spontaneous polarization is also changed. The effect of changing the spontaneous polarization of crystals, linked to the uniform change of their temperature, is called pyroelectric effect or pyro effect.

A small change in temperature ΔT, which is the same at all points in the crystal (uniform heating), the change in the spontaneous polarization vector $\Delta\vec{P}^s$ is proportional to the change in temperature, i.e.,

$$\Delta\vec{P}^s = \vec{\gamma}\Delta T \equiv \Delta P_i^s = \gamma_i \Delta T \tag{5.23}$$

where $\vec{\gamma}(\gamma_1, \gamma_2, \gamma_3)$ is the vector of pyroelectric coefficients or, simply, pyroelectric coefficient, analytically it is defined as follows:

$$\gamma_i = \frac{dP_i^s}{dT} \tag{5.24}$$

The pyroelectric coefficient $\vec{\gamma}$ is considered positive if the increase in temperature results in an increase in spontaneous polarization of the crystal.

Crystal symmetries impose certain restrictions. The electric polarization vector $\vec{P}^s$ can only be in the crystal along the unitary polar direction. Therefore, as already mentioned, according to Neumann's principle, the pyro effect can be revealed by crystals belonging to the 10 polar symmetry classes: 1,2,3,4,6, *m*, *mm*, 3*m*, 4*mm*, 6*mm*.

Furthermore, the ∞ and ∞m symmetry textures can also reveal the pyroelectric effect. The choice of the coordinate system is determined by the crystal symmetry and the direction of the $\vec{\gamma}$ vector is linked to the symmetry elements. Thus, in each of the 10 polar classes, this vector will have well-defined, nonzero components. Thus, in the 3*m* point symmetry group (such symmetry is possessed by tourmaline, in which the pyro effect was first observed), the direction of the 3 axis of symmetry is the unitary polar direction.

Thus, along this axis lie the polarization vector $\vec{P}^s$ and the vector $\vec{\gamma}$. In the coordinate system from the crystal, the symmetry axis 3 coincides with the coordinate X_3; the vector $\vec{\gamma}$ is directed along the X_3 axis and its components $\gamma_1 = \gamma_2 = 0$, $\gamma_3 \neq 0$, i.e., $\vec{\gamma}$ has only one independent component.

Let us now review a technical matter concerning tourmaline. A pyroelectric receiver consisting of a 1*mm* thick tourmaline sheet, whose sensitivity is 10^{-3} V/step, what temperature changes could this receiver measure? What is the best way to cut a sheet for this purpose? What value could be used to characterize the sensitivity of the receiver? How many times would the sensitivity of the receiver increase if, instead of a tourmaline sheet, we used a lithium sulphate sheet of the same thickness cut perpendicular to the polar axis?

The first thing to notice is that the only pyroelectric coefficient γ_3 of tourmaline is bound to the crystal in the polar direction of the crystal, that is with the 3 axis. According to equation (5.22), the most convenient way to orient the tourmaline sheet is to cut it perpendicularly to the 3 axis, that is, along the (0001) plane. The temperature change that this sheet can "feel" can be found by equation (5.23), in this way:

$$\Delta P_3^s = \gamma_3 \Delta T \tag{5.25}$$

Considering that $\Delta P^s = |\Delta\sigma|$, $\Delta V = \Delta Q/c \cong 10^{-3}$ V/step, $\Delta Q = \Delta\sigma S$ and $c = \epsilon_{33} S/d$; from equation (5.25), it is found that the minimum temperature change that can be measured, using a millivoltmeter and a 1 mm thick tourmaline sheet will be:

$$\Delta T = \frac{\epsilon_{33} \Delta V * 8.85 * 10^{-12}}{d\gamma_3} = 10^{-5}\, grad \tag{5.26}$$

As for the sensitivity of the pyroelectric stress receiver for the idling regime, it can be characterized by the relationship

$$\frac{\Delta V}{\Delta T} = A \frac{\gamma_i}{\epsilon_{ii}} \tag{5.27}$$

where A is a constant equal to $d/(8.85 * 10^{-12})$; γ_i and ϵ_{ii} are values of the pyroelectric coefficient and the dielectric constant along the polar axis X_i. Using the values of γ_i and ϵ_{ii} for tourmaline and lithium sulfate (for lithium sulfate, as for all class 2 crystals, the polar axis in a standardized system is the X_2 axis), we obtain that the sensitivity of the receiver with the lithium sulfate foil is approximately 27 times higher than that of the receiver with the tourmaline foil.

5.3 Electrocaloric effect

From thermodynamic considerations, it follows that there is an effect opposite to the pyroelectric effect. This effect consists of the change in temperature of the pyroelectric material when an electric field acts on it and is called the electrocaloric effect. The equation for the electrocaloric effect is mathematically written as:

$$\Delta T = q \Delta E \tag{5.28}$$

where q is the electrocaloric coefficient.

On the other hand, the variation of the energy of the elemental volume of the crystal, which accompanies small changes in the electrical and thermal conditions, is the total differential, that is:

$$dU = EdP + TdS \equiv d(U - EP) = -PdE + TdS \tag{5.29}$$

where dS is the change in entropy; dP is the change in polarization of the crystal due to the application of the E field. Therefore, let us write the expression for T explicitly and take the derivative dT/dE, using the equation of the electrocaloric effect (5.28) we obtain:

$$q = \left(\frac{dT}{dE}\right)_S = -\left(\frac{dP}{dS}\right)_E = -\frac{\partial P}{\partial T}\frac{\partial T}{\partial S} \tag{5.30}$$

Since $\partial P/\partial T = \gamma$, $q = -\gamma\ \partial T/\partial S$. Using the thermodynamic correlations $dS = Q/T$, $dQ = dT\rho cJ$, we obtain equation (5.31):

$$q = -\frac{\gamma T}{\rho cJ} \tag{5.31}$$

where ρ and c are, respectively, the density and heat capacity of the crystal, J is the mechanical equivalent of heat, which is equal to $4.18 * 10^7$ erg/cal. Plugging in the tourmaline data, we get $q_3 = 1.5 * 10^3$ CGSE units.

Replacing equation (5.31) in equation (5.28) we obtain the equation of the electrocaloric effect

$$\Delta T = -\frac{\gamma T}{\rho cJ}\Delta E \tag{5.32}$$

It follows that if the coefficient γ is positive and the electric field intensity vector $\vec{E}$ coincides with the direction of the vector $\vec{P}$, the coefficient q will be negative, that is, in this case, the increase in the polarization of the crystal at the expense of the applied field lowers its temperature.

Bibliography

Aggarwal, R.L. and Ramdas, A.K. (2019). Physical Properties of Diamond and Sapphire. CRC Press.

Helliwell, J.R. (2017). Science of crystal structures: highlights in crystallography.

Jerphagnon, J., Chemla, D. and Bonneville, R. (1978). The description of the physical properties of condensed matter using irreducible tensors. Adv. Phys. 27: 609–650.

Lovett, D. (2018). Tensor Properties of Crystals. CRC Press.

CHAPTER 6

Determination of Crystal Structure and Fundamentals of Rietveld Refinement

The X-ray diffraction (XRD) technique of polycrystalline samples remains an internationally useful analytical tool for industry and modern research in materials science, which, for the development of many products and processes from certain raw materials, requires knowledge of their chemical composition and the content of their crystalline phases, including their amount of amorphous material. Using this technique, it is not only possible to obtain qualitative and quantitative information on the crystalline phases present in the sample, but also to determine the amount of amorphous material if present. Currently, there are two widely used quantitative methods to carry out this type of analysis, which are:

1. Reference Intensity Ratio method (RIR)
2. Rietveld method

In this chapter, a brief description of the RIR method and a more in-depth analysis of the Rietveld method will be given.

6.1 Scattering by a single electron

In electron X-ray scattering, the scattered X-rays have the same frequency (wavelength) as the incident beam and are "coherent" with the incident X-rays. If the incident X-ray beam traveling along the X-axis encounters a single electron with mass m (kg) and charge e coulombs (C) located at the origin O, the intensity I of the scattered X-rays at the position of some point P located in the $X-Z$ plane at a distance r (m) from the origin can be expressed by the "Thomson equation":

$$I = I_0\left(\frac{\mu_0}{4\pi}\right)^2\left(\frac{e^4}{m^2r^2}\right)\sin^2\alpha = I_0\frac{K}{r^2}\left(\frac{1+\cos^2 2\theta}{2}\right) \tag{6.1}$$

where I_0 is the intensity of incident X-ray, $\mu_0 = 4\pi \times 10^{-7}$ (mKgC^{-2}), and α is the angle between the scattering direction and that of the acceleration of electron, 2θ is the angle

DOI: 10.1201/9781003344438-6

between the line of *OP* connecting from the origin *O* to the point of measurement *P*, y, and the X-axis the direction of the X-ray beam. $K = (2.8179 \times 10^{-15})^2\ m^2$ is equivalent to the square of the electron radius r_e in classical electromagnetic theory. The term in (6.1)

$$\left(\frac{1+\cos^2 2\theta}{2}\right) \tag{6.2}$$

is called the "polarization factor."

For example, the ratio of (I/I_0) at the position of 0.01 m away from an electron at origin is extremely small, at about 7.94×10^{-26}. Nevertheless, the intensity of scattered X-rays can be amplified for detection without any difficulty, because the number of electrons contained in 1 mg of substance is of the order of $10^{20} \sim 10^{21}$ and such many electrons interfere with each other.

Finding the absolute value of the electron scattering intensity using (6.1) is complex either by measurement or by calculation. However, for most applications, only the relative values of the X-ray scattering and X-ray diffraction intensities are required.

It can be assumed that all terms in (6.1) except the polarization factor can be considered as a constant. There is another form of X-ray scattering by an electron and it is characterized by a mechanism quite different from Thomson scattering. Compton scattering occurs when X-rays encounter free or loosely bound electrons, this type of scattering was already discussed in Chapter 1.

6.2 Scattering by a single atom

If an X-ray beam interacts with an atom consisting of a nucleus and a certain number of electrons, each electron produces a scattering intensity generated by constructive interference between the waves scattered by the atom after the interaction of the incident waves with this atom. The interaction can occur through two different processes: incoherent and coherent scattering by an electron.

The first method occurs when the X-ray beam hits the atom, and its oscillating electric field component interacts with its electrons and makes them oscillate around the nucleus. The dipole formed by the oscillation of the electrons becomes a source of radiation of the same frequency according to classical electromagnetic theory. This type of scattering is commonly called coherent scattering. Analogously as in the case of scattering by an electron, when there is a loss of energy during the interaction of the X-ray beam with an atom which manifests itself with a change in the wavelength of the scattered radiation, this is still the phenomenon of Compton scattering and is called incoherent scattering.

In coherent scattering arising from the scattering of an X-ray beam by an atom consisting of the nucleus and several electrons, each electron produces a coherent scattering intensity described by the Thomson equation (6.1). Since the mass of the nucleus is much greater than that of the electron, the X-ray beam cannot oscillate the nucleus to any appreciable extent. The acceleration and deceleration of the nucleus to emit X-rays are not appreciable, therefore, in considering the net effect of X-ray scattering from an atom, it is necessary to consider only the scattering from the electrons associated with the atom. This is also evident from (6.1), since the square of the mass of the scattering particle appears in the denominator.

The scattering amplitude of a single atom with atomic number Z containing Z electrons is equal to Z times the scattering amplitude of an electron in the forward direction. Because in the forward direction the scattering angle is zero (2θ), the phases of the X-rays scattered by all electrons in an atom are completely coincident, so that the amplitude of the scattered X-rays can be simply summed. However, when the scattering angle has nonzero values, there is variation in the phase of the scattered X-rays from individual electrons in an atom, that is, when the scattering angle is nonzero, the X-rays scattered by electrons located, for example, at a point *A* and at a point *B* in space, have phase difference due to the difference in the optical path.

This implies that the scattering amplitude from a single atom decreases with increasing θ. The scattering amplitude from an atom also depends on the wavelength of the incident X-rays. For example, at the same scattering angles 2θ, the shorter the wavelength, the larger the phase difference. Therefore, the scattering amplitude becomes relatively small when the shortest wavelength is used.

To calculate the scattering amplitude of X-rays for atoms containing more than two electrons, it is necessary to consider that the charge of the electrons is not concentrated at fixed points but is distributed in space like a cloud. The electron density function $\rho(r)$ as a function of distance r from the nucleus at the origin is useful to describe the distribution of electrons. Let us consider that the wave vectors of the incident X-rays and the scattered X-rays are given by ($\vec{l}_0$) and ($\vec{l}$), respectively. The scattered X-rays at the distance r will produce the optical path difference $(\vec{l} - \vec{l}_0) \cdot \vec{r}$ compared to the scattered X-rays at the origin. If the wavelength of the incident X-rays is given by, the scattering amplitude of the X-rays irradiated in the direction of $\vec{l}$ can be described by:

$$\rho \exp\left[\frac{2\pi i}{\lambda}(\vec{l} - \vec{l}_0)\cdot\vec{r}\right] dV \tag{6.3}$$

The amplitude of coherent single-electron scattering can be obtained by integrating over the volumes occupied by the electron with the help of the phases relevant to ρdV. Then, the scattering factor f_e per electron in the electron unit is obtained, given by the expression:

$$f_e = \int \exp\left[\frac{2\pi i}{\lambda}(\vec{l} - \vec{l}_0)\cdot\vec{r}\right] \rho dV \tag{6.4}$$

From expression (6.4), the value of f_e corresponds to the ratio of the amplitude of coherent scattering by a single electron having electric charge distribution to that of X-rays scattered by a single electron whose location is fixed at a point according to the classical view. Furthermore, the wave vectors of ($\vec{l}_0$) and ($\vec{l}$) and s are related as follows:

$$\vec{q} = \vec{l} - \vec{l}_0 \Rightarrow |\vec{q}| = q = \frac{2\sin\theta}{\lambda} \tag{6.5}$$

The vector $\vec{q}$ is called the scattering vector or wave vector and represents the amount necessary to rotate the direction of the incident X-rays towards that of the scattered radiation by an angle 2θ, therefore, $\vec{l} = \vec{q} + \vec{l}_0$.

The hypothesis of assuming spherical symmetry for the inner shell electron density distribution to more easily calculate the scattering factors for atoms containing more than two electrons has shown an accurate fitting with experimental results. For example, if the electron density distribution around the nucleus located at the origin is given by $\rho = \rho(r)$ as a function of the distance r, the scattering factor f_e can be obtained as:

$$f_e = \int_0^{\infty} 4\pi r^2 \rho(r) \frac{\sin 2\theta qr}{2\pi qr} dr \tag{6.6}$$

To calculate the coherent scattering amplitude per atom involving n electrons, it is calculated by summing all the f_e contributions with respect to all electrons using the following equation:

$$f = \sum_i f_{ei} = \int_0^{\infty} 4\pi r^2 \rho_i(r) \frac{\sin 2\theta qr}{2\pi qr} dr \tag{6.7}$$

The quantity f given by expression (6.7) is usually called the "atomic X-ray scattering factor" or simply "atomic scattering factor", sometimes also called the "shape factor", because it depends on the distribution of electrons around the nucleus. The quantity f gives the coherent scattering efficiency or capacity per atom and is defined as a ratio of the amplitude of the wave scattered from an atom to that scattered from an electron under the same conditions. Therefore, $f = Z$ for any atom scattering in the forward direction.

Equation (6.7) implies that f is a function of ($\sin\theta/\lambda$) for any atom. Electron density distributions in atoms have been given from the electron wave functions by using various techniques such as the Hartree–Fock and Fermi–Thomas–Dirac approximations, and several theoretical calculations have been performed for the atomic dispersion factors of the elements as a function of ($\sin\theta/\lambda$). Such results are compiled in the International Tables for X-ray Crystallography (Al-Tabbakh 2017, Clegg 2023).

We can define the variable Q as:

$$Q = 2\pi q = 4\pi \times \frac{\sin\theta}{\lambda} \tag{6.8}$$

To simplify the writing, so using the expression (6.8) we can write (6.7) as:

$$f = \sum_i f_{ei} = \int_0^{\infty} 4\pi r^2 \rho_i(r) \frac{\sin Qr}{Qr} dr \tag{6.9}$$

It is useful to summarize some important points regarding single-atom X-ray scattering. When a monochromatic X-ray beam encounters an atom, two scattering processes, coherent and incoherent, occur simultaneously in all directions. It has been found that the intensity of incoherent (Compton) scattering for light elements (with a small atomic number) increases with decreasing atomic number Z (see Chapter 1). Furthermore, as the amount of $\frac{\sin\theta}{\lambda}$ increases, the intensity of incoherent (Compton) scattering increases.

Therefore, the intensities of unmodified (coherent) scattering and modified (Compton) scattering change oppositely with Z and $\frac{\sin\theta}{\lambda}$. Furthermore, the sum of the coherent and incoherent scattering intensities is equal to the classical scattering intensity per electron. If i_e represents the incoherent scattering intensity in electron units for the completely unpolarized incident X-ray beam, the following equation can be obtained:

$$\begin{aligned} & I_0 \frac{e^4}{m^2c^4r^2}\left(\frac{1+\cos^2 2\theta}{2}\right)f_e^2 + I_0 \frac{e^4}{m^2c^4r^2}\left(\frac{1+\cos^2 2\theta}{2}\right)i_e \\ & = I_0 \frac{e^4}{m^2c^4r^2}\left(\frac{1+\cos^2 2\theta}{2}\right) \end{aligned} \tag{6.10}$$

or simply

$$i_e = 1 - f_e^2 \tag{6.11}$$

the phase of the radiation attributed to the Compton scattering effect has no fixed relationship with that of the incident beam, due to the variation in wavelength after the collision, suggesting that the interference effect cannot be produced by modified Compton radiation. Therefore, the incoherent scattering intensity $i(M)$ per atom can be given by the simple sum of the incoherent scattering intensities of the respective electrons (in the case of crystalline structures it is usually neglected).

6.3 Scattering by a unit cell

Recall that X-ray diffraction is described by Bragg's law (1.22) (see Chapter 1), which in terms of the Miller indices for the first order of diffraction (n = 1) can be written as (see Figure 2.1):

$$2d_{hkl} \sin\theta = \lambda \tag{6.12}$$

where $d_{hkl} = d'/n$ is the interplanar distance for families (hkl) in the seven crystal systems and can be calculated from the expressions given in Table 2.3 (see Chapter 2).

As mentioned in Chapter 2, a crystal is defined as a solid consisting of atoms arranged in a periodic pattern defined by the unit cell. Therefore, when considering the scattering intensity of crystals, it is important to obtain information from phase differences based on the relationships between scattering intensity and atomic positions in a unit cell.

The phase difference of X-rays whose path difference is one full wavelength is 2 radians (360°). Consider an atom A located at the origin (000) and an atom B in real

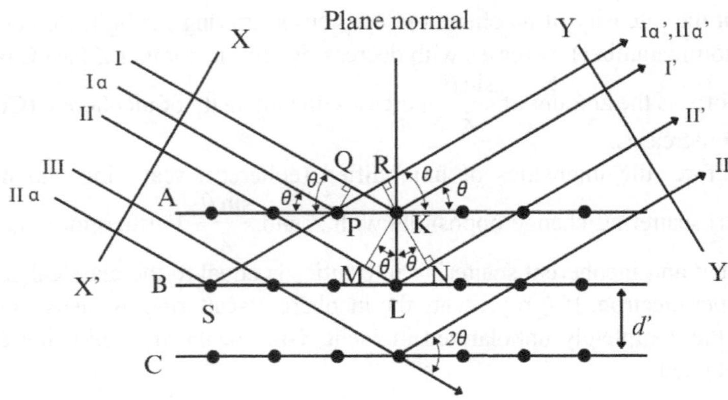

Figure 6.1. Schematic diagram of diffraction of X-rays by a crystal (Bragg condition).

coordinates (x y z) that can be expressed by fractional coordinates (u v w) where $u = x/a$, $v = y/b$ and $w = z/c$, where a, b and c are the lattice parameters along each axis. Under these conditions, the phase difference between the X-rays scattered by atom B and those scattered by atom A at the origin (see Figure 6.1) is given by the following equation for the reflection (hkl):

$$\phi = 2\pi(hu + kv + lw) \tag{6.13}$$

This relationship applies to a unit cell of any shape. If atoms A and B are of different elements, there will be differences not only in phase but also in amplitude. By summing the waves scattered by all the atoms in a unit cell, including the atom at the origin, the resulting wave can be obtained. Each wave can be described by a vector whose length is equal to the amplitude and this vector is inclined to the horizontal axis by the phase angle.

The amplitude and phase of the resulting wave are then obtained by simply adding two vectors using the parallelogram law. Such a geometric solution can be simplified if we use complex numbers to represent the vectors. The complex exponential function can also be used. The variation of the electric field intensity E with time t is often expressed in the form:

$$E = A\,\exp(2\pi\nu ti) = A\cos(2\pi\nu t) + iA\sin(2\pi\nu t) \tag{6.14}$$

where A is the amplitude and ν is frequency. On the other hand, the amplitude of the scattered wave from each atom in a unit cell is given by the atomic dispersion factor f of the atom and the value of $\frac{\sin\theta}{\lambda}$ relevant to diffraction.

Since the phase of each wave is given by (6.13) for the reflection (*h k l*) using fractional coordinates (*u v w*) of the atom, any scattered wave can be expressed in the complex exponential form.

$$Ae^{i\phi} = fe^{2\pi i(hu + kv + lw)} \tag{6.15}$$

Based on this relationship, the sum of scattered waves from atoms in a unit cell can be calculated using summation over all atoms:

$$F_{hkl} = \sum_{j=1}^{N} f_j e^{2\pi i(hu_j + kv_j + lw_j)} \tag{6.16}$$

where N is the total number of atoms involved in the unit cell. F, defined by (6.16), is called the structure factor or geometrical structure factor and one can obtain the value of F only if the position and the type of atoms in the unit cell are given. For this reason, (6.12) as well as the Bragg condition given by (6.16) are considered very important relationships for analyzing the structure of crystals by X-ray diffraction.

The structure factor F is usually given as a complex number and represents both the amplitude and the phase of the scattered wave obtained by summing over all the atoms in the unit cell. Furthermore, the absolute value of F represents the amplitude of the scattered wave resulting from summing the waves scattered by the individual atoms in the unit cell, based on that of the wave scattered by a single electron. Therefore, the intensity of the resulting wave scattered from all the atoms in the unit cell in the direction satisfying Bragg's law is simply proportional to $|F|^2$. The value of $|F|^2$ can be obtained by multiplying F from (6.16) by its complex conjugate F^*.

Example 1. In the case of body-centered cell containing two atoms of the same type (position 000 and ½ ½ ½ (in (6.16))) in the unit cell, the structure factor is given as follows:

$$uvw = 000,\ \tfrac{1}{2}\ \tfrac{1}{2}\ \tfrac{1}{2} \tag{6.17}$$

$$F_{hkl} = fe^{2\pi i * 0} + fe^{2\pi i * \left(\frac{h}{2} + \frac{k}{2} + \frac{l}{2}\right)} = f[1 + e^{\pi i \times *(h+k+l)}] \tag{6.18}$$

When the number of $h - k + l$ is even: $F = 2f$, $F^2 = 4f^2$. When the number of $h + k + l$ is odd: $F = 0$, $F^2 = 0$. The results show that the reflections will be observed for planes such as (110), (200), and (211) whose indices satisfy the condition ($h + k + l$) is even, but not for the planes (111), (210), (300), etc., because the waves cancel each other being out of phase.

Example 2. Consider a hexagonal close-packed cell where two atoms of the same type are located at positions (000) and (1/3, 2/3, 1/2) and calculate the structure factor F_{hkl}.

$$F_{hkl} = \sum_{j=1}^{2} f_j e^{2\pi i(hu_j + kv_j + lw_j)} \tag{6.19}$$

Expanding the summation (6.19)

$$F_{hkl} = fe^{2\pi i \times 0} + fe^{2\pi i * \left(\frac{h}{3}+\frac{2k}{3}+\frac{l}{2}\right)} = f\left[1+e^{2\pi i * \left(\frac{h+2k}{3}+\frac{l}{2}\right)}\right] \tag{6.20}$$

Let $p = \dfrac{h+2k}{3} + \dfrac{l}{2}$ then the expression (6.20) becomes:

$$F_{hkl} = fe^{2\pi i * 0} + fe^{2\pi i * \left(\frac{h}{3}+\frac{2k}{3}+\frac{l}{2}\right)} = f\left[1+e^{2\pi i p}\right] \tag{6.21}$$

The intensity of the waves scattered from the plane $(h\,k\,l)$ is simply proportional to the square of the absolute value of the structure factor and is obtained by multiplying the F value by its complex conjugate

$$|F_{hkl}|^2 = f(1 + e^{2\pi ip}) * f(1 + e^{-2\pi ip}) = f^2 (2 + e^{2\pi ip} + e^{-2\pi ip}) \tag{6.22}$$

Let us remember from basic mathematics that $e^{ix} + e^{-ix} = 2\,cos\,x$ y $cos\,2x = 2cos^2 x - 1$ and let us use these results in expression (6.22):

$$|F_{hkl}|^2 = f^2 (2 + 2\cos 2\pi p) = f^2 [2 + 2(2cos^2 \pi p - 1)] = f^2 (4cos^2 \pi p) \tag{6.23}$$

which implies that:

$$|F_{hkl}|^2 = 4f^2 \cos^2 \pi\left(\frac{h+2k}{3} + \frac{l}{2}\right) \tag{6.24}$$

It is easy to check that $cos^2 \pi n = 1$ (n is an integer). On the other hand, when $(h + 2k)$ is a multiple of 3, while l is an odd number as in (001) and (111), the term $((h + 2k)/3 + l/2)$ gives $\pi/2$ and $3\pi/2$. In these cases, the intensity of the scattering is not detected. When l is an even number, while when $(h + 2k)$ is a multiple of 3, such as (002) and (112), the value $p = (h + 2k)/3 + l/2$ is an integer and $cos\,\pi p = \pm 1$ then $|F_{hkl}|^2 = 4f^2$. This implies that the intensity of the scattering is observed in these cases. However, it should be kept in mind that all structural factors of such plans have not always the same value as in the following examples.

$$|F_{101}|^2 = 4f^2 \cos^2 \pi\left(\frac{h+2k}{3} + \frac{l}{2}\right) = 4f^2 \cos^2 \pi\left(\frac{1}{3} + \frac{1}{2}\right) = 3f^2 \tag{6.25}$$

$$|F_{102}|^2 = 4f^2 \cos^2 \pi\left(\frac{h+2k}{3} + \frac{l}{2}\right) = 4f^2 \cos^2 \pi\left(\frac{1}{3} + \frac{2}{2}\right) = f^2 \tag{6.26}$$

The results for the structure factor of a hexagonal close-packed cell are summarized in Table 6.1.

Example 3: the case of a unit cell of diamond structure belongs to the cubic crystal system and contains eight atoms of the same type, and their positions are as follows.

$$(0,0,0)\left(\frac{1}{2},\frac{1}{2},0\right)\left(\frac{1}{2},0,\frac{1}{2}\right)\left(0,\frac{1}{2},\frac{1}{2}\right)\left(\frac{1}{4},\frac{1}{4},\frac{1}{4}\right)\left(\frac{3}{4},\frac{3}{4},\frac{1}{4}\right)\left(\frac{3}{4},\frac{1}{4},\frac{3}{4}\right)\left(\frac{1}{4},\frac{3}{4},\frac{3}{4}\right)$$

Table 6.1. Summary of structure factor of a hexagonal close-packed cell.

$H + 2K$	l	$\|F\|^2$	**Examples**
$3n$	Odd number	0	001 111 221
$3n$	Even number	$4f^2$	002 110 112
$3n \pm 1$	Odd number	$3f^2$	101 103 201
$3n \pm 1$	Even number	f^2	100 102 200

n : integer

Compute the structure factors F_{hkl} and $|F|^2$.

$$F = f\left[e^{2\pi i(0+0+0)} + e^{2\pi i\left(\frac{h}{2}+\frac{k}{2}+0\right)} + e^{2\pi i\left(\frac{h}{2}+0+\frac{l}{2}\right)} + e^{2\pi i\left(0+\frac{k}{2}+\frac{l}{2}\right)}\right]$$

$$+ f\left[e^{2\pi i\left(\frac{h}{4}+\frac{k}{4}+\frac{l}{4}\right)} + e^{2\pi i\left(\frac{3h}{4}+\frac{3k}{4}+\frac{l}{4}\right)} + e^{2\pi i\left(\frac{3h}{4}+\frac{k}{4}+\frac{3l}{4}\right)} + e^{2\pi i\left(\frac{h}{4}+\frac{3k}{4}+\frac{3l}{4}\right)}\right]$$

After a short algebra we can obtain:

$$F = f\left[1 + e^{\pi i\frac{(h+k+l)}{2}}\right]\left[1 + e^{\pi i(h+k)} + e^{\pi i(h+l)} + e^{\pi i(k+l)}\right]$$

To analyze the restrictions imposed by this expression, let us first consider the second term. We find that the second term will be zero, if $h\ k\ l$ is a mixture of even and odd numbers and considering the fact $e^{n\pi i} = (-1)^n$, where n is an integer. The relations $e^{\pi i} = -1$, and $e^{2\pi i} = +1$, are also useful in computing the structure factor. For example, the following results are obtained with respect to the planes at (100), (110), and (211), respectively.

$$1 + e^{\pi i(1+0)} + e^{\pi i(1+0)} + e^{\pi i(0+0)} = 1 - 1 - 1 + 1 = 0$$

$$1 + e^{\pi i(1+1)} + e^{\pi i(1+0)} + e^{\pi i(1+0)} = 1 + 1 - 1 - 1 = 0$$

$$1 + e^{\pi i(2+0)} + e^{\pi i(1+0)} + e^{\pi i(1+1)} = 1 - 1 - 1 + 1 = 0$$

On the other hand, the second term is found to be 4 when $h\ k\ l$ is not a mixture of even and odd numbers and then, the structure factor F can be expressed in the following form.

$$F = 4f\left[1 + e^{\pi i\frac{(h+k+l)}{2}}\right]$$

Multiplying member by member by the complex conjugate of this expression, we obtain:

$$F^2 = 16f^2\left[1 + e^{\pi i\frac{(h+k+l)}{2}}\right]\left[1 + e^{-\pi i\frac{(h+k+l)}{2}}\right] = 32f^2\left[1 + \cos\frac{\pi}{2}(h+k+l)\right]$$

Note that the cosine function gives –1 when the argument is an odd multiple of π, +1 for an even multiple of π, and zero for an odd multiple of $2/\pi$. Therefore, the following results hold for nonmixed $h\ k\ l$ planes. If $(h + k + l)$ are odd integers such as (111), (311), then $\cos x = 0 \rightarrow |F|^2 = 32f$. If $(h + k + l)$ is given by odd multiple of 2 such as (110), (200), then $\cos x = -1 \rightarrow |F|^2 = 0$. If $(h + k + l)$ is given by the even multiple of 2 such as (220), (400), then $\cos x = +1 \rightarrow |F|^2 = 64f^2$. In fact, $|F|^2 = 0$ when *hkl* is a mixture of even and odd numbers.

Example 4. Sodium chloride ($NaCl$) has a cubic lattice with four Na^+ ions and four Cl^- ions per unit cell as shown below:

$$Na^+(000)\left(\frac{1}{2}\frac{1}{2}0\right)\left(\frac{1}{2}0\frac{1}{2}\right)\left(0\frac{1}{2}\frac{1}{2}\right)$$

$$Cl^-(000)\left(\frac{1}{2}00\right)\left(00\frac{1}{2}\right)\left(\frac{1}{2}\frac{1}{2}\frac{1}{2}\right)$$

Let us calculate the structure factor F_{hkl}, if the dispersion factors of Na and Cl are given by the following expressions, f_{Na} and f_{Cl}, respectively and do the calculation for the (111) and (200) planes in particular, then again using expression (6.16) we obtain:

$$F = f_{Na}\left[e^{2\pi i(0+0+0)} + e^{2\pi i\left(\frac{h}{2}+\frac{k}{2}+0\right)} + e^{2\pi i\left(\frac{h}{2}+0+\frac{l}{2}\right)} + e^{2\pi i\left(0+\frac{k}{2}+\frac{l}{2}\right)}\right]$$

$$+ f_{Cl}\left[e^{2\pi i\left(0+\frac{k}{2}+0\right)} + e^{2\pi i\left(\frac{h}{2}+0+0\right)} + e^{2\pi i\left(0+0+\frac{l}{2}\right)} + e^{2\pi i\left(\frac{h}{2}+\frac{k}{2}+\frac{l}{2}\right)}\right]$$

After developing a simple algebra, we arrive at

$$F = [f_{Na} + f_{Cl}e^{\pi i(h+k+l)}][1 + e^{\pi i(h+k)} + e^{\pi i(h+l)} + e^{\pi i(k+l)}]$$

The first term shows the basis of the unit cell, Na at (000) and Cl at (1/2, 1/2, 1/2), respectively, and the second term indicates that this structure can be reproduced by face-centered translations. Note that the term for the translation is the same as the description of F for an FCC lattice. Since the sums $(h + k)$, $(h + l)$ and $(k + l)$ are even integers if h, k and l are all even or all odd (not mixed), the value of all three exponentials has the value 1.

On the contrary, the sum of the three exponentials will be 1 for mixed indices. Specifically, the scattered intensity of the (111) and (200) planes is observed, but the intensity of the (100) and (210) planes is not observed. However, even if h, k and l are not mixed, we must include the atomic dispersion factors for the different atoms in the present case. The results from an FCC lattice cannot be directly applied to this case. For the (111) plane when h, k and l are not mixed and $(h + k + l)$ is an odd number, then

$$F = 4(f_{Na} - f_{Cl}) \rightarrow |F|^2 = 16(f_{Na} - f_{Cl})^2$$

$$F_{111} = 16(f_{Na} - f_{Cl}) \text{ decrease in intensity due to } (f_{Na} - f_{Cl})$$

(200) plane: when h, k and l are unmixed and $(h + k + l)$ is even number

$$F = 4(f_{Na} + f_{Cl}) \rightarrow |F|^2 = 16(f_{Na} + f_{Cl})^2$$

$$F_{200} = 16(f_{Na} + f_{Cl}) \text{ decrease in intensity due to } (f_{Na} + f_{Cl})$$

To calculate the structure factor F of ionic crystals, the scattering factor of the ions must be used instead of the atoms. However, this does not make a big difference, since the scattering factors of the atoms are almost identical to those of the ions at larger angles, greater than ($\sin\theta/\lambda = 0.3$). Only small differences are found at small angles.

Example 5. Structural similarity between an element and a compound is common, as demonstrated for example by the case of diamond and zinc stearate ($\beta - ZnS$). Let us show some points in these two structures, when the atomic positions of Zn and S in the unit cell of $\beta - ZnS$ are as follows:

$$Zn(000)\left(0\frac{1}{2}\frac{1}{2}\right)\left(\frac{1}{2}0\frac{1}{2}\right)\left(\frac{1}{2}\frac{1}{2}0\right)$$

$$S\left(\frac{1}{4}\frac{1}{4}\frac{1}{4}\right)\left(\frac{1}{4}\frac{3}{4}\frac{3}{4}\right)\left(\frac{3}{4}\frac{1}{4}\frac{3}{4}\right)\left(\frac{3}{4}\frac{3}{4}\frac{1}{4}\right)$$

The unit cell of both diamond and $\beta - ZnS$ is expressed by a face-centered cubic lattice. In diamond, eight atoms per unit cell occupy (000), $\left(\frac{1}{4}\frac{1}{4}\frac{1}{4}\right)$, and the translational positions of its face center. These structural fundamentals do not change in ZnS, but Zn occupies the translational position of its face center, while S occupies $\left(\frac{1}{4}\frac{1}{4}\frac{1}{4}\right)$ and the translational position of its face center (or the reverse).

The structure of zinc blende can be characterized as follows. One of the two elements (e.g., Zn) occupies the corner of the unit cell as well as the face center position, and another element (e.g., S) occupies the tetrahedral positions as in diamond. These structural features can also be expressed as superposition of the fcc lattice of Zn with the fcc lattice of S after translating the position by (1/4 1/4 1/4) from the lattice of Zn.

Let us now determine the structure factor of $\beta - ZnS$ and show the cases in which intensity can be observed by X-ray scattering. In this case we have:

$$F = f_{Zn}\left[e^{2\pi i(0+0+0)} + e^{2\pi i\left(0+\frac{k}{2}+\frac{l}{2}\right)} + e^{2\pi i\left(\frac{h}{2}+0+\frac{l}{2}\right)} + e^{2\pi i\left(\frac{h}{2}+\frac{k}{2}+0\right)}\right]$$

$$+ f_S\left[e^{2\pi i\left(\frac{h}{4}+\frac{k}{4}+\frac{l}{4}\right)} + e^{2\pi i\left(\frac{h}{4}+\frac{3k}{4}+\frac{3l}{4}\right)} + e^{2\pi i\left(\frac{3h}{4}+\frac{k}{4}+\frac{3l}{4}\right)} + e^{2\pi i\left(\frac{3h}{4}+\frac{3k}{4}+\frac{l}{4}\right)}\right]$$

After developing a simple algebra, we obtain

$$F = [1 + e^{\pi i(h+k)} + e^{\pi i(h+l)} + e^{\pi i(k+l)}]\left[f_{Zn} + f_S e^{\pi i\left(\frac{h+k+l}{2}\right)} \right]$$

The second term represents the basis of the unit cell where Zn is located at (000) and S at $\left(\frac{1}{4}\frac{1}{4}\frac{1}{4}\right)$, respectively. On the other hand, the first term is a relationship observed in the description of the structure factor for an FCC lattice. That is, if $h\ k\ l$ is a mixture of even and odd numbers, the first term will be zero, while it becomes four for unmixed indices. Therefore, we can examine the case where $h\ k\ l$ is unmixed using the following equation:

$$F = 4\left[f_{Zn} + f_S e^{\pi i\left(\frac{h+k+l}{2}\right)} \right]$$

F^2, which is proportional to the intensity, can be calculated by multiplying by its member-wise complex conjugate to obtain the following:

$$|F|^2 = 4\left[f_{Zn} + f_S e^{\pi i\left(\frac{h+k+l}{2}\right)} \right] * 4\left[f_{Zn} + f_S e^{-\pi i\left(\frac{h+k+l}{2}\right)} \right]$$

$$= 4\left[f_{Zn}^2 + f_S^2 + 2 f_{Zn} f_S \cos\frac{\pi}{2}(h+k+l) \right]$$

When the cosine argument is an even multiple of π, it gives +1; and when it is an odd multiple of π, it gives –1. Taking this into account, the following results are obtained for the sum of the indices $h + k + l$ recorded in Table 6.2.

When h, k, and l are mixed, some are odd and some are even numbers, the intensity is not observed because $|F|^2 = 0$.

In the calculation of the structure factor, no information about the crystal systems is required. In Example 1, the information given is only about the body-centered cell containing two atoms. No information about the shape of the unit cell, such as cubic, tetragonal, and orthorhombic, was used. The structure factor is completely independent of the shape and size of the unit cell. This also illustrates an important point; for any body-centered cell, there are reflections that do not occur when $(h + k + l)$ is an odd number without reference to the specific crystal system. In other words, the information about the missing reflections provides a clue about the actual

Table 6.2. Relationship between $h + k + l$ and $|F|^2$ for $\beta - ZnS$.

$h+k+l$	$\cos x$	$\|F\|^2$
Odd numbers	0	$4(f_{Zn}^2+f_S^2)$
Even multiple of 2	+1	$16(f_{Zn}+f_S)^2$
Odd multiple of 2	–1	$16(f_{Zn}-f_S)^2$

Table 6.3. Relationship between the Bravais network and allowed and disallowed reflections.

Crystal Type	Bravais Lattice Type	Reflections possible present	Reflections necessarily absent
Simple	Primitive	Any h, k, l	None
Body-centered	Body-centered	$h + k + l$ even	$h + k + l$ odd
Face-centered	Face-centered	h, k and l unmixed	h, k and l unmixed
Diamond cubic	Face-centered cubic	As FCC, but if all even and $h + k + l \neq 4N$; the absent	h, k and l mixed and if all even and $h + k + l \neq 4N$
Base-centered	Base-centered	h and k both even or both odd* $h + 2k = 3N$ with l even	h and k mixed*
Hexagonal close-packed	Hexagonal	$h + 2k = 3N \pm 1$ with l odd $h + 2k = 3N \pm 1$ with l even	$h + 2k = 3N \pm 1$ with l odd

* These relationships apply to a cell centered on the C face. If reflections are present only when h and k are unmixed, or when k and l are unmixed, then the cell is centered on the B or A face, respectively.

atomic arrangement in the crystals. The relationship between the Bravais lattice and diffraction behavior is summarized in Table 6.3.

6.4 Estimation of X-ray diffraction intensity from a polycrystalline sample

Suppose we have N atoms, the scattering amplitude is A, and the scattered X-rays satisfy the so-called Bragg condition, then they are said to be completely in phase. In this case, the amplitude of the diffracted X-rays will be given by NA, so that the total diffraction intensity is $(NA)^2$, that is, the intensity of the diffracted X-rays that are in phase when satisfying the Bragg condition will be N times the diffraction intensity given by $(NA)^2$, corresponding to the case where the Bragg condition is not met and no interference is recognized.

The value of N is said to be about 10^{22} per gram even in a very small crystal, so that diffracted X-rays from a (crystalline) powder sample can be measured with sufficient reliability. Note, however, that the intensity of the diffracted X-rays is considerably weak compared to the intensity of the incident X-rays. Some selected examples for the "structure factor" required to estimate the intensity of diffracted X-rays are described in section 6.3. This structure factor $|F|^2$ is a way of knowing the relationship between the crystal structure and the intensity of diffracted X-rays from each possibly measured crystal plane. In addition to the structure factor, the measured intensity of diffracted X-rays from powder samples contains several components, such as the polarization factor, the multiplicity factor, the Lorentz factor, the absorption factor, and the temperature factor. These factors are described in more detail below.

6.4.1 Structure Factor (see also section 6.3)

The structure factor for the (*hkl*) reflection is given by (6.16)

$$F_{hkl} = \sum_{j=1}^{N} f_j e^{2\pi i(hu_j + kv_j + lw_j)}$$

where N represents the total number of atoms in a unit cell, f_j represents the atomic scattering factor of the j-th atom, and $u_j v_j w_j$ are the fractional coordinates for the position of the j-th atom in the corresponding unit cell. Since reflections where $F = 0$ will be zero intensity, we will not observe the diffraction peak, and these reflections are called forbidden.

6.4.2 Polarization factor

The Thomson equation described in Section 6.1 (see equation (6.2)) is estimated from the assumption of a completely unpolarized incident X-ray beam, such as that emitted by an X-ray tube. However, it is impossible for a portion of the X-rays diffracted from crystals to be unrelated to the polarization property. The polarization factor, P, is given in the following form as a function of the diffraction angle 2θ for three typical X-ray diffraction experiments with characteristic X-rays of constant wavelength:

I. Using a filter only:

$$P = \frac{1 + \cos^2 2\theta}{2} \tag{6.27}$$

II. Using a crystal monochromator in the incident X-ray path:

$$P = \frac{1 + x\cos^2 2\theta}{1 + x} \tag{6.28}$$

III. Using a crystal monochromator in the diffracted X-ray path:

$$P = \frac{1 + x\cos^2 2\theta}{2} \tag{6.29}$$

In cases II and III $x = \cos^2 2\theta_M$ is used for an ideally mosaic monochromator crystal or $x = \cos 2\theta_M$ for an ideally perfect monochromator crystal if $2\theta_M$ is twice the Bragg angle of the monochromator crystal.

6.4.3 Multiplicity factor

The multiplicity factor represents the number of crystal planes that have the same spacing and structure factors but different orientations. For example, in the case of the {100} planes for a cubic crystal, it is six, denoted by (100), (010), (001), $(\bar{1}00)$, $(0\bar{1}0)$ and $(00\bar{1})$, and in the case of the {111} planes, it is eight. The multiplicity factors in various crystal systems are summarized in Table 6.4. When the crystal orientations in powder samples are completely random, the probability of crystal

Table 6.4. Multiplicity factors for crystalline powder samples.

Cubic	hkl	hkk	$hk0$	$hh0$	hhh	$h00$		
	48*	24	24*	12	8	6		
Hexagonal	$hk \cdot l$	$hh \cdot l$	$h0 \cdot l$	$hk \cdot 0$	$hh \cdot 0$	$h0 \cdot 0$	$00 \cdot 1$	
	24*	12	12	12*	6	6	2	
Trigonal referred to Rhombohedral axes	hkl	$\bar{k}kl$	hkk	$hk0$	$\bar{k}hh$	hhh	$hh0$	$h00$
	12*	12*	6	12*	6	2	6	6
Trigonal referred to hexagonal axes	$hk \cdot l$	$hh \cdot l$	$h0 \cdot l$	$hk \cdot 0$	$hh \cdot 0$	$0h \cdot 0$	$00 \cdot l$	
	12*	12*	6	12*	6	6	2	
Tetragonal	hkl	hhl	$hh0$	$hk0$	$h0l$	$h00$	$00l$	
	16*	8	4	8*	7	4	2	
Orthorhombic	hkl	$hk0$	$h00$	$0k0$	$00l$	$h0l$	$0kl$	
	8	4	2	2	2	4	4	
Monoclinic	hkl	$hk0$	$0kl$	$h0l$	$h00$	$0k0$	$00l$	
Orthogonal axis: b	4	4	4	2	2	2	2	
Triclinic	hkl	$hk0$	$0kl$	$h0l$	$h00$	$0k0$	$00l$	
	2	2	2	2	2	2	2	

* In some crystals, planes having these indices comprise of two forms with the same spacing but different structure factor. In such case, the multiplicity factor for each form is half the value given here.

orientations satisfying the so-called Bragg condition is given by the ratio of the multiplicity factors, for example, 8–6 for the {111} to {100} condition.

6.4.4 Lorentz Factor (LF)

In experimental X-ray measurements using a diffractometer, the sample is usually rotated during the measurement. On the other hand, we have already seen that the intensity of the diffracted X-rays is maximum at diffraction angles that satisfy Bragg's law; however, the intensity of the diffracted X-rays is usually noticeable at angles that deviate slightly from the Bragg angle, so that the curve of measured intensity as a function of 2θ is given in the form of Figure 6.2.

The diffracted X-rays produced from a crystal sample, as it rotates through the diffraction angle, are collected by a detector, and then the total energy of the diffracted X-rays can be measured. This energy is the integrated intensity of the corresponding peak and can also be obtained from the area under the curve of the corresponding peak. That is, measuring the intensity of the diffracted X-ray beam from a crystal sample means measuring the integrated values of the diffracted intensities produced from volumes with a certain limited size.

The attributed volume of the integrated intensity depends on the diffraction angle, so this must be considered when comparing diffracted intensities from different crystal planes. Such effects are represented by the so-called Lorentz factor.

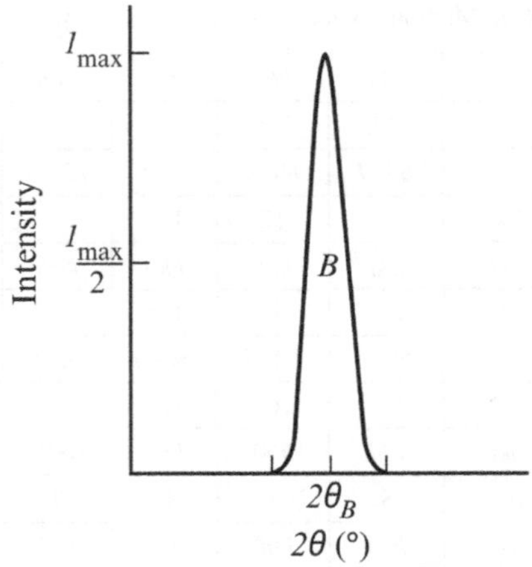

Figure 6.2. Intensity diffracted from a crystalline sample rotated through the Bragg angle.

For example, the Lorentz factor [LF] for powder samples is given by the following equation as a function of the diffraction angle.

$$LF = \frac{1}{\sin^2\theta\cos\theta} \tag{6.30}$$

The Lorentz factor can be used as a Lorentz polarization factor by coupling with the polarization factor, which is also a function of the diffraction angle, or simply used independently.

6.4.5 Absorption Factor (AF)

Finely ground powder samples are placed on a sample holder to make a flat plate shape for measuring diffracted X-rays using a diffractometer. Let us consider the case when the incident X-ray beam with its intensity I_0 per unit cross-sectional area is incident on the powder sample with the angle γ from the sample surface, and we measure the intensity dI diffracted by an infinitesimally thin layer characterized by thickness dx located at a depth x below the top surface of the sample at the angle β from the sample surface. The resulting integrated intensity diffracted from such a small volume is given in the following equation:

$$dI = \frac{I_0}{\sin\gamma}\exp - \mu x\left(\frac{1}{\sin\gamma} + \frac{1}{\sin\beta}\right)dx \tag{6.31}$$

where μ is the linear absorption coefficient of sample materials.

The total intensity is obtained by integrating the expression (6.31) over an infinitely thick sample (t is the thickness), and we obtain the following simple form for the case of $\gamma = \beta = \theta$ since this configuration is allowed in measurements using a diffractometer.

$$I = \frac{I_0}{\sin\theta}\int_0^t \exp\left[-\left(\frac{2\mu x}{\sin\theta}\right)\right]dx = \frac{I_0}{2\mu}\left[1-\exp\left[-\left(\frac{2\mu t}{\sin\theta}\right)\right]\right] = (AF) * I_0 \quad (6.32)$$

the absorption factor is represented by

$$AF = \frac{1}{2\mu}\left[1-\exp\left[-\left(\frac{2\mu t}{\sin\theta}\right)\right]\right] \quad (6.33)$$

Since a sample is generally considered to be of infinite thickness, it becomes $t \to \infty$, and we conclude that the absorption factor is simply stated as $\frac{1}{2\mu}$. Therefore, the absorption factor for measurements with a commonly used diffractometer can be negligible as far as handling the relative amounts of intensities is concerned, because the absorption factor in cases where the flat plate-shaped sample shows infinite thickness is constant and independent of the diffraction angle. The judgment criterion that a sample can be of infinite thickness can be calculated using the following method:

In the diffractometer with the condition $\gamma = \beta = \theta$, the relationship between the intensity of the diffracted X-ray beam for the sample with its thickness t and that of the sample for the semi-infinite plate is given by the following equation with reference to (6.31).

$$G_t = \frac{\int_0^t \frac{I_0}{\sin\theta}\exp\left[-\left(\frac{2\mu x}{\sin\theta}\right)\right]dx}{\int_0^\infty \frac{I_0}{\sin\theta}\exp\left[-\left(\frac{2\mu x}{\sin\theta}\right)\right]dx} = 1-\exp\left[-\left(\frac{2\mu t}{\sin\theta}\right)\right] \quad (6.34)$$

$$t = -\frac{\sin\theta}{2\mu}\ln(1-G_t) \quad (6.35)$$

The value of G_t is correlated as a function of the sample thickness t.

6.4.6 Temperature Factor

Atoms in a crystal are not held at fixed points but move around their mean positions by thermal vibration and the amplitude of such vibration increases with increasing temperature. For example, an atom is estimated to shift by 5–6% from its mean position at room temperature. Due to this thermal vibration effect of atoms, the reduction of the intensity of diffracted X-rays from a crystal sample is detected. The effect of atomic vibrations on the intensity of diffracted X-rays is commonly considered as a decreasing factor. In practice, for the effect of temperature represented as $\exp[-2M_T]$ on the intensity of diffracted X-rays, the quantity M is calculated in the following equations with the atomic dispersion factor f.

$$f = f_0 e^{-M_T} \tag{6.36}$$

M_T is obtained from the following expression

$$M_T = 8\pi^2 \langle \mu^2 \rangle \left(\frac{\sin\theta}{\lambda} \right)^2 = B_T \left(\frac{\sin\theta}{\lambda} \right)^2 \tag{6.37}$$

and $\langle \mu^2 \rangle$ is the mean square displacement of the atom in a direction normal to the diffraction planes.

It should be noted that the exact calculation of M_T as a function of temperature is complex. For this reason, the coefficient B_T in (6.38) is estimated from experimental intensity data measured at different temperatures and the results are applied in different cases.

6.5 General formula of the intensity of diffracted x-rays for powder crystalline samples

A general formula for calculating the intensity of diffracted X-rays for crystalline powder samples can be obtained by considering the factors mentioned above together. For example, the intensity *I* measured by a diffractometer with characteristic radiation monochromated by a filter is given in the following equation:

$$I = |F|^2 \, p \left(\frac{1+\cos^2 2\theta}{2\sin^2\theta\cos\theta} \right) \frac{1}{2\mu} \left(1 - \exp\left[-\left(\frac{2\mu t}{\sin\theta} \right) \right] \right) e^{-2M_T} \tag{6.38}$$

where *F* represents the structure factor, p is the multiplicity factor, and the parentheses in the third term correspond to the Lorentz polarization (*LP*) factor.

Regarding the *LP*, note that another expression without the numerical value 2, which appears in the denominator, is also frequently used. This difference is considered negligible in the calculation of the intensity, because it is a constant independent of the angle. If a sample of infinite thickness is considered, the absorption factor is independent of the diffraction angle, being a constant with the value $\frac{1}{2\mu}$. Although the temperature factor e^{-2M_T} affects the atomic dispersion factor *f*, it is usually omitted in the calculation of the desired intensities, as far as the handling of the relative amounts of intensities is concerned.

6.6 Crystal structure determination

6.6.1 Cubic systems

The analysis of diffraction data is, so to speak, an "index pattern analysis". The crystal structure of metallic elements, which account for about 70% of the periodic table, shows the "cubic systems" characterized by a relatively simple atomic arrangement, such as FCC, BCC, HCP, and diamond. Therefore, the analysis of diffraction data of metallic elements is not such a difficult task. This is also supported by the following reasons. The first step of structural analysis for the measured X-ray diffraction data is to calculate the position of the scattering angle 2θ, corresponding to the location of the diffraction peaks.

Since current X-ray diffraction experiments are performed under computer control, the results usually automatically provide the values of 2θ, d, and ($I = I_1$), which are the plane spacing calculated from the Bragg condition using the given wavelength (λ) of the X-rays and the relative intensity ratio of the detected peaks with reference to the first peak intensity I_1, respectively.

By combining the Bragg condition (see (6.12)), with the plane spacing for a cubic system (see Table 2.3 (Chapter 2)), the diffraction peaks with the values of $\sin^2\theta$ satisfy the following equation:

$$\frac{\sin^2\theta}{(h^2+k^2+l^2)} = \frac{\sin^2\theta}{S} = \frac{\lambda^2}{4a^2} \tag{6.39}$$

where $S = h^2 + k^2 + l^2$.

From expression (6.40), it is seen that the sum of the square of the S-plane indices, corresponding to the measured diffraction peaks, is always an integer and $\frac{\lambda^2}{4a^2}$ is taken as a constant for any X-ray diffraction pattern. There are only four possibilities in cubic systems, including the diamond lattice, and the Miller indices are given as shown in Table 6.5. Therefore, the easiest way is by trial and error, as follows.

A crystal lattice is assumed to estimate the sum of the square of the plane indices (S). Then, from expression (6.40), the value of $\frac{\sin^2\theta}{S}$ is calculated. When the values are found to be constant, it can be said that the assumed crystal lattice is well accepted. Also, the lattice parameter (a) can be estimated from $\frac{\lambda^2}{4a^2}$, applying the wavelength (λ) of the X-rays used. Of course, when a constant value is not obtained, another crystal lattice is assumed, and the same calculation is carried out.

Certain integers such as 7, 15, and 23 will never appear as the value of $S = h^2 + k^2 + l^2$. This implies that when a number such as 7 or 15 appears, there is an error in the analysis. It is customary to use expression (6.40) in another form to analyze diffraction data from cubic systems, as follows:

$$\frac{4\sin^2\theta}{\lambda^2} = \frac{1}{d^2} = \frac{(h^2+k^2+l^2)}{a^2} \tag{6.40}$$

6.6.2 Tetragonal and hexagonal Systems

In non-cubic systems, the analysis of indexing patterns becomes more difficult because the number of unknown parameters increases. With respect to tetragonal or hexagonal systems whose plane spacing can be represented by axial ratios (c/a), a special graphical method called the "Hull–Davey chart" has been proposed and has been used in past decades. For example, the relationship between the plane spacing of tetragonal systems and the Bragg condition is given in the following form with two unknown parameters, a and c.

Table 6.5. Characteristic sequence of diffraction peaks of four common cubic lattice types together with a hexagonal close-packed lattice.

		Cubic					Hexagonal	
$2\theta = 0° \rightarrow$		Simple		BCC	FCC	Diamond	HPC	
$s \rightarrow$			*hkl*					
	1	▬▬▬	100					$hk \cdot l$
	2	▬▬▬	110	▬▬▬				
	3	▬▬▬			▬▬▬	▬▬▬	▬▬▬	10.0
			111				▬▬▬	00.2
							▬▬▬	10.1
	4	▬▬▬	200	▬▬▬	▬▬▬			
	5	▬▬▬	210					
	6	▬▬▬	211	▬▬▬			▬▬▬	10.2
	8	▬▬▬	220	▬▬▬	▬▬▬	▬▬▬	▬▬▬	11.0
	9	▬▬▬	300, 221				▬▬▬	10.3
	10	▬▬▬	310	▬▬▬				
	11	▬▬▬	311		▬▬▬	▬▬▬	▬▬▬	20.0
							▬▬▬	11.2
							▬▬▬	20.1
	12	▬▬▬	222	▬▬▬	▬▬▬		▬▬▬	00.4
	13	▬▬▬	320				▬▬▬	20.2
	14	▬▬▬	321	▬▬▬			▬▬▬	10.4
	16	▬▬▬	400	▬▬▬	▬▬▬	▬▬▬		
	17	▬▬▬	410, 322				▬▬▬	20.3
	18	▬▬▬	411, 330	▬▬▬			▬▬▬	21.0
	19	▬▬▬	331		▬▬▬	▬▬▬	▬▬▬	21.1
	20	▬▬▬	420	▬▬▬			▬▬▬	11.4
$2\theta = 180° \rightarrow$								

$$\frac{1}{d^2} = \frac{4\sin^2\theta}{\lambda^2} = \frac{h^2 + k^2}{a^2} + \frac{l^2}{c^2} \tag{6.41}$$

Taking logarithms on both sides of the expression (6.41) after a short algebra we obtain the following expressions:

$$2\log d = 2\log a - \log\left[(h^2 + k^2) + \frac{l^2}{(c/a)^2}\right] \tag{6.42}$$

$$\log\sin^2\theta = \log\frac{\lambda^2}{4a^2} + \log\left[(h^2 + k^2) + \frac{l^2}{(c/a)^2}\right] = -2\log d + \log\frac{\lambda^2}{4} \tag{6.43}$$

From expression (6.42), for two different distances and subtracted member by member, we obtain:

$$2\log d_1 - 2\log d_2$$

$$= 2\log a - \log\left[(h_1^2 + k_1^2) + \frac{l_1^2}{\left(\frac{c}{a}\right)^2}\right] \tag{6.44}$$

$$+ \log\left[(h_2^2 + k_2^2) + \frac{l_2^2}{\left(\frac{c}{a}\right)^2}\right]$$

This equation suggests that the difference between the values of 2 log d for any two planes depends only on the axial ratio c/a and the indices hkl. Using this fact, Hull and Davy (1921) developed a method built from the variation of the quantity

$$\left[(h_1^2 + k_1^2) + \frac{l_1^2}{\left(\frac{c}{a}\right)^2}\right]$$

with c/a. This method was frequently used when we did not have electronic calculators and only manuals with mathematical tables were available. Nowadays, more analytical methods are being developed that make numerical calculations easier. Let us show with an example the use of an analytical method in the case of a hexagonal system.

The relationship between the plane spacing of the hexagonal system (see Table 2.3 (Chapter 2)) and the Bragg condition (6.12), is given in the following form:

$$\frac{4\sin^2\theta}{\lambda^2} = \frac{1}{d^2} = \left(\frac{4}{3}\right)\frac{h^2 + hk + k^2}{a^2} + \frac{l^2}{c^2} \tag{6.45}$$

This expression can be written as:

$$\sin^2\theta = \frac{\lambda^2}{4a^2}\left[\left(\frac{4}{3}\right)(h^2 + hk + k^2)\right] + \frac{\lambda^2 l^2}{4c^2} \tag{6.46}$$

The lattice parameter a and the axial ratio c/a are unique values for a substance of interest. Therefore, when $X = \frac{\lambda^2}{3a^2}$ and $= \frac{\lambda^2 l^2}{4c^2}$, expression (6.46) can be rewritten as follows:

$$\sin^2\theta = X(h^2 + hk + k^2) + Yl^2 \tag{6.47}$$

The hkl plane indices are always integers; therefore, the value of the sum $(h^2 + hk + k^2)$ must be 0, 1, 3, 4, 7, etc. and the value of l^2 is 0, 1, 4, 9, etc. Considering these features, we divide the $\sin^2\theta$ values for the measured diffraction peaks by the integers, 3, 4, 7, etc. and tabulate the results. Then, we examine the numerical results and find the common quotient, which is equal to another or equal to one of the measured $\sin^2\theta$ values. We tentatively obtain the possible value of X by finding the diffraction peaks corresponding to ($hk0$) with $l = 0$. The value of X obtained in this process satisfies the following relationship for the general indices h, k, and l.

$$Yl^2 = \sin^2\theta - X(h^2 + hk + k^2) \tag{6.48}$$

To calculate the value of Y, we re-tabulate the results by subtracting the value of X, $3X$, and $4X$ from each $\sin^2\theta$ and look for the ones in the proportion of 1, 4, 9, etc. because the peaks of ($00l$) will have $l^2 = 1, 4, 9, \ldots$ only.

Note that the remaining peaks which are neither ($hk0$) nor ($00l$) should be assigned as peaks belonging to the hexagonal system. According to this procedure, a time indexing can be established for several peaks involving the X and Y values. Once these results are obtained, all measured diffraction peaks can be recalculated, and a comparison is made between the calculated $\sin^2\theta$ values and the experimental data.

When no inconsistent points can be found, the analysis is said to be "complete". At the same time, the network parameters are calculated from the relations of $X = \dfrac{\lambda^2}{3a^2}$ and $Y = \dfrac{\lambda^2 l^2}{4c^2}$.

6.7 Data processing (Profiling reduction)

In this section we will give the fundamental concepts involved in data processing. In practice, this processing is done with the help of computational tools that already have the algorithms and statistical analysis programmed. In the appendix we will show explicitly how this is done with the PANalytical X'Pert HighScore software.

A diffraction pattern is made up of contributions which include diffraction of desired wavelengths, and dispersion and fluorescence from the sample and the sample holder, respectively. The purpose of data processing is to extract the necessary information from the experimental pattern to process it in a quantitative analysis.

For this purpose, the following steps are proposed:

1. Smoothing
2. Background noise subtraction (background)
3. $K_{\alpha 2}$ peak removal

6.7.1 Smoothing

Since the production of X-rays is a random process, any intensity count data will have an associated error, the magnitude of which depends on the number of counts taken (i.e., the product of the counting rate and time). This statistical counting process introduces random fluctuations into the raw data as well. Such fluctuations can be partially removed by the data smoothing process. Figure 6.3 illustrates the basic idea of smoothing: an odd number of contiguous data points (collected points that form

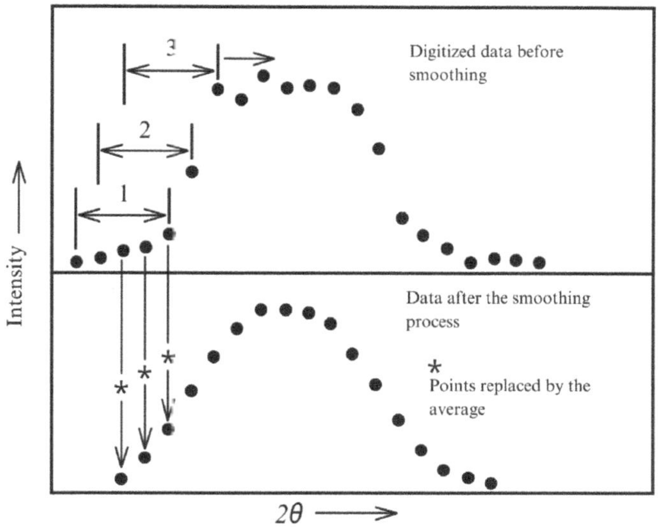

Figure 6.3. Statistical process of data smoothing.

a peak in the diffraction profile) are averaged and the central data point is replaced by the average, the starting data point is then incremented by one, and the process is repeated until the entire pattern is smoothed.

However, the smoothing process and the degree of smoothing is determined by the statistical count that the detector associates with each peak, the sharpness or definition of each peak, the uniqueness of each peak, and the number of data selected to smooth each peak. In quantitative analysis by the RIR and Rietveld method, it is important to perform a good smoothing of the diffraction profile, since it is necessary to extract the areas of the highest intensity peaks. Therefore, all profile fitting is performed by using quadratic or cubic polynomials that act as digital filters that perform the interpolation between the selected points.

6.7.2 Background noise subtraction (background)

The radiation incident on the sample or the support containing it can be diffracted, producing secondary fluorescence radiation in addition to the white radiation coming from the radiation source. Therefore, the X-ray reaching the detector is additionally modified regardless of the type of monochromator used. Thus, the ease of interpretation of the resulting diffraction pattern is invariably dependent on the resolution of the profile and the overall signal-to-noise ratio of that pattern. Thus, a diffraction profile with broad peaks is easier to process than a pattern with well-defined sharp peaks. In general, among the most common factors that can produce significant background noise are:

- Scattering generated by the sample holder, usually seen at low values of 2θ, where a region that is too wide is produced.
- The fluorescence of the sample, which can be controlled to a certain extent with the use of a monochromator on the diffracted beam.

- The presence of significant amounts of amorphous material in the sample.
- Air scattering, which has a greater effect at low values of 2θ.

In some cases, depending on the quantitative analysis method to be used, the background may or may not be eliminated. In the case of the RIR and Rietveld methods, it is always possible to do so to more accurately estimate the area under the diffraction peak of interest.

For the subtraction of the background noise, several points are specified that are considered to be background noise at low and high angles of each of the peaks of the diffraction profile. An average of the intensities of these points is obtained and used as the intensity of the background noise, creating an interpolation range at each end of the peak and this gives us the information of the straight line from which the intensities below this are considered background noise and therefore are eliminated, as can be seen in Figure 6.4.

This process is known as the Sonneveld and Visser method, which uses a least squares polynomial. For a large data set, the number of sampling points is an integer in the range 150–250, and the repetition time can be an integer between 20 and 50. If the background noise is also flat, the sampling points can be increased or the repetition time reduced, or vice-versa.

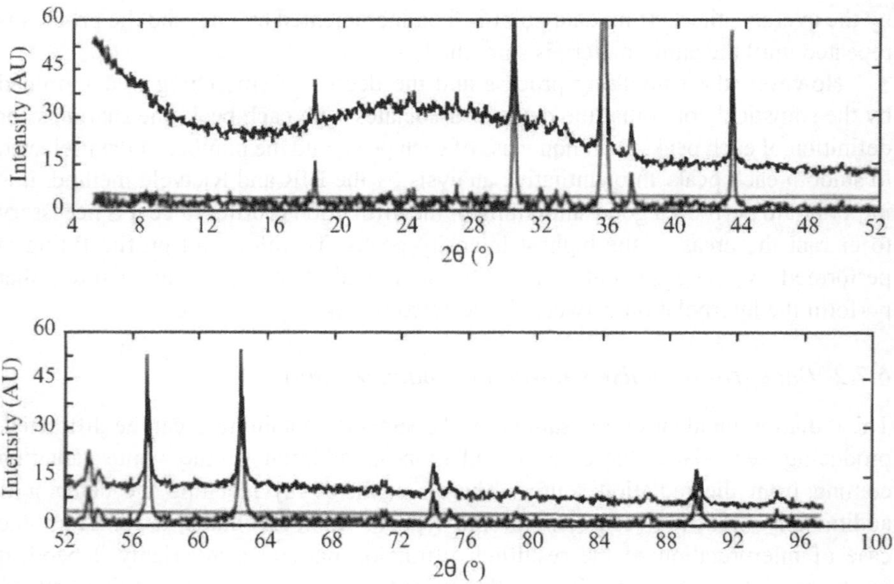

Figure 6.4. Removing background noise in a DRX profile.

6.7.3 $K_{\alpha 2}$ peak removal

Many experimental X-ray diffraction tests are performed using the *Cu* $K_{\alpha 1}$ and $K_{\alpha 2}$ doublets, certain peak distortions can occur due to partial separation of the $K_{\alpha 1}$ and $K_{\alpha 2}$ doublets over some values in certain ranges of 2θ, for example, at low values of 2θ the $K_{\alpha 1}$ and $K_{\alpha 2}$ doublet is not resolved and at high values of 2θ it is completely

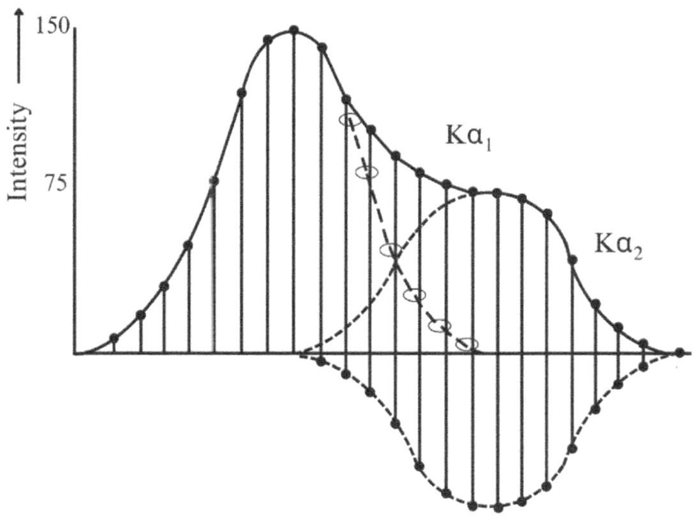

Figure 6.5. Rachinger method for removing the $K_{\alpha 2}$.

separated and in the intermediate ranges the doublet is partially resolved. In a quantitative analysis where the integrated intensities of a few reflections are used, for example, in RIR and Rietveld methods, it is important to have well resolved diffraction peaks; therefore, it is necessary to eliminate the $K_{\alpha 2}$ peak.

Methods to remove the $K_{\alpha 2}$ peak are generally based either on the Rachinger technique (Rachinger 1948), which is applied to the 2θ values or on a Fourier transform convolution of the powder pattern. The Rachinger method uses the wavelength values of the $K_{\alpha 1}$ and $K_{\alpha 2}$ lines and the ratios of their intensities. And since the ratio of the $K_{\alpha 1}$:$K_{\alpha 2}$ intensities is 2:1, half of the intensity measured at low 2θ angles can be subtracted from the intensity value at high angles where the $K_{\alpha 2}$ peak diffracts (see Figure 6.5).

In this way the negative peak is determined and subtracted from the originally measured data, obtaining a pure $K_{\alpha 1}$ peak profile.

6.8 Quantitative analysis methods by XRD

This section introduces quantitative analysis using X-ray diffraction methods. Quantitative analysis by XRD is usually concerned with the determination of amounts of different phases in multiphase samples. It can also be considered in terms of the determination of characteristics of single phases, such as crystal structure, crystal size, and shape. In quantitative analysis, the goal is to determine phase proportions with reliable numerical accuracy from experimental data. Although standard patterns and structural data are often used as a starting point, the bulk of successful quantitative analysis usually involves modeling of the diffraction pattern.

All quantitative analyses require a precise and accurate determination of the sample diffraction pattern in terms of peak positions and intensities, since their success depends on careful sample preparation, good data quality, complete information on the working material, and possible sources of error in the experiments. This is because diffraction data are generally dependent on systematic errors of the diffractometer and its data collection system.

The most effective quantitative methods are those that particularly include a modeling of the total pattern, which is generated by applying some specialized analytical software.

6.8.1 Intensity equation

The X-ray diffraction pattern of a sample contains information about the positions of the peaks and their respective intensities. The positions are indicative of the crystal structure and symmetry of the contributing phase. The intensities reflect the total dispersion of each of the planes in the phase and are directly dependent on the distribution of atoms in the crystal structure. Ultimately, the intensities are closely related to the structure and composition of the phase. The intensity equation for an *hkl* reflection of an α phase is given by:

$$I_{(hkl)\alpha} = \frac{K_{(hkl)\alpha} X_\alpha}{\rho_\alpha \left(\frac{\mu}{\rho}\right)_m} \tag{6.49}$$

where

X_α: weight fraction of the phase α

ρ_α: α pase density

$\left(\frac{\mu}{\rho}\right)_m$: mass absorption coefficient of the sample

$K_{(hkl)\alpha}$: constant for each *hkl* reflection of a crystal structure *a* in a given mixture and experimental conditions.

This constant is calculated with the following expression:

$$K_{(hkl)\alpha} = \frac{I_0 \lambda^3 e^4 M_{(hkl)}}{64\pi r m_e^2 c^4 V_\alpha^2} \left|F_{(hkl)\alpha}\right|^2 \left(\frac{1+\cos^2 2\theta \cos^2 2\theta_m}{\sin^2\theta \cos\theta}\right) \tag{6.50}$$

where

I_0: Incident ray intensity

r: Diffractometer radius (distance from sample to detector)

λ: X-ray wavelength

c: Speed of light in vacuum

$2\theta_m$: Monochromator angle

e: Electric charge of the electron

m_e: electron mass

$M_{(hkl)}$: Multicity factor for *hkl* reflection phase *a*

V_α: volume of the unit cell of phase a

$\left(\dfrac{1+\cos^2 2\theta}{\sin^2\theta\cos\theta}\right)$: Lorentz-Polarization correction factor for the diffractometer

$F_{(hkl)\alpha}$: Fase structure a for hkl reflection

6.8.2 Quantitative methods based on the relationship of intensities

There are several methods for quantitative analysis of multiphase mixtures that use the intensity ratio between one phase of interest and another, possibly a standard (Seeck and Murphy 2015, Bunaciu et al. 2015). Some of these methods are of a special nature, for example, to determine mixtures of polymorphic phases with the same mass absorption coefficients and mixtures whose chemical composition is known.

Most of these general-use methods involve the addition of a known quantity as an internal standard to relate the diffraction peaks of the standard quantity and that of the phase being determined. Other methods do not require the use of diffraction line intensities, such as Vegard's law, used to determine the composition of substitutional solid solutions.

A substitutional solid solution is one in which one element is replaced by another in the entire range of possible compositions determined by the compositions of two end members. This law provides a direct method for theoretically determining the composition of a solid solution, by simply measuring the value of the unit cell parameter precisely and referring it to a calibration curve.

The following describes in general terms the quantitative multiphase analysis methods by XRD that involve the use of the intensity ratio.

6.8.2.1 Diffraction-absorption method

This method is based fundamentally on the relationship between the intensity of a reflection *hkl* of a phase a in the mixture under study and the intensity of a reflection *hkl* of phase α in its pure form. To carry out the quantitative analysis using this method, the corresponding mass absorption coefficients of the phases present must be known in advance or, if possible, calculated. Although for many materials this value may remain undetermined. The calculation expression for the quantitative analysis in this method is

$$\frac{I_{(hkl)\alpha}}{I^0_{(hkl)\alpha}} = \frac{\left(\mu/\rho\right)_\alpha}{\left(\mu/\rho\right)_m} X_\alpha \tag{6.51}$$

where

$I^0_{(hkl)\alpha}$: Intensity of the *hkl* reflection in the pure α phase.

X_α: Mass fraction of the pure α phase.

$\left(\mu/\rho\right)_\alpha$: Mass absorption coefficient of the α phase.

$\left(\mu/\rho\right)_m$: Mass absorption coefficient of the total sample.

The mass absorption coefficient of the total sample can be calculated from the following equation:

$$\left(\mu/\rho\right)_m = \sum_i \left(\mu/\rho\right)_i X_i \tag{6.52}$$

For this method, there are several application cases, which are dependent only on the validity of equation (6.51). The first case is when the quantitative analysis is carried out on samples whose chemical composition is known. In this case, the mass absorption coefficients of the pure phase to be analyzed and that of the mixture can be calculated from equation (6.52) and likewise the intensities $I_{(hkl)\alpha}$ and $I^0_{(hkl)\alpha}$ are measured from the unknown phase a in the mixture and from a pure sample of α, respectively, thus the weight fraction of phase a (X_a) can be calculated using equation (6.51). The second case is when the mass absorption coefficients of the mixture and the phase are equal, therefore, equation (6.51) is reduced to:

$$\frac{I_{(hkl)a}}{I^0_{(hkl)a}} = X_a \tag{6.53}$$

This case is particularly present in the mixture of polymorphs. Common examples are the analysis of cristobalite in a quartz and an amorphous SiO_2 matrix, or the analysis of the cubic, tetragonal, and monoclinic forms of zirconium oxide in pure ZrO_2 samples. There is also the determination of anatase in rutile, as a quality control in the processing of paints and glazed paper. In this way, since the two oxides have the same mass absorption coefficient, the same density and the diffraction lines are not excessively wide, the concentrations of rutile and anatase can be related and determined in a direct way using the equation

$$\frac{X_{rutile}}{X_{anatase}} = \frac{I^0_{(101)anatase}}{I^0_{(110)rutile}} \times \frac{I_{rutile}}{I^0_{anatase}} \tag{6.54}$$

A special case for binary mixtures is when the absorption coefficients for each pure phase are known, thus allowing the calculation of the amounts of the two phases without the requirement of $\left(\mu/\rho\right)_m$ using the following equation:

$$X_\alpha = \frac{\left(\dfrac{I_{(hkl)\alpha}}{I^0_{(hkl)\alpha}}\right)\left(\mu/\rho\right)_\beta}{\left(\mu/\rho\right)_\beta - \left(\dfrac{I_{(hkl)\alpha}}{I^0_{(hkl)\alpha}}\right)\left[\left(\mu/\rho\right)_\alpha - \left(\mu/\rho\right)_\beta\right]} \tag{6.55}$$

This equation is known as the Klug equation (Alexander and Klug 1948), since it was first formulated by H. P. Klug. As a test of the accuracy of the analysis, it is possible to perform the calculation for each independent phase and then compare the results. The general case of the diffraction-absorption method requires that the mass absorption coefficient of the sample be known. This quantity can be experimentally determined by several methods (Seeck and Murphy 2015; Bunaciu et al. 2015) and used in respective calculations.

Although it is theoretically possible to obtain it, the errors generally involved in this measurement are too large to be practically used in a quantitative analysis by XRD. Since the mass absorption is largely a function of atomic dispersion, the $\left(\mu/\rho\right)_m$ can be estimated from the atomic dispersion factors, provided the elemental composition of the sample is known.

Elemental mass attenuation coefficients are published in the International Conference on X-ray Crystallography, and using them, the value of the coefficient of the sample can be estimated with reasonable accuracy. In this way, the diffraction-absorption method can be generalized for direct use.

6.8.2.2 Standard additions method

Lennox (1957) was one of the first to develop this method, which was also known as the Spiking method. This method, like the internal standard method, is equally applicable to any α phase in a mixture. The only requirement is that the mixture must contain, among other phases, a β phase, which has a reflection that is not overlapped by any reflection from the α phase and which also serves as a reference. In a sample containing both a and β, the ratio of the intensities of a reflection from each phase can be obtained from the equation:

$$\frac{I_{(hkl)\alpha}}{I_{(hkl)'\beta}} = \left(\frac{K_{(hkl)\alpha}}{K_{(hkl)'\beta}}\right)\left(\frac{\rho_\beta}{\rho_\alpha}\right)\left(\frac{X_\alpha}{X_\beta}\right) \tag{6.56}$$

where

X_α: Initial weight fraction of the α phase

X_β: Initial weight fraction of the β phase.

The other terms are already defined in equation (6.49) and (6.50).

Using the method of standard additions, a quantity of the pure phase α is added to the mixture containing the unknown concentration of α. After Y_a grams of the α phase have been added, per gram of the unknown phase, equation (6.56) becomes:

$$\frac{I_{(hkl)\alpha}}{I_{(hkl)'\beta}} = \left(\frac{K_{(hkl)\alpha}}{K_{(hkl)'\beta}}\right)\left(\frac{\rho_\beta}{\rho_\alpha}\right)\left(\frac{X_\alpha + Y_a}{X_\beta}\right) \tag{6.57}$$

In general, the intensity relationship is given by

$$\frac{I_{(hkl)\alpha}}{I_{(hkl)'\beta}} = K(X_\alpha + Y_a) \tag{6.58}$$

where K is the slope of the graph of $\frac{I_{(hkl)\alpha}}{I_{(hkl)'\beta}}$ vs Y_a in units of grams of α per gram of sample. Thus, to determine the desired concentration of the α phase in the original sample, several additions of Y_a must be performed to obtain a graph like that in Figure 6.6, in which the negative intercept with the X axis will be sought X_a.

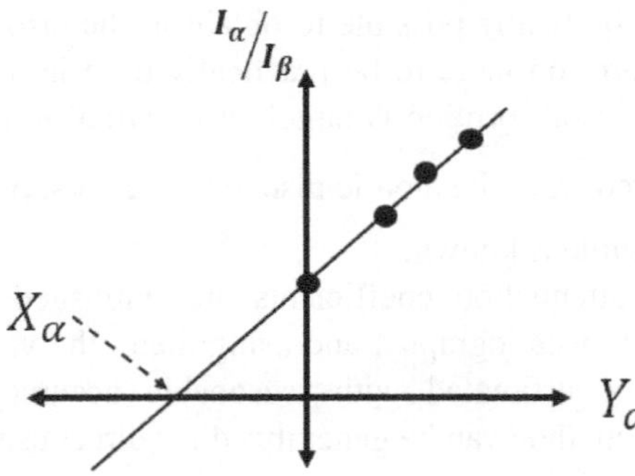

Figure 6.6. Analysis of the Spiking Method.

Due to tedious sample preparation and errors encountered at low concentrations of the two phases, this method is rarely applied in X-ray diffraction analysis, instead it is widely used for X-ray fluorescence elemental analysis.

6.8.2.3 Internal standard method

This method is the most frequently used method among the methods used for quantitative analysis by XRD in polycrystalline samples. It is based on the elimination of the mass absorption factor $\left(\mu/\rho\right)_m$ of the sample, by the relation of two expressions of the same type of equation (6.49), from which another expression of the form is obtained:

$$\frac{I_{(hkl)\alpha}}{I_{(hkl)'\beta}} = k\frac{X_\alpha}{X_\beta} \tag{6.59}$$

This equation represents the basis of quantitative analysis by the internal standard method, which is a straight line as a function of the weight fraction of the α phase, and whose slope *k* of the calibration curve is obtained from a graph of $\frac{I_{(hkl)\alpha}}{I_{(hkl)'\beta}}$ vs $\frac{X_\alpha}{X_\beta}$.

Thus, to quantify each of the components of the mixture, the values of *k* must first be established for each of the phases from the standards of known concentration, X_β, of a standard mineral to the mixture of phases (which may include amorphous material).

This allows to pre-establish a calibration curve with the defined value of the slope *k* and thus, to calculate the weight fraction of any phase α in the original mixture. It should be noted that *k* is a function of *(hkl)*, *(hkl)'*, *α* and *β* for a constant weight fraction of the internal standard.

The method requires careful selection of the internal standard for a given type of analysis. It is evident that the addition of a new phase to the analysis mixture will increase the complexity of the diffraction pattern, causing problems in the measurement of peak intensities. In general, any pure phase can act as an internal standard; but in practice it must have one or more completely resolved reflections that do not overlap with any reflections from the phases of interest, and it must also be free from the effects of micro absorption (this could be a small particle size or a similar micro absorption value to the sample) and extinction.

Typically, face-centered cubic (*F*) crystal system materials with small cell values are chosen because of their simple diffraction pattern. Sometimes the analysis of a single phase in a rock or mineral may have a large particle size and a strong orientational bias. In this case, a common mistake is to add an internal standard selected from a chemical reagent pantry, where the compound has been made by chemical precipitation and therefore has a very fine particle size, which during the mixing process will layer over the large sample particles, creating local heterogeneity and thus much worse problems than sample absorption. This is why it is essential to reduce the sample particle size to the 1 μm range.

This method lends itself to the generalization of the RIR method in a system of linear equations where the use of overlapping reflections and chemical constraint analysis are considered.

6.8.2.4 RIR (Reference Intensity Ratio) method and the $I/I_{corundum}$

From equation (6.59) we can obtain the following graph:

$$X_\beta \frac{I_{(hkl)\alpha}}{I_{(hkl)'\beta}} \; vs \; X_\alpha \tag{6.60}$$

which is a straight line with slope *k*, hence, *k* is an inherent measure of the diffracted intensities of the two phases. Visser and De Wolff 1964 were the first to propose that *k* values could be published as constants for each material type, defining the reference phase β as corundum in a 50:50 mixture by weight with the α phase and using the *hkl* of the 100% intensity reflections. This *k* value, known as (I/I_c) or RIR has been widely accepted and is now published in the ICCD phases of the Power Diffraction File.

The procedure for measuring the two reflections in the calculation of I/I_c is rapid but has several disadvantages: the preferred orientation often affects the observed intensities unless careful sample preparation methods are employed; other problems are extinction, mixing inhomogeneity, and crystal variability.

All these effects suggest that published I/I_c values are subject to substantial error. For best accuracy, multiple reflections of the α phase in the sample and the reference standard (corundum) should be used, which may somehow reduce or eliminate some of these effects. Multiple reflections allow the calculation of an average RIR value with its respective standard deviation. Such an approach often reveals whether the preferred orientation is present or not and provides true measures of the reproducibility in the calculation of the I/I_c value. The average $RIR_{\alpha,c}$ and the estimated standard errors are calculated from the following mathematical models:

$$RIR_{\alpha,c} = \left(\sum_{i=1}^{n} \left(\frac{I}{I_c} \right)_{ij} \right) / n \tag{6.61}$$

$$\sigma_{\alpha,c} = \left(\left(\sum_{i=1}^{n} \left(\frac{I}{I_c} \right)_{ij} - RIR_{\alpha,c} \right) / (n-1) \right)^{1/2} \tag{6.62}$$

where n is the total number of I/I_c measurements from all selected reflection pairs.

6.8.2.5 Generalized RIR method

In practice, RIRs can be defined for any reference phase using any of the diffraction reflections, that is, not only corundum is used as a reference standard. So, a more general definition of RIR for a phase α to a reference phase β is:

$$RIR_{\alpha,\beta} = \left(\frac{I_{(hkl)\alpha}}{I_{(hkl)'\beta}} \right) \left(\frac{I^{rel}_{(hkl)'\beta}}{I^{rel}_{(hkl)'\alpha}} \right) \left(\frac{X_\beta}{X_\alpha} \right) \tag{6.63}$$

This equation allows the calculation of an RIR from any combination of reflections in a two-phase mixture of any composition, with *hkl*, (*hkl*)′, and X_β as variables. The I^{rel} term relates the relative intensities of the peaks used; if the peaks of the two phases are 100%, the value of this term is 1. In this way, the RIR values for any phase can be determined experimentally using some material as a standard.

Commonly used standards are Al_2O_3 (corundum), SiO_2 (quartz) and ZnO (zincite) as these compounds exhibit very good diffraction peaks. Several RIR values can be calculated for different peaks in the same phases as a test of the accuracy of the determinations. RIR carefully determined in the same laboratory under the same conditions can be used to produce good repeatable analyses. Furthermore, having good RIR values allows one to select the best standard to use with minimal peak overlap for a given sample.

6.8.2.6 RIR standardized method

Chung (1974a; 1974b) found that if all phases in a mixture along with the RIR values are known for all of them, then the sum of all the weight fractions of the phases would be equal to 1. This allowed a system of *n* equations to be written to solve with n weight fractions, using the following mathematical model:

$$X_\alpha = \frac{I_{(hkl)\alpha}}{RIR_\alpha I^{rel}_{(hkl)\alpha}} \left[\frac{1}{\sum_{j=1}^{\#\,phases} \dfrac{I_{(hkl)'j}}{RIR_j I^{rel}_{(hkl)j}}} \right] \tag{6.64}$$

Chung called this method the adiabatic principle, but it is now also universally known as the standardized RIR method, which allows quantitative calculations to be made without the presence of an internal standard. But it is also limited by the presence of some unidentified or amorphous phase. In fact, it should also be noted that in practically all samples, for example, in rocks, phases will be found that are undetectable and thus the method will not work satisfactorily either.

When RIR values are known from another source, for example, published I/I_c values, these may be tentatively used in conjunction with I^{rel} values. These values are generally accurate enough to be used directly in quantitative analysis, but it is up to the analyst to accurately determine the relative intensities and RIR values for his analysis by careful calibration of his measurements. In fact, the word "standardless" is a misnomer. Standards are always requirements, however, the RIR method allows data from the literature to be used for semi-quantitative analysis.

6.8.3 Rietveld method: generalities

The Rietveld method was originally developed for the refinement of magnetic and nuclear structures (Rietveld 1969) and was later applied to quantitative phase analysis by XRD (Hill and Howard 1987). This type of analysis is mainly used when the number of phases to be quantified in a sample increases considerably, because the diffraction profile produced by the mixture contains many superimposed peaks and therefore the application of a quantitative method involving a simple measurement of the intensities of certain reflections is not possible in these cases.

This is why Rietveld analysis has gained so much popularity in recent years as a fast and accurate quantitative technique. This method is based on minimizing the difference between a theoretical diffraction pattern calculated with structural models for each of the phases present and the corresponding experimental pattern, which makes use of the complete point-by-point diffraction pattern, which is why the superposition of reflections can be resolved by means of this methodology.

The minimized quantity in the Rietveld refinement is the residual parameter obtained by the least squares fitting, expressed by:

$$S_y = \sum_i w_i (y_{0i} - y_{ci})^2 \tag{6.65}$$

where y_{0i} and y_{ci} are the observed and calculated intensities, respectively, at the i-th data, and w_i is the weight assigned to each observed intensity. On the other hand, the preferred orientation (*PO*), very frequent in the direction of polycrystalline samples, can be refined, which is an advantage in phase quantification, since the *PO* modifies the relative intensities of certain reflections and thus produces erroneous results.

Rietveld analysis uses a model that includes the crystal system and unit cell constants for each phase, a contribution to the background, parameters that define the shape and width of the diffraction lines, and a scale factor to adjust the relative intensities of the contribution each makes to the diffraction pattern. It is precisely from these scale factors that the composition of the sample can be quantified. The main fitting indicators for a Rietveld refinement are the R_{wp} and R_{Bragg} values (Young 1993). These values are calculated with a refinement program using the equations:

$$R_{wp} = \left[\frac{\sum_i w_i (y_{0i} - y_{ci})^2}{\sum_i w_i y_{0i}^2} \right]^{1/2} \tag{6.66}$$

$$R_{Bragg} = \frac{\sum_i |I_{0i} - I_{ci}|}{\sum_i |I_{0i}|} \tag{6.67}$$

where I_{0i} and I_{ci} correspond to the observed and calculated integrated intensities I_{hkl} respectively, y_{0i} and y_{ci} to the observed and calculated point-by-point intensities obtained for phase i. In this way, the value of R_{wp}, which is unique for a refined diffraction pattern, is an indication of how good the refinement of the diffraction pattern is. The value of R_{Bragg} is determined for each crystallographic phase and tells us whether the structural model has been chosen correctly. On the other hand, if an amorphous phase were to be present in the sample, the weight fraction values could be overestimated. The problem is usually solved by using an internal standard, which is added in a known quantity to the sample and considered as another component of the sample. In this way, the percentage of the amorphous phase X_a in the original mixture can be calculated directly from the weight of the added internal standard X_s and the value obtained from the standard in the refinement $X_{s,c}$.

Although the Rietveld method is not a method of structure determination but of refinement, it is necessary to have an approximately correct structural model to make the fitting and complete the structure.

6.8.4 The mathematical model of the Rietveld method

The Rietveld method proposes to represent the observed intensities of the reflections in a diffraction profile, by means of a theoretical model that aims to include all the effects that influence the obtaining of said profile. In particular, the physical model aims to include the effects that determine the position, intensity, and shape of the diffraction maxima of the profile, leaving the modelling of the background noise to an empirical fitting or directly performing the subtraction of the background.

The position and intensity of the maxima are determined by the crystal structure of the compound, except in the presence of preferential orientation of the sample, where the intensities cannot be represented only by the structure, and the shape of the peaks by instrumental and intrinsic factors of the sample that should be included in the model in such a way that they can represent the observed intensities. Considering that there is a set of data $(2\theta_i, y_{0i})$ where $2\theta_i$ represents the position in units of 2θ. The mathematical model is implemented to determine the calculated intensity (y_{ci}) at the same points $2\theta_i$ determined according to the following mathematical model:

$$y_{ci} = S\sum_i L_{(hkl)} \left|F_{(hkl)}\right|^2 \phi(2\theta_i - 2\theta_{(hkl)}) P_{(hkl)} A + y_{bi} \tag{6.68}$$

where

y_{bi}: Background intensity at point $2\theta i$.

S: Scale factor.

$L_{(hkl)}$: Factor containing the Lorentz, polarization, and multiplicity factors.

$P_{(hkl)}$: Preferential orientation function of the sample.

$\phi(2\theta_i - 2\theta_{(hkl)})$: Function representing the shape of the diffraction peak.

A: Absorption factor and associated phenomena.

$F_{(hkl)}$: Structure factor corresponding to the peak with the atomic dispersion vector f which in turn has the temperature factor, see equations (6.16), (6.36), and (6.37), that is.:

$$F_{(hkl)} = \sum_{j=1}^{N} N_j f_j e^{2\pi i(hu_j + kv_j + lw_j)} \exp\left[-M_T \frac{\sin^2\theta}{\lambda^2}\right] \tag{6.69}$$

where

N_j: is a multiplier of the occupancy factor (considers atoms in their special positions, atomic disorder, or both)

f_j: Atomic dispersion factor

M_T: Atomic temperature factor defined by equation (6.37)

Each of the terms in equation (6.69) may contain within itself one or more adjustable parameters x_j ($j = 1,\ldots,m$) such as the atomic positions r_j in $F_{(hkl)}$ or the peak widths $H_{(hkl)}$ in $\phi(2\theta_i - 2\theta_{(hkl)})$. We are then faced with the problem of finding the set of m parameters ($x_1,\ldots,x_m$) that allow us to adjust y_{ci} to y_{0i} for all i.

This problem generally lacks an exact solution, because the experimental values of y_{0i} are affected by random errors (the presence of uncorrected or modeled systematic errors invalidates the analysis). One of the most widely used methods to solve this problem is linear least squares. Using this method, the function is minimized:

$$S_y = \sum_i w_i (y_{0i} - y_{ci})^2 \tag{6.70}$$

where

$w_i = \dfrac{1}{y_{0i}}$: It is a weight factor assigned to the observed intensity y_{0i}

y_{0i}: Observed intensity with background noise correction at step i.

y_{ci}: Calculated intensity, formed by the sum of the contributions of all the Bragg reflections that overlap in step i.

Proceeding according to the method, it is found that the solution to the problem requires the inversion of a matrix of size $m \times m$, called a normal matrix, which has the form:

$$M_{jk} = -2\sum_i w_i \left[(y_{0i} - y_{ci}) \frac{\partial^2 y_{ci}}{\partial x_j x_k} - \left(\frac{\partial y_{ci}}{\partial x_j}\right)\left(\frac{\partial y_{ci}}{\partial x_k}\right) \right] \tag{6.71}$$

The complexity of the normal matrix is therefore greater as the model includes a greater number of variables (more atomic parameters), whether of the same or different type (more complex peak functions with more adjustable parameters). To simplify the calculation of the normal matrix somewhat (and make its programming possible), it is assumed that the first term within the bracket is null. This term is only null if the relationship between y_{ci} and each of the x_j is linear.

This does not happen even in the simplest structural models, since the relationship between the structural factor $F_{(hkl)}$ and the vector r_j that is inevitably present in the structural part of the model is of the exponential type. This simplification allows the calculation and the inversion process of the normal matrix to be carried out, but this makes a system that is not linear. By linearizing the system, the calculation of the best values of x_j cannot be carried out directly from the observed values, and an iteration process must be used where it is necessary to include starting values of x_j for all j. By inverting the normal matrix, it is possible to determine the changes in x_j according to the model:

$$\Delta x_j = \sum_k M_{jk}^{-1} \frac{\partial S_y}{\partial x_k} \tag{6.72}$$

Repeating the process of calculating M_{jk} with the new values $x_j = x_{j0} + \Delta x_j$, the inversion of M and the calculation of Δx_j until reaching a set of values x_j that do not change from one cycle to the next, that is, they present a minimum of the function S_y. Note that M_{jk}^{-1} is not the inverse of M_{jk} but of the element jk of the inverse matrix of M.

Unfortunately, an inherent condition of the linear least squares method is that if the starting values $(x_1,\ldots,x_m)$ are not close to the final values, the refinement process may converge to a relative minimum or may not converge at all. This condition prevents the resolution of structures by this mechanism, since the need to start from a set of structural parameters r_j close to the final ones is to introduce an adequate structural model into the model, and therefore, requires having at least partially solved the structure.

Generally, the practical procedure used to fitting the model to the data and avoid false minima requires working with increasingly complex partial models where a series of reasonable values $(x_1,\ldots,x_m)_0$ of the initial parameters are left fixed, allowing the subset $(x_{m-n+1},\ldots,x_m)$ to vary until reaching the minimum.

The model is expanded as the fitting improves and the minimization function approaches the absolute minimum until all the model parameters are reasonably free to vary simultaneously in a stable manner. The selection of the different structural models of the respective phases to be refined by the Rietveld method can be taken from the different crystal structure databases, such as the ICSD (International Centre Structural Database), among others.

6.8.5 Analysis for numerical fitting

To achieve an accurate fitting of the measured intensities, the model used must consider all the factors involved in the formation of the diffraction profile. The use of an incomplete model will inevitably lead to incorrect parameters $(x_1,\ldots,x_m)$. However, even with an adequate model, experimental errors in the measurement produce mismatches that cannot be avoided. To assess the degree of fitting of the model to the data, it is necessary to use a series of numbers that indicate different relationships between the observed intensities (y_{0i}) and the calculated ones (y_{ci}). This series of numbers is called refinement fitting criteria, and is calculated from the residual values R_p (R-pattern), R_{wp} (R-heavy pattern), and R_e (R-experimental) and the χ^2 whose expressions are shown below:

$$R_p = \sum_i |y_{0i} - y_{ci}| / \sum_i |y_{0i}| \tag{6.73}$$

$$R_{wp} = \left[\sum_i w_i (y_{0i} - y_{ci})^2\right]^{1/2} / \left[\sum_i w_i y_{0i}^{\ 2}\right]^{1/2} \tag{6.74}$$

$$R_e = [(N-P)]^{1/2} / \left[\sum_i w_i y_{0i}^{\ 2}\right]^{1/2} \tag{6.75}$$

$$\chi^2 = GOF = \left[\frac{S_y}{(N-P)}\right]^{1/2} = \frac{R_{wp}}{R_e} \tag{6.76}$$

R_p represents the direct relationship between the observed and calculated intensities without any kind of weight between them. It is a very reliable parameter to observe during the minimization process since it responds to small changes in the degree of adjustment both in the background and in the maxima. However, its value is always low since, as it is not weighted, it usually represents more the background (most of the points in a diffraction profile) than the diffraction maxima. It is enough for a diffraction profile to have a high background and be well adjusted for this parameter to be very low, since the relative difference between intensities becomes very low if y_{0i} is always high.

R_{wp} is the fitting parameter to be observed since its numerator contains the function that is minimized by the least squares method. Unlike R_p, it includes a weighting function on the intensities that gives greater importance to the mismatches in the high intensities (the diffraction peaks) than in the low ones (the background). It is the most significant parameter that is usually reported when publishing results, since it gives us an indication of how good the refinement of the diffraction pattern is.

R_e is the so-called expected R, it is the quotient between the degrees of freedom of the system and the sum of the standard deviations of the intensities. It is a lower bound of R_{wp} when the quality of the diffraction data is good.

χ^2 is the parameter used to perform the goodness of fit test (*GOF*). Numerically it is equal to the square root of S_y divided by the difference between the number of observations (N) and the number of fitted parameters (P). Its expression is related to parameters R_{wp} and R_e. This value tends to 1 when the fit is perfect, but when it takes values less than 1 it can be concluded that the quality of the data (count statistics at each point, number of points, etc.) is insufficient to describe the number of parameters that are being fitted. A simpler model should then be chosen or a set of data more appropriate to the model should be obtained.

In addition to the above-mentioned fitting parameters, there are others such as R_B (*R*-Bragg Factor) and R_F (*R*-structure Factor) that are calculated to assist the experimenter in fitting the calculated intensities I_c (by integrating the curve fitted to the experimental maxima) to the observed intensities I_0 (determined by numerical integration of the peaks in the region determined by the model).

$$R_B = \sum_i \left| I_{0i} - I_{ci} \right| / \sum_i \left| I_{0i} \right| \tag{6.77}$$

$$R_F = \sum_i \left| I_{0i}^{1/2} - I_{ci}^{1/2} \right| / \sum_i \left| I_{0i}^{1/2} \right| \tag{6.78}$$

These two fitting parameters do not consider the intensities of the diffraction pattern regions where there are no diffraction maxima, and are therefore biased in favour of the structure, that is, they give us an indication of whether the chosen model has been correct. When the fit of the maxima is bad but the measured and calculated intensities coincide, the value of these parameters is low, however, the other diffraction profile fitting parameters are high. Therefore, the latter can only be used to compare the quality of the structure fitting to a data set if the fitting parameters of the complete profile are comparable.

Although six numerical parameters have been mentioned that guide the quality of the fit, the best indication of this can be obtained during the experiment, that is, by visual inspection of the y_{0i} vs 2θ curve when it is represented together with the y_{ci} vs 2θ curve, which also allows to analyze details of this in specific regions of the diffractogram. Additionally, it is useful to also represent in the same figure, the intensity difference curve $(y_{0i} - y_{ci})$ vs 2θ. In the latter, it is easy to detect errors in the fit of the intensity, the shape of the peak, the background, etc., in specific regions of the diffraction profile that have little effect on the global parameters of the fit. Therefore, all Rietveld refinement programs have a graphical program that updates the result of each refinement cycle automatically.

6.8.6 Fitting parameter

The theoretical model that fits the diffraction profile is composed of several functions, each of which contains specific information and must be analyzed independently when selecting the fitting parameters. But the model must also be considered globally to analyze the correlations between parameters. Next, we return to the description of the parameters of interest given in section 6.4, but now in terms of their contribution to the diffraction profile.

1. Global parameters of the diffraction profile

   The scale factor, wavelength, polarization state of the incident wave, background, 2θ zero error, and sample absorption (A) are the most important non-structural global parameters in the diffraction profile. The misalignment of any of them (except absorption) with respect to the real value usually prevents the refinement from being stable, and they must be set to values very close to the correct ones.

   - **Scale factor** is refined to adjust the intensity calculated by the model in relation to the observed intensity. It has no physical meaning, except in the case where multiple phases are being refined in the same diffraction profile, that is, during a quantitative analysis. In these cases, the scale factor is a fundamental component to determine the proportion of each of the phase's present in a sample. It is usually the first parameter to be refined in a diffraction profile, but its refinement does not make sense if the cell parameters and the 2θ zero are misaligned. The correlation between the 2θ zero and the cell parameters, as they are the three parameters that determine the position of the maxima in the diffraction profile, is very large, so that the misalignment of one of them necessarily implies the misalignment of the others.

- **Polarization factor** significantly determines the diffraction intensity at angles close to $2\theta = 90°$, since it is in this direction that the component normal to the sample of the electric field of the incident rays is not diffracted, leaving only the component parallel to the sample available. Conventional X-ray instruments that work with filtered radiation emit non-polarized radiation, so 50% of the incident wave is not available at $2\theta = 90°$. For this case, the polarization correlation takes the form of equation (6.27).

$$P = \frac{1 + \cos^2 2\theta}{2}$$

The use of direct beam monochromators (in X-rays and neutrons) and diffracted beam monochromators affect the polarization of the beam, mainly if it is not linear and parallel to the sample, so in these cases special polarization factors must be included.

In powder diffraction equipment, the system setup is such that the polarization is linear and parallel to the plane of the sample and therefore $P \cong 1$ for all angles. However, a value of 1 is never really obtained and it is possible to refine the parameter P although it presents a high correlation with profile shape parameters, and temperature factors of the atoms. The polarization factor appears explicitly among the refinable parameters, but combined with the Lorentz factor (6.30) in the structural model as follows:

$$LFP = \frac{1}{\sin^2 \theta \cos \theta} * P$$

- **2θ zero:** shifts the calculated maxima by a constant amount in 2θ and is often very small. Values of 2θ zero can result in distortion of the diffraction maxima and loss of instrument resolution. It is important to refine it in diffraction profiles with very narrow maxima, although under normal conditions it should have a value of the same order as the step used for data acquisition.
- **Background radiation** reaches the detector generating a uniform background throughout the diffraction profile. However, frequent scattering of X-rays by air generates a background dependent on 2θ which is characteristic for rising for angles close to 0° and 180°. When the sample has an amorphous component, a scattering of radiation is generated by this material which produces a characteristic rise in the background at low angles (normally less than 50° in 2θ). Correct representation of the background is essential to correctly describe the smallest diffraction maxima in the diffractogram and

the peak shape of the Lorentzian-type maxima which have long tails around the maximum.

All programs offer the ability to model the background using a function with refined parameters, while some also offer the ability to set background points and interpolate linearly or polynomial to describe the total background. Careful attention should be paid to the peak shape when the latter mode is selected.

Typical functions used to refine the background in 2θ ranges between 10° and 170° are orthogonalized Chebyshev polynomials and Fourier cosine series. Other special functions to describe typical background phenomena such as diffuse thermal scattering or air scattering (at angles less than 5°) have also been implemented. Depending on the shape of the background and the 2θ scanning range of the diffraction profile, 3 to 10 parameters are often refined simultaneously. It is common to observe whether the value of R_{wp} changes when adding or removing a background parameter. It is always advisable to use as few background parameters as possible.

- **Absorption:** The absorption of the sample depends on the experimental arrangement used to obtain the diffraction profile. The absorption problem can be very serious in transmission geometry for X-ray diffraction, due to the dependence of the correction factor on 2θ and the impossibility of deriving an exact theoretical model.

  In reflection geometry (Bragg – Brentano) absorption produces a constant intensity drop along the diffraction profile and is therefore included in the scale factor. In this geometry, the effect of low absorption (sample transparency) causes fundamental distortions in the peak shape. Surface roughness can also cause absorption-related problems, but this can be avoided by proper sample-space treatment.

2. Peak parameters

   Peak shape is an important factor when studying a powder diffraction sample. Correct refinement of the peak shape is essential not only for a reasonable visual fit between the calculated and observed intensity, but also for the values of the sample's structural parameters, dependent on the integrated intensity, to make physical sense.

   Although in a diffraction profile it is observed that H (the maximum width at the height of the peak) is a function of the modulus of the scattering vector, in the absence of anisotropy due to crystal size or lattice stresses, the function H can be modeled empirically. This makes the parameter H of each peak not one of the variables in the fit, but rather coefficients of the equations that describe the variation of the maximum width at the average height of the peak with 2θ are fitted. An example of this is the equation of Cagliotti, Pauletti, and Ricci that describes the variation of H with θ for Gaussian profiles:

$$(FWHM)_G = H_k^2 = U\tan^2(\theta) + V\tan(\theta) + W \qquad (6.79)$$

Or Lorenzian profiles:

$$(FWHM)_L = H_k = X \tan \theta_k + Y/\cos \theta_k \tag{6.80}$$

where U, V, W and/or X, Y are the half-width resolution parameters, which are independent of θ, and are adjusted during refinement. Accordingly, there are several functions to analyze the peak shape, some of them are:

Gaussian peak shape

$$G_{ik} = \frac{2}{H_k}\sqrt{\frac{\ln 2}{\pi}}\exp\left[-4\ln 2\frac{(\Delta 2\theta_{ik})^2}{H_k^2}\right]$$

Lorentzian peak shape

$$G_{ik} = \frac{2}{\pi H_k}\left[1+4\left(\frac{\Delta 2\theta_{ik}}{H_k}\right)^2\right]^{-1}$$

Pseudo-Voight beak shape

$$G_{ik} = \frac{2\eta}{\pi H_k}\left[1+4\left(\frac{\Delta 2\theta_{ik}}{H_k}\right)^2\right]^{-1} + (1-\eta)\frac{2\sqrt{\ln 2}}{\sqrt{\pi H_k}}\exp\left[-4\ln 2\left(\frac{\Delta 2\theta_{ik}}{H_k}\right)^2\right]$$

Pearson Function VII

$$G_{ik} = \frac{\Gamma(n)}{\Gamma(n-0.5)}\times\frac{(2^{1/n}-1)}{\pi}\times\frac{2}{H_k}\left[1+4(2^{1/n}-1)\left(\frac{\Delta 2\theta_{ik}}{H_k}\right)^2\right]^{-n}$$

where

$\Delta 2\theta_{ik} = 2\theta_i - 2\theta_{(hkl)}$: indicates the calculated position for the Bragg reflection corrected for the change in the zero point (Zero shift) of the detector.

$2\theta_i$: sweep angle at the i-th point.

η: is a parameter that defines the fraction of Lorentzian character in the Pseudo-Voight profile

$\Gamma(n)$: is the gamma function, which when it takes the value of $n = 1$, the Pearson VII function becomes Lorentzian and when $n = 8$ it becomes Gaussian.

Finally, although not strictly related to the peak shape, there is the preferential orientation of the sample. All polycrystalline samples exhibit this effect to a greater or lesser degree, but it is particularly pronounced in materials that have cleavage planes, or a laminar (plates) or acicular (needles) morphology.

In these cases, the crystals tend to orient themselves in a preferred way, introducing changes in the relative intensities of the diffraction pattern, and often alerting the experimenter from the moment of mounting it in the sample holder, so that it is feasible to eliminate or reduce its effects if sufficient care is taken. These orientations can be modeled analogously to the structure, introducing functions with refinable parameters that prefer the preferred directions of the sample with respect to that of the scattering vector.

3. Structural model

   The structural model is what determines the positions and intensities of the diffraction maxima. The first structural parameters that must be carefully selected in an initial model and refined are those of the unit cell. The symmetry of the lattice limits the number of them, varying from one in cubic structures to six in triclinic structures with the consequent degree of difficulty associated with the number of refinable parameters. In cases where the diffraction maxima are narrow and well-defined (there is not much overlap), the refinement of the cell converges quickly to reasonable values. The peak shape and preferred orientation parameters can be correlated with those of the unit cell, mainly the asymmetry and anisotropy parameters that allow some distorted maxima to fit as a function of *hkl*.

   The structural model itself, containing structural variables such as position, temperature factor (isotropic or anisotropic), occupation factor of one or more atoms, provides many variables to be refined, originating from non-linear equations and often correlated with each other (usually the occupation factor and the temperature factors of the same atom). These parameters must be carefully varied to ensure that the convergence produces a minimum that corresponds to a realistic chemical model.

   It is common to observe that temperature factors take unrealistic, negative, or very large values, due to mismatches in the peak shape, or the polarization of the incident beam, etc. It is often stated that temperature parameters act as a garbage dump for mismatches, being almost always affected by any poorly adjusted variable. Corrections for anomalous absorption and dispersion usually improve the values obtained.

   The type of data used for refinement must be considered when varying a very large number of structural and temperature parameters. The presence of atoms with very different dispersion powers can affect the correct determination of the structural parameters. There are many factors that must be considered when performing a Rietveld refinement, paying special attention to the structural parameters, since the stress and grain size calculations may have little physical value if the crystallographic structure obtained at the end of the refinement is not reasonable.

4. Quantitative analysis by the Rietveld method

   In a quantitative analysis carried out by the Rietveld method, several phases in the same sample can be refined simultaneously and the respective scale factors for each of them can also be obtained. It is precisely from these refined scale factors that the quantitative relationships of each of the phases that make up the sample are obtained. It has been shown that, by this method, more precise quantitative results are obtained, compared to other methods based on X-ray diffraction.

The information about the relative weight fraction W by this Rietveld method for a phase p present in a mixture of n phases is obtained using the following expression:

$$W_p = S_p(ZMV)_p / \sum_{i=1}^{n} S_i(ZMV)_i \tag{6.81}$$

where S is the scale factor obtained in the last cycle of the Rietveld refinement, Z is the number of (formula) units per cell, M is the mass of the formula unit (in atomic mass units), and V is the volume of the unit cell in (Å^3). The summation in i extends to all phases included in the model. Now, if an amorphous phase were to be present in the sample, the weight fraction values could be overestimated.

The problem is normally solved by using an internal standard, s, which is added in a known quantity to the sample and considered as another component of the sample. In this way, the percentage of the amorphous phase X_a in the original sample can be calculated directly from the weight of the added internal standard X_s and the value obtained from the standard in the refinement $X_{s,c}$ by means of the following expression:

$$X_a = \frac{100}{(100 - X_s)}\left(1 - \frac{X_s}{X_{s,c}}\right) \tag{6.82}$$

The advantages of the Rietveld method for making a quantitative phase analysis have several advantages over other methods, among which the following stand out:

- The total diffraction profile fitting minimizes the problem of overlapping between diffraction peaks. Indeed, all reflections corresponding to the different phases are explicitly included in the refinement, thus reducing the uncertainty in the weight fraction values.
- The fitting process allows for the consideration of the existence of preferred orientation effects.
- The amount of information needed for quantitative analysis is small and is related to the crystalline structure of the phases.

In summary, quantitative analysis by the Rietveld method consists of four stages:

✓ Generation of a calculated XRD pattern using structural data from a model.

✓ Point-to-point comparison of the experimentally obtained XRD pattern with the calculated XRD pattern

✓ Minimization of the differences between the two diffraction patterns by least squares fitting.

✓ Quantitative determination of each of the phase's present from the refined scale factors.

The parameters used in the Rietveld model can be divided into four main groups:

i. **Crystal structure parameters:** These describe the dimensions of the unit cell and its contents, namely the atom type, atomic coordinates, thermal vibration parameters and occupancy factors. These parameters are used to calculate peak intensities and positions and are the most widely used.

ii. **Peak profile parameters:** Describe the width and shape of the diffraction peaks and are used to confine the peak intensities of numeral *i.* in the observed diffraction profile.

iii. **Scale parameters:** are used to provide the overall scale of the calculated pattern to the observed pattern and to model the effect of preferred orientation.

iv. **Background parameters and zero 2θ diffraction angle:** unlike *i.* to *iii.* These parameters apply to the total pattern, rather than to a particular phase.

Bibliography

Al-Tabbakh, A.A. (2017). Symmetry in Crystallography: Understanding The International Tables, by Paolo G. Radaelli: Scope: Review. Level: Advanced Undergraduate, Postgraduate, Early Career Researcher, Researcher, Teacher, Specialist, Scientist, Engineers. Taylor & Francis.

Alexander, L. and Klug, H. P. (1948). Basic aspects of X-ray absorption in quantitative diffraction analysis of powder mixtures. Analytical Chemistry 20(10): 886–889.

Bunaciu, A.A., UdriŞTioiu, E.G. and Aboul-Enein, H.Y. (2015). X-ray diffraction: instrumentation and applications. Critical Reviews in Analytical Chemistry 45(4): 289–299.

Chung, F.H. (1974a). Quantitative interpretation of X-ray diffraction patterns of mixtures. I. Matrix-flushing method for quantitative multicomponent analysis. Journal of Applied Crystallography 7(6): 519–525.

Chung, F.H. (1974b). Quantitative interpretation of X-ray diffraction patterns of mixtures. II. Adiabatic principle of X-ray diffraction analysis of mixtures. Journal of Applied Crystallography 7(6): 526–531.

Clegg, W. (2023). Space groups–the final frontier: a tutorial guided tour of some entries in International Tables for Crystallography Volume A. Crystallography Reviews 29(4): 228–246.

Hill, R.J. and Howard, C.J. (1987). Quantitative phase analysis from neutron powder diffraction data using the Rietveld method. Journal of Applied Crystallography 20(6): 467–474.

Hull, A.W. and Davey, W.P. (1921). Graphical determination of hexagonal and tetragonal crystal structures from X-ray data. Physical Review 17(5): 549.

Lennox, D.H. (1957). Monochromatic diffraction-absorption technique for direct quantitative X-ray analysis. Analytical Chemistry 29(5): 766–770.

Rachinger, W.A. (1948). A correction for the α1 α2 doublet in the measurement of widths of X-ray diffraction lines. Journal of Scientific Instruments 25(7): 254.

Rietveld, H.M. (1969). A profile refinement method for nuclear and magnetic structures. Applied Crystallography 2(2): 65–71.

Seeck, O.H. and Murphy, B. (2015). X-ray Diffraction: Modern Experimental Techniques. CRC Press.

Visser, J.W. and De Wolff, P.M. (1964). Absolute intensities. Rep. 641: 202–204.

Young, R.A. (1993). Introduction to the Rietveld method. The Rietveld Method 1–38.

APPENDIX 1

X-Ray Diffraction Applications Using PANalytical X'Pert HighScore

1. Introduction

The X-ray diffraction technique of powder samples has become one of the most useful tools in the international arena for the crystallographic analysis of materials. Based on this technique, various methods have been developed with which it is not only possible to obtain qualitative and quantitative information on the crystalline phases in a material, but also on the amorphous content if the material is semi-crystalline.

The most widely used methods of this type are the Rietveld phase quantification method and the internal standard method for quantifying the amorphous content. In the Rietveld method, the entire observed diffraction profile is modeled from structural parameters of the constituent phases, which allows refining parameters of an instrumental and crystallographic nature, the calculated and observed diffractograms are compared, the differences are reduced through the least squares method and quantitative results are obtained from this. In the internal standard method, an estimate of the amorphous content is obtained by mixing a known quantity of an appropriate internal standard with the sample and based on this, the phase content in the mixture is corrected, quantified by the Rietveld method.

XRD testing can be used to determine

- ✓ Phase composition of a sample
  - – Quantitative phase analysis determines the relative amounts of phases in a mixture by reference to relative peak intensities
- ✓ Unit cell lattice parameters and Bravais lattice symmetry
  - – Index peak positions
  - – Lattice parameters can vary depending on, and therefore give you information about, alloying, doping, solid solutions, strains, etc.

✓ Residual strain (macro formation) | Crystalline structure
  - By Rietveld refinement of the entire diffraction pattern
  - Epitaxy/Texture/Orientation
✓ Crystallite size and micro strain
  - Indicated by peak broadening
  - Other defects (stacking faults, etc.) can be measured by analyzing peak shapes and widths
✓ XRD also has in situ capabilities, to evaluate all of the above properties as a function of time, temperature, and gas environment.
✓ Preferred orientation.
✓ The group and space of symmetries and from this information extract physical and chemical reactivity properties.
✓ Amount of amorphous phase, when this phase is a minority in relation to the crystalline phases.

X'Pert High Score Plus® v3.0e (HSP) software, developed by PANalytical (now part of Malvern Panalytical), is a tool designed for qualitative and quantitative phase analysis by X-ray diffraction (XRD). This software is widely used in materials research, mineral sciences, and industrial applications due to its ability to provide accurate and detailed analysis of the crystal structure of materials.

X'Pert High Score Plus® v3.0e is equipped with an intuitive and robust user interface that allows users to efficiently process and analyze XRD data. One of its most notable features is the implementation of the Rietveld method for the quantification of identified phases.

This appendix will introduce basic functions of HSP, v3.0. There is much more that can be done with the program than will be said here. It will hint at things that can be done with HighScore Plus (HSP) without going into detail. In such cases, this is because a quick read of the relevant help section in (HSP) will tell you clearly how to accomplish such a task.

This appendix provides an introduction to the use of HSP only for phase identification and quantification in crystalline materials with very little amorphous phase. The software is too robust to attempt to cover all or more topics than those discussed here, without going beyond the scope of this book. The quantification of amorphous phases in semicrystalline materials, by the internal standard method, or by the modified Rietveld method using information obtained through the Warren-Averbach model, will not be discussed in this appendix. The Warren-Averbach model uses the Fourier transform of diffraction profiles to analyze the size of crystalline domains and defects in the material. Although it was developed to analyze the width of diffraction peaks in crystalline phases, it can be modified to include the amorphous contribution.

For materials that are mostly amorphous, the use of several techniques is required to quantify their phases. The most recommended is to use FTIR, Raman Spectroscopy, DSC, TGA complemented with XRD for this purpose, but this, due to its importance and complexity, will be the subject of another book.

2. Preliminaries

When you open the program, the Main and initial screen of the HSP software, v3.0 is displayed (see Figure 1).

Figure 1. Main and initial screen of the HSP software, v3.0.

At the top are all the main fields (File, Edit, View, Reference Patterns, Analysis, Reports, Tools, Customize; Window, Help) that contain the program instructions, as shown in Figure 2.

Figure 2. Main fields of the HSP.

To open the data file to be analyzed, click on the File field and the menu shown in Figure 3 will be displayed.

Then click "open". Browse to your file in the location you have on your PC and then click "Open" to open the file. Now you have the file open, and you will see a window like the one shown in Figure 4.

You may see a window with some differences to this one, don't worry, it won't be a problem for all the subsequent processing. These differences have to do with

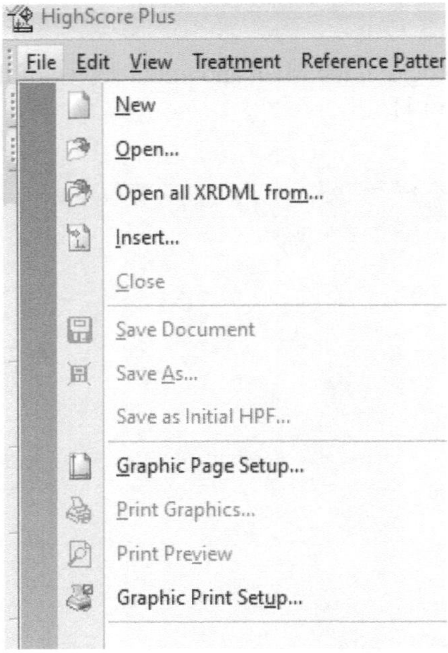

Figure 3. File field window displayed.

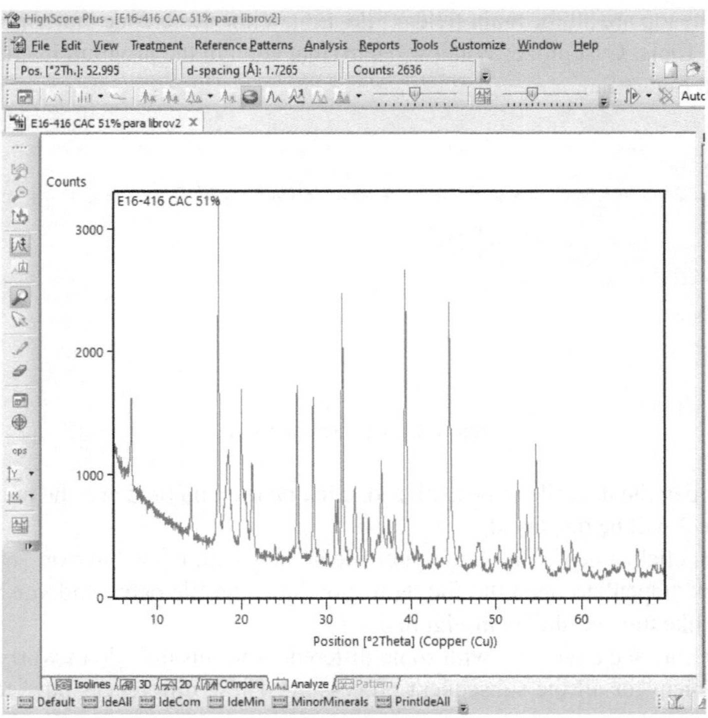

Figure 4. Window with the diffractogram file open.

versions of the operating system, or of this software itself and if the file has had any previous processing. Obviously, you will see your diffractogram, not the one in Figure 4.

"Desktops" are predefined arrangements of windows and toolbars in HSP. You can create your own desktop or use one of the default desktops provided with the program. Default desktops are designed to be optimal for specific tasks, such as Phase ID or Profile Fitting. If you begin using HSP and the desktop is arranged in a way that you are not familiar with, you can change the desktop to one that you find more useful.

In the View field click to display the window and then click on desktop (see Figure 5).

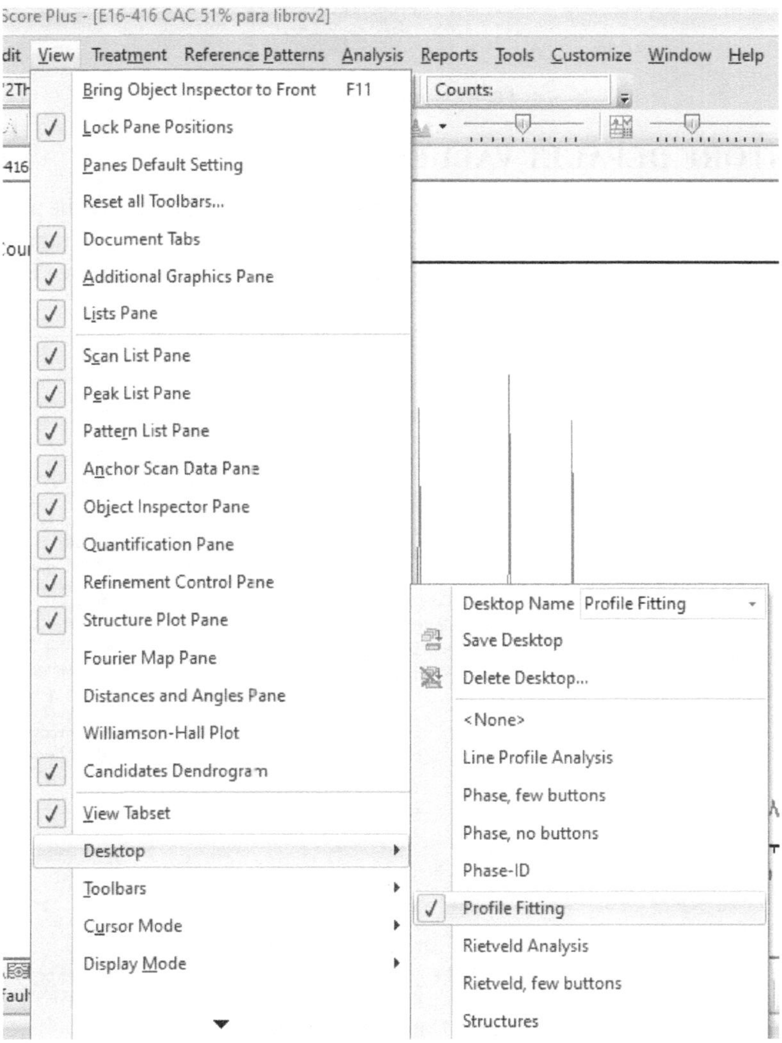

Figure 5. List of possible work desks.

Choose Profile Fitting (see Figure 6)

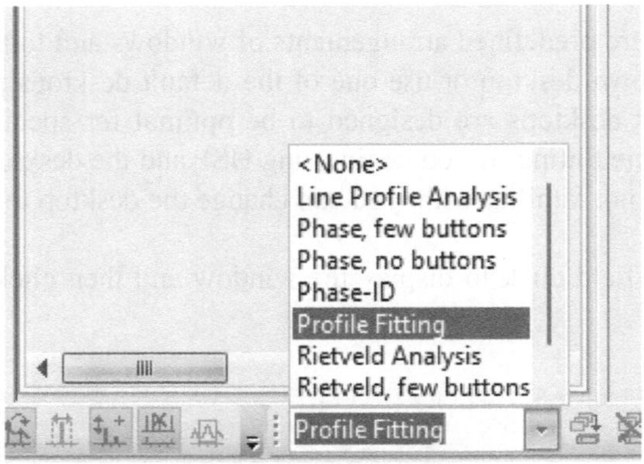

Figure 6. Profile Fitting Desk.

RESTORE DEFAULT VALUES

If you start using HSP and it behaves unusually, you may want to restore the program's behavior to its factory defaults. To do this: Go to the Customize > Program Settings field and click Reset All to Default and click OK to close the window (see Figure 7).

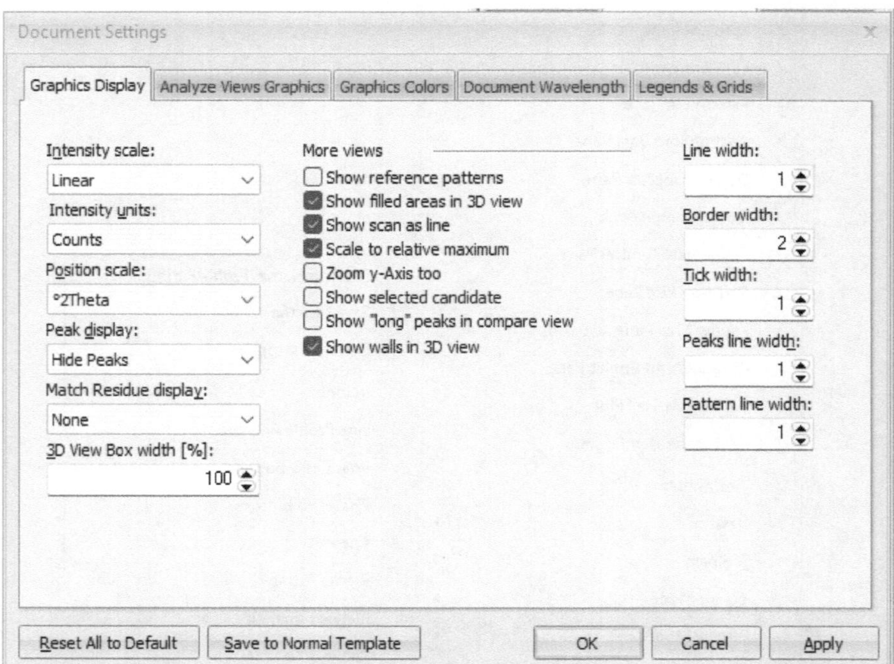

Figure 7. Reset All to Default.

You can also explore the Document Settings to change options for the graphic display, Analyze Views Graphics, Graphics, Document Wavelength, Legends & Grids (see Figure 7). We will return to these fields later.

By default, fields only show the most recently used commands. To disable this feature and be able to see all commands in the menus: Select the View > Toolbars > Customize field. From the three tabs, select the Options tab and then Uncheck the box next to "Menus show recently used commands first". Click Close (see Figure 8).

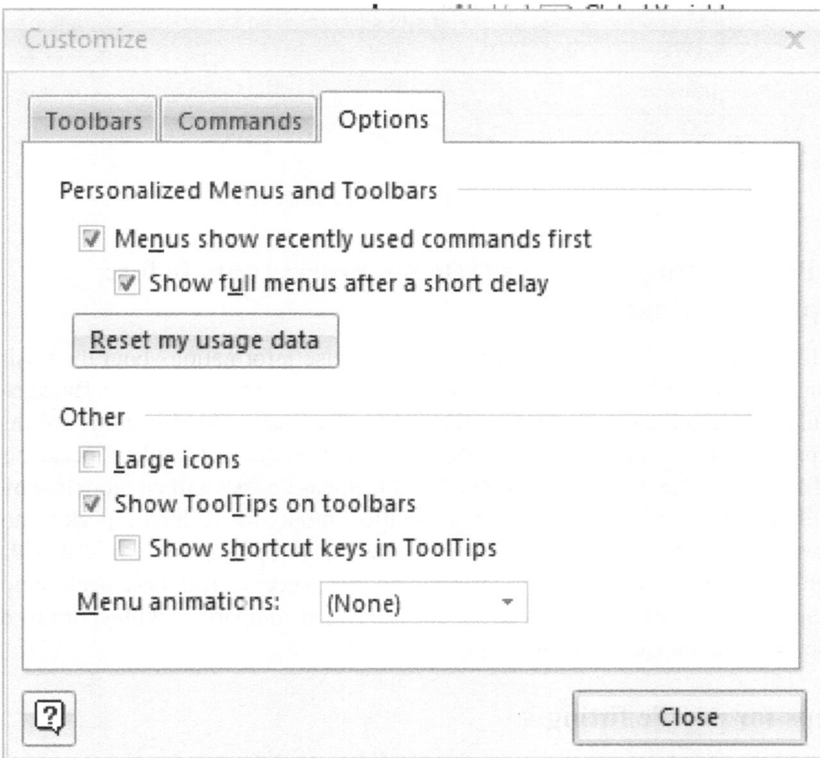

Figure 8. Personalized Menus and Toolbars.

Go to the View > Panes Default Setting field to get the Additional Graphics window. The main window is divided into three large blocks, which we will be talking about throughout the appendix, Main Graphics, Lists pane and Object Inspector (see Figure 9).

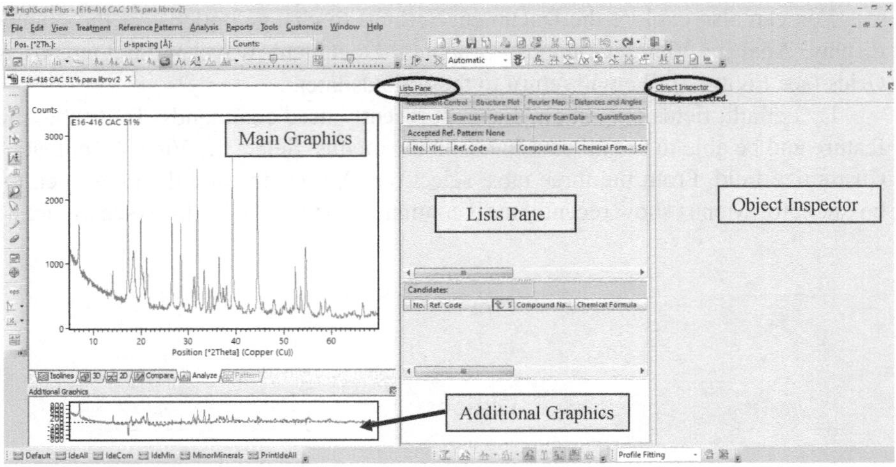

Figure 9. Additional Graphics and main windows.

3. Profile fitting procedure (Do not process data before profile fitting)

Profile fitting is a technique used to extract precise information about the position, intensity, width and shape of each individual peak in a diffraction pattern. Background should not be subtracted before profile fitting. The background is modelled during the profile fitting process. It is also not correct to remove the K alpha-2 peaks before profile fitting. The doublet of Kalpha-1 and Kalpha-2 peaks will be modelled by the profile function and changing the mathematical model for Kalpha-2 peak removal, if desired, is done after profile fitting; by default the peaks are removed with the Rachinger model. For further information on this aspect of Kalpha-2 peak removal, see appendix 2. Profile fitting is based on a statistical comparison with your raw data, which will not be valid if your data is smoothed.

Steps for profile fitting

The steps to adjust the profile are as follows and are given in the logical order.

1. Fit the background.
2. Identify the peaks (i.e., peak search).
3. Profile fit the data.
4. Evaluate the quality of the profile fit.
5. Do calculations based on the result.

3.1 Step 1. Fit the background

Go to the Treatment > Determine Background field (see Figure 10).

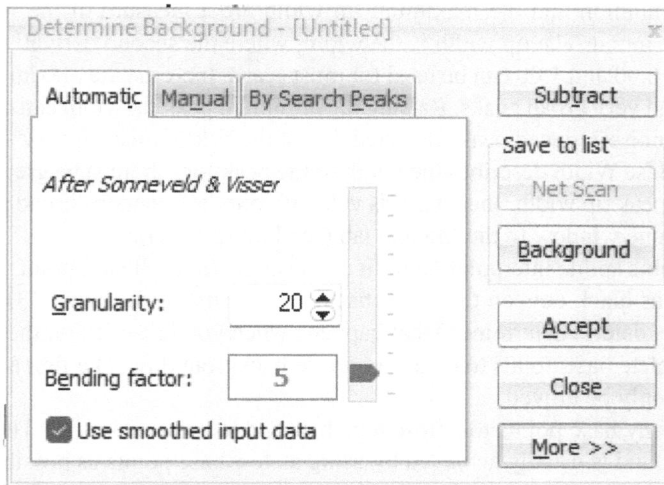

Figure 10. Determine Background.

Use a small bending factor (0–2) and a granularity between 15 to 30. The granularity parameter changes the number of intervals used for background determination. The bending factor slider adjusts the curvature of the background. When the background is good, Accept the background. The background is good when it passes as low as possible on the base (see Figure 11).

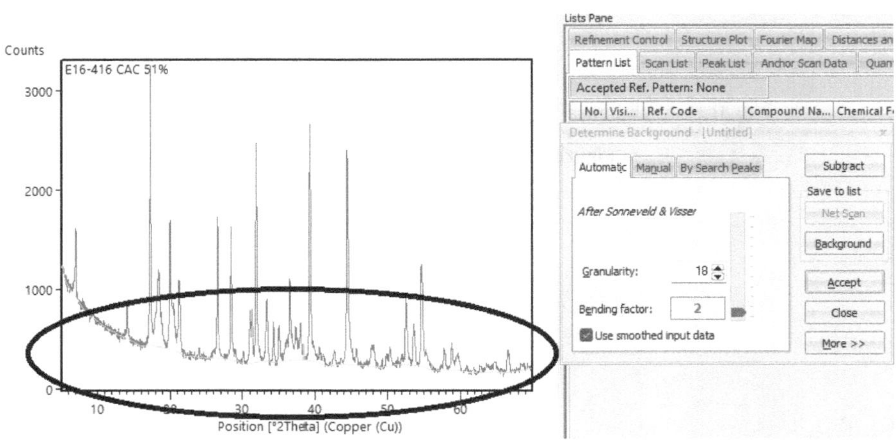

Figure 11. Background Fitting.

In the same window, click on the tab > By Search Peaks (see Figure 12(a)). 'Significance' refers to an estimate of the probability that a possible peak is not induced by noise. Only peaks with a significance above the Minimum Significance are accepted by the peak search procedure. A large Minimum Significance (2 or more) is useful for noisy data—however, some very small peaks could be missed.

The minimum and the maximum tip width are a measure of the width of the peaks to be found. The tip width is the width, where the second derivative is < zero. The values 0.00 and 1.00 can be used for most scans. Increase the maximum peak tip width to find very broad peaks. Raising the minimum peak tip width can decrease the number of noise-induced peaks detected, but at the risk of missing very small peaks. The Peak Base Width describes the width of the peak foot. It must be greater than the maximum peak tip width. Increase this value if your data contains broad peaks. Now go in the same window to the Manual tab (see Figure 12(b)).

Use cubic spline interpolation to fit curves between each base point. Base points are shown as black dots on the green background curve (see Figure 13). The mode (add, move, delete) determines what happens when you left-click on the base point. Add and delete base points to create a precise background fit. The first and last base points can only be moved.

Too many base points too close together tend to create poor fits. The best fit to the background is usually achieved by using as few base points as possible.

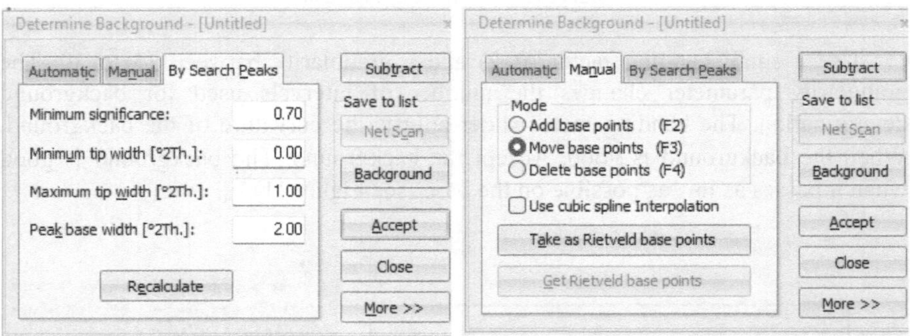

(a) By Search Peaks (b) Manual Option

Figure 12. Background fitting options.

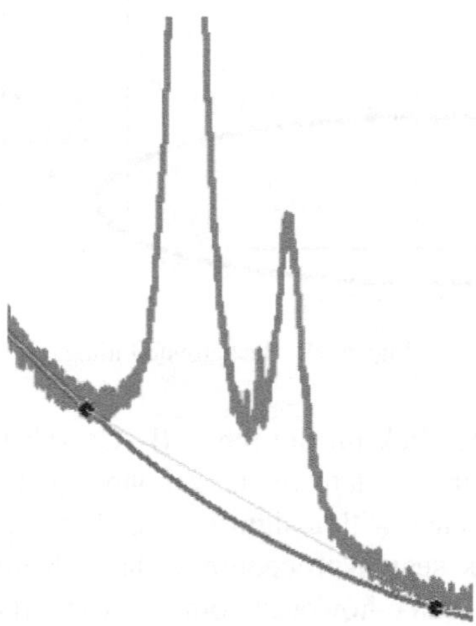

Figure 13. Cubic spline interpolation.

3.2 Step 2. Identify the peaks (peak search)

Go to the Treatment > Search Peaks field (see Figure 14)

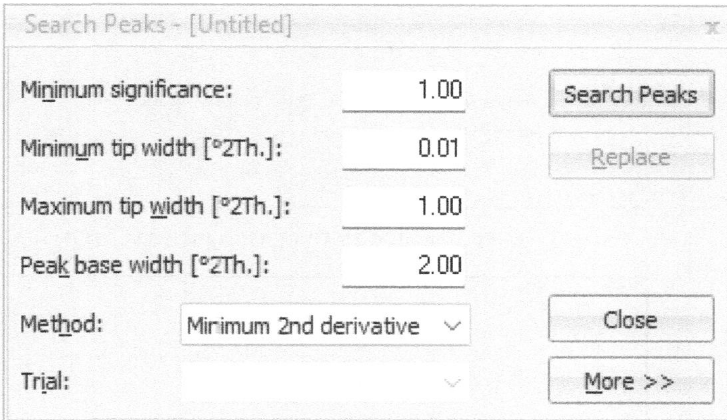

Figure 14. Search Peaks.

PEAK SEARCH PARAMETERS

There are two methods available to locate the peak:

- Minimum of Second Derivative, or
- Top of Smoothed Peak.

Use Minimum of Second Derivative for scans with overlapping peaks to get optimal results. The significance of a peak is a measure of the probability that a peak found is not noise-induced. Only peaks with a significance above the Minimum Significance are accepted by the peak search procedure. A typical value for the minimum significance is 0.75 for 2θ-based raw scans. Setting the minimum significance to 0.75 ensures that all peaks found are reliable, however, some very small peaks could be missed.

Often higher significances are used for phase identification to ensure that only clear, strong peaks are taken into account.

If you subtracted the background or manipulated the anchor pattern in any other way, much higher significances could be required to suppress noise-induced peaks.

The minimum and the maximum tip width are a measure of the width of the peaks to be found. The tip width is the width, where the 2nd derivative is < zero. The values 0.00 and 1.00 can be used for most scans. Increase the maximum peak tip width to find very broad peaks. Raising the minimum peak tip width can decrease the number of noise-induced peaks detected, but at the risk of missing very small peaks.

The Peak Base Width describes the width of the peak foot. It must be greater than the maximum peak tip width.

This value is only used to determine the background during peak search. If background data are available, the peak base width is not used at all.

If you previously subtracted a background, then no background values are present. In this case, peak search generates a new background.

In the More tab (see Figure 14) >> click and choose the default parameter: set Identify, printIdeAll, Untittled. Click on the Search peaks button and then on Accept (see Figure 15).

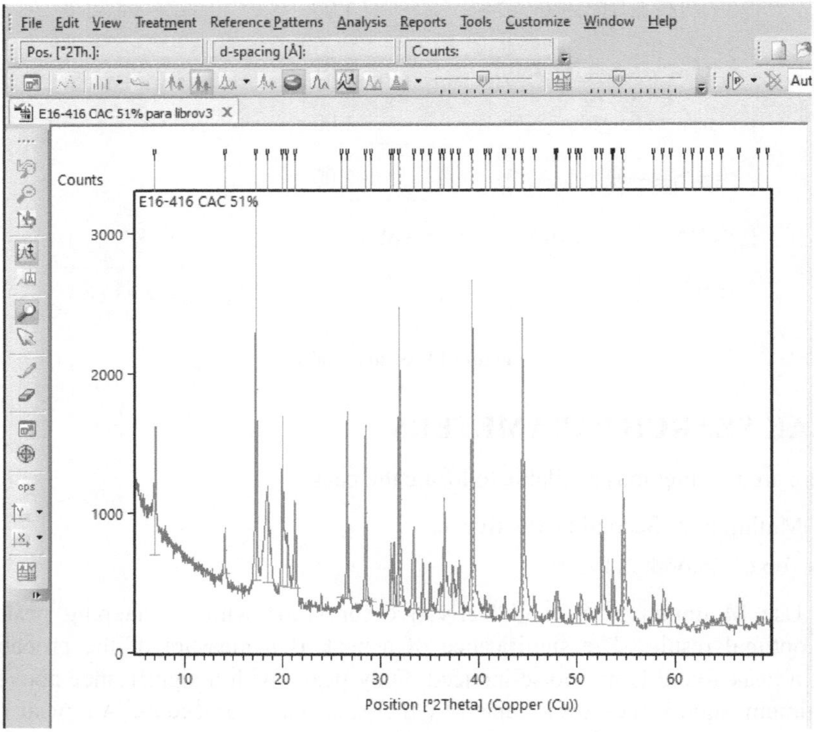

Figure 15. Peaks identified.

PARAMETER SETS

Most functions in HSP have a parameter set, which contains all parameters required for this task. Parameter sets are stored per user. The concept of parameter sets allows you to work quickly using the available defaults, but it also supports you when you want to try out, save and restore your own preferred choices.

All parameter sets share a common look and feel. The name of the parameter set in use is always shown in the title bar of the corresponding dialog. You can edit parameter sets using edit parameter set even, when there is no open diffraction document, or if a certain function is not available for the actual document. All possible parameter sets are shown in the overview.

There are four different types of parameter sets:

- ✓ Default parameter sets
- ✓ Untitled parameter sets
- ✓ Named parameter sets
- ✓ Public parameter sets

Usually, parameter sets are stored per windows user. But every user can designate any parameter set to become a public parameter set. These parameter sets are available to all users running HSP on a certain PC.

PARAMETER SET MANIPULATION

To select, save, delete or print parameter sets, you must first expand the dialog using the More (see Figure 14) >> button. Use the drop-down list box to select a parameter set. Five buttons allow you to:

Save As...:

Save the actual parameter set with the entered name. If the name already exists, then it is overwritten by the new parameter set.
You need to explicitly **name and save** each parameter set. The question "Do you want to save...?" is **not shown** in HSP.

Delete:

Delete the selected parameter set.

Copy:

Copy a description of the selected parameter set to the clipboard.

Print:

Print a description of the selected parameter set (to the text printer).

Set Public:

Switched ON, the selected parameter set becomes publicly available for all users.
Sometimes additional (less frequently used) parameters are also available under the More >> button.

Default:

A default parameter set is always present. It represents the factory default setting and cannot be changed. Under general circumstances the default parameter set is a good starting point, but not necessarily the optimum for the actual analytical problem. At the very moment you change a single parameter, the default (or any other) parameter set turns into "Untitled".

Untitled:

Untitled parameter sets are created when you edit a default, or a named parameter set of a specific function. Any change immediately results in an untitled parameter set. These parameter sets are meant to be used for a trial. They are easily overwritten by any subsequent editing. There is only one Untitled parameter set available for each function.

Changing any parameter set will overwrite the untitled parameter set of the corresponding function!

An Untitled parameter set is used in the ordinary way, but it cannot be part of a user batch program.

Named:

Any named parameter set has been saved on purpose under its own specific name. It is kept until it is deleted on purpose. Each user can have as many named parameters sets as he wants to.

Parameter set names are not case-sensitive. "HELLO", "Hello", and "hello" point to one single parameter set. A maximum length of 255 characters is possible for a parameter set name.

If you edit a named parameter set, its name is automatically changed to untitled. You must save it under its old, existing name in order to change the contents and keep its name.

Note:

If you delete or change parameter sets which are used in another function (like in Reporting or in Search-Match), you first get a warning. After you deleted it, the default parameter set is used in place of the deleted parameter set.

Missing parameter sets still used in a batch program result in non-execution of the corresponding steps of the batch program.

Public:

Usually all named parameter sets are only available for the user, who made and stored them. But in some situations (many users, one supervisor) it is desired to make certain tasks (requiring a set of parameters) available for many users, bypassing the export and import of parameter sets, which is possible too.

The property public is used to make a parameter set available for all windows users of the software on one PC. To make a parameter set public, click the button when you save or edit the parameter set. Click the button once more to de-activate it and to make the parameter set private again.

Note:

There are no user rights. Any user can make any parameter set public or de-publicize it.

It is possible, that another user makes a parameter set public, which has the same name then one of your private parameter sets. If this happens, the public parameter set is leading and overrules the private parameter set.

Note the notches above, each one represents a peak, if they do not appear go to the button shown in Figure 16 and click to bring up the menu and select show peaks outside.

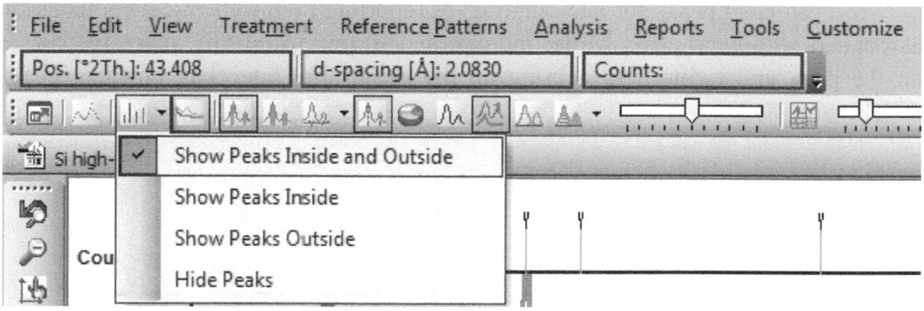

Figure 16. Visualization of identified peaks.

You can long click and sweep some peaks you want to enlarge in the graph and release-click on the closed box (see Figure 17) to observe the identified peaks in relation to the experimental diffractogram.

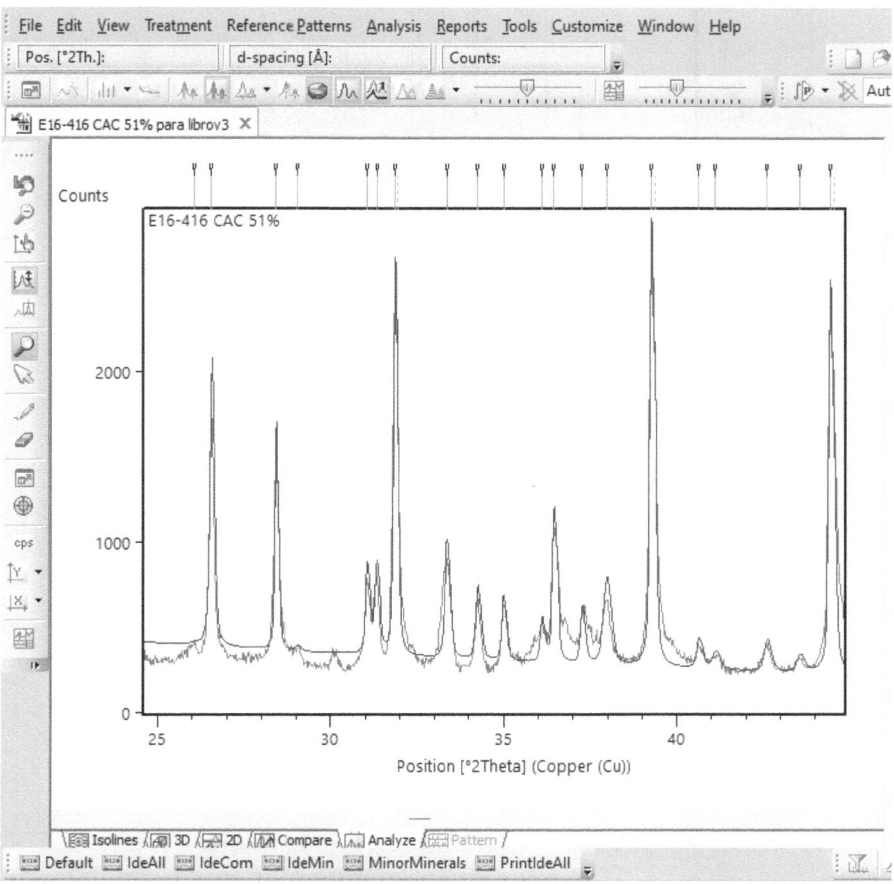

Figure 17. Visualization of identified peaks with experimental diffractogram relation.

You can label peaks for later analysis or reporting by the chemical composition associated with each peak, Miller indices, and so on. To do this, place the cursor on the peak and right-click to bring up the menu and then click on label peaks (see Figure 18).

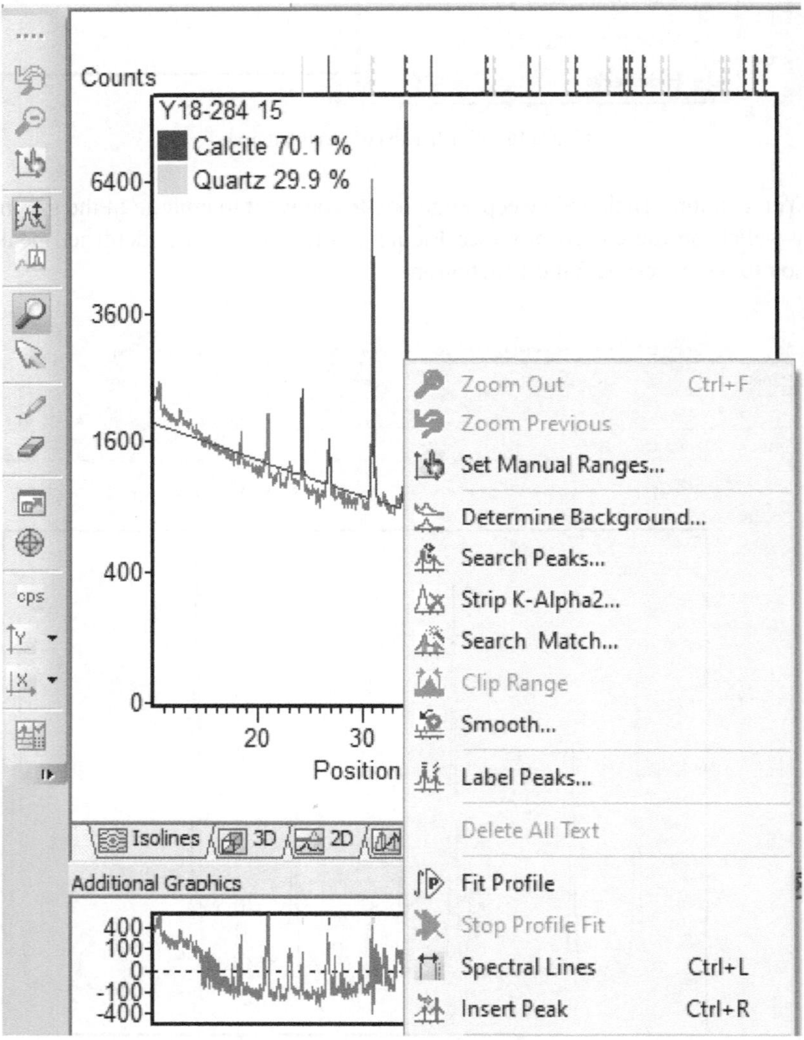

Figure 18. Labels the identified peaks.

You can then label the peaks with whatever labels you want (see Figure 19).

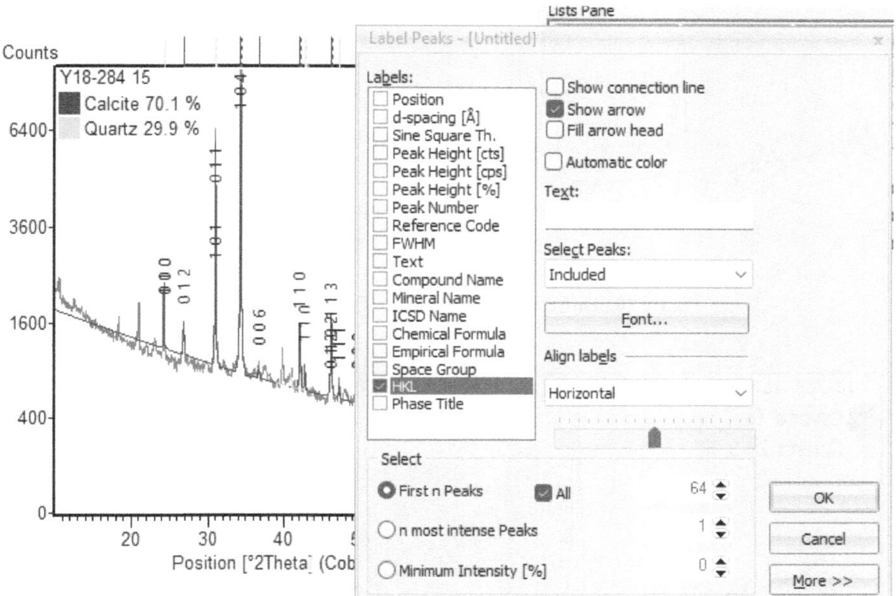

Figure 19. Peaks labeled with Miller indices.

You should not yet see any identified phase, much less quantified, such as those observed in the images of Calcite (70.1%) and Quartz (29.9%). We will see this in detail later.

Profile fitting will use the available peaks to fit the data as best as possible. You can include peaks. When peaks are included correctly, profile fitting accurately separates contributions from overlapping peaks.

To delete a peak: Select its line in the Peak List and do one of the following: Press the Delete key on your keyboard or right-click and select Delete Peak—if you move the mouse cursor over a peak line marker, that peak will automatically be highlighted in the Peak List. Remember that the grey-shaded lines in the Peak List are K-alpha2 peaks. They are shown as dotted lines in the Main Graphs. There is no need to edit them—they will be automatically corrected during profile fitting.

To insert a peak: Right-click in the main graphics window (see Figure 20) and select Insert Peak (see Figure 21). Position the cursor to match the height and position of the peak, then left-click to insert the peak. Right-click in the main graphics window and select Insert Peak to disable the peak insertion cursor (see Figure 22).

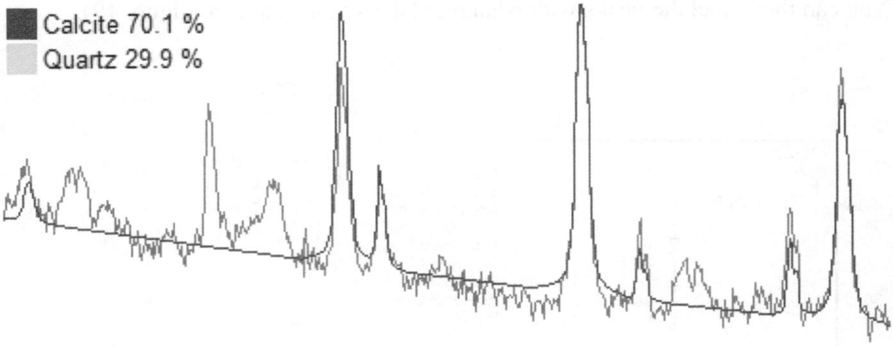

Figure 20. Diffractogram without peak insertion correction.

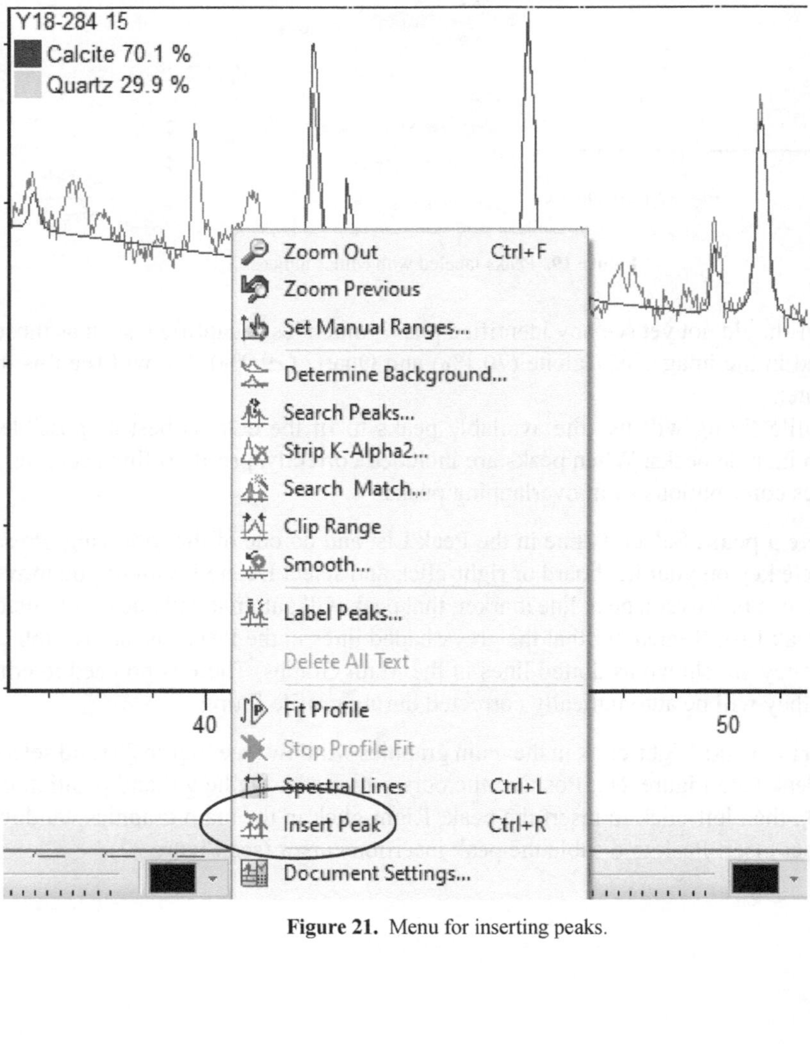

Figure 21. Menu for inserting peaks.

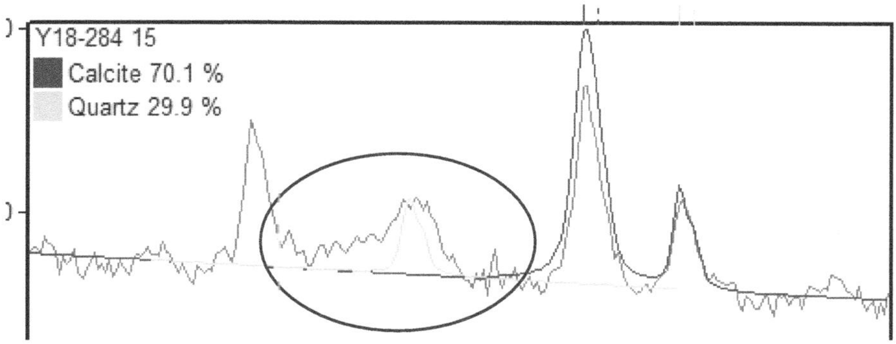

Figure 22. Peak insertion-corrected diffractogram.

3.3 Step 3. Profile fit the data

Inspect the program settings for profile fitting. Select the Refinement Control tab in the Lists panel. Left-click on the Global Variables item in the Refinement Control panel. Locate the global settings in the Object Inspector panel (see Figure 23).

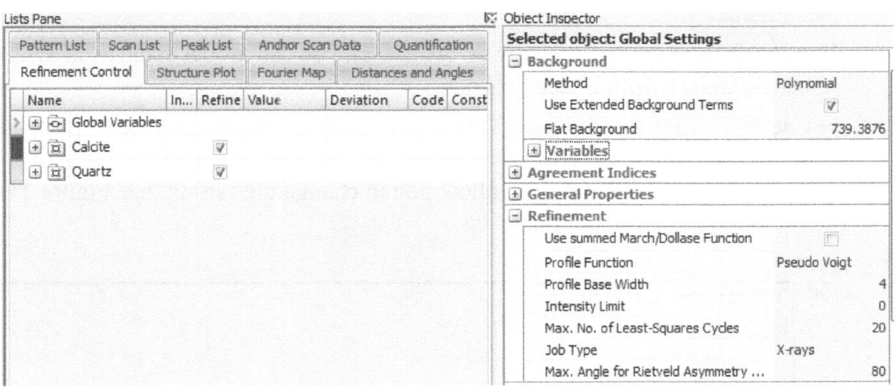

Figure 23. Profile fit the data.

Inspect the most relevant profile setting preferences.

Background:

The Method can be changed using the drop-down menu.
The most common settings are:

- ✓ **Polynomial:** fit the background using a polynomial curve.
- ✓ **Shifted Chebyshev:** fit the background using a curve that is useful for samples with amorphous content.
- ✓ **Use Available Background:** fix the background to the curve created in step 1 and do not refine it during profile fitting. This is a useful selection when the background is complex or when the pattern features many overlapping peaks.

Profile Function:

- ✓ The function can be changed using the drop-down menu
- ✓ This determines the equation used to fit each diffraction peak
- ✓ Pseudo Voigt and Pearson VII are the most common

For settings with a text-based value, you click on the word describing what is being used (such as the "Polynomial" for Background Method or "Pseudo Voigt" for Profile Function) and a drop-down menu will appear to allow you to change the setting.

Profile Fit Peak Base Width

- ✓ This number determines the cut-off where the peak intensity is no longer calculated
- ✓ If the profile fit does not extend for the full length of the tails of the peak, then increase this number

Asymmetry Type

- ✓ When set to "No Asymmetry Function", the peaks will be modeled as symmetric (see Figure 24)
- ✓ If peaks are asymmetric (usually at low angle), you can model using Split Width (most common), Split Shape, or both.
- ✓ For settings with a text-based value, you click on the word describing what is being used (such as "No Asymmetry Function" for Asymmetry type) and a drop-down menu will appear to allow you to change the setting (see Figure 25).

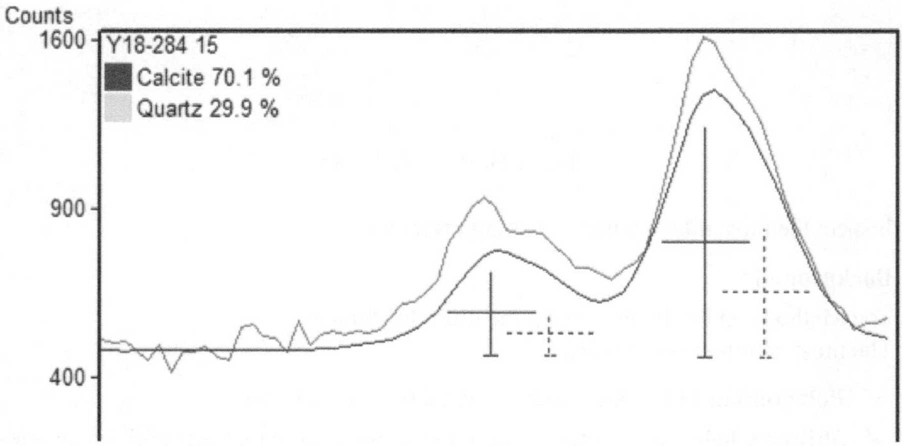

Figure 24. Symmetry of peaks.

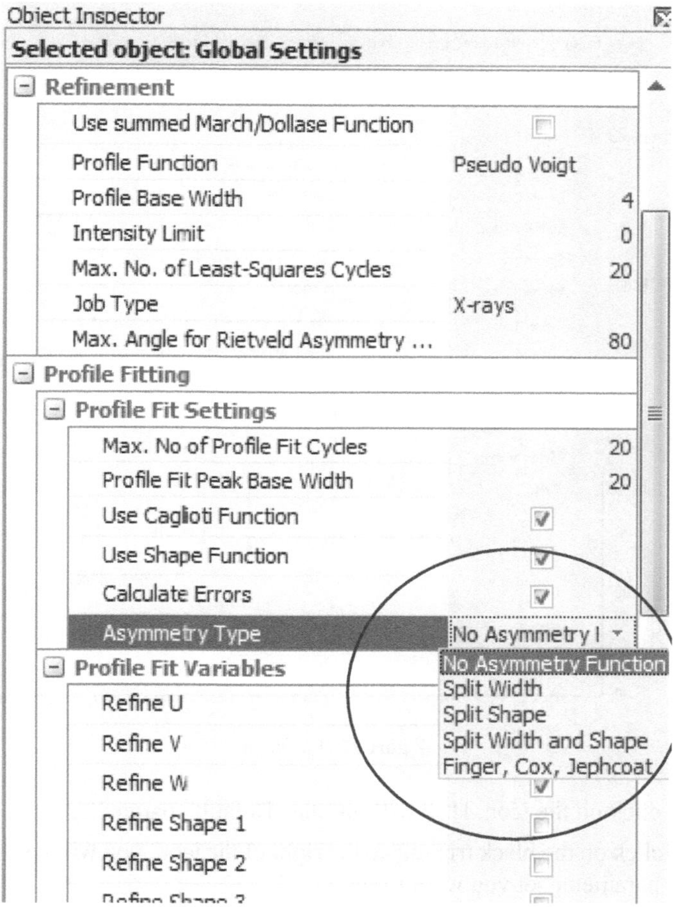

Figure 25. Asymmetry type.

You can customize or restore default values for profile setting preferences
Go to the menu Customize > Defaults

- ✓ Select the Default Global Settings tab.
- ✓ The Profile Fitting Preferences, such as background method and asymmetry type, are listed here.
- ✓ Changes made to these preferences will change the default starting values for all new profile fitting analyses.

ADJUSTING AUTOMATIC PROFILE FITTING PARAMETERS

Many items for profile fitting are found in the menu Treatment > Profile Fit
You can execute different profile fitting treatments from the menu Treatment > Profile Fit > Fit Profile or by using the toolbar (see Figure 26).

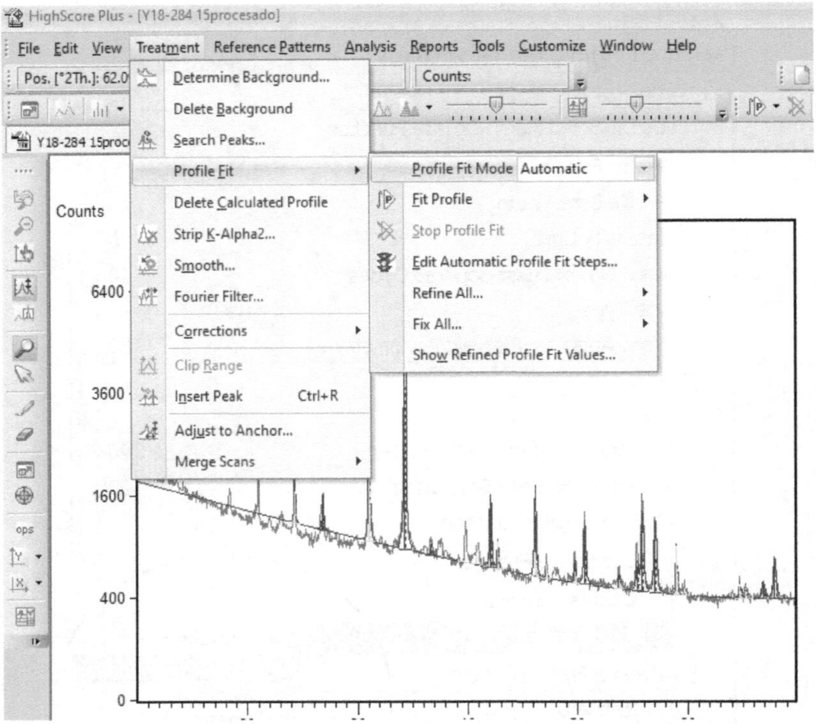

Figure 26. Profile fit.

- ✓ If you click on the icon, HSP will run the "Default" parameter set.
- ✓ If you click on the black triangle to the right of the icon, you will be able to select which parameter set you want to run.

You can further review and set the parameters relevant to your problem and thus have a much more accurate and reliable profile fitting (see Figure 27 and Figure 28).

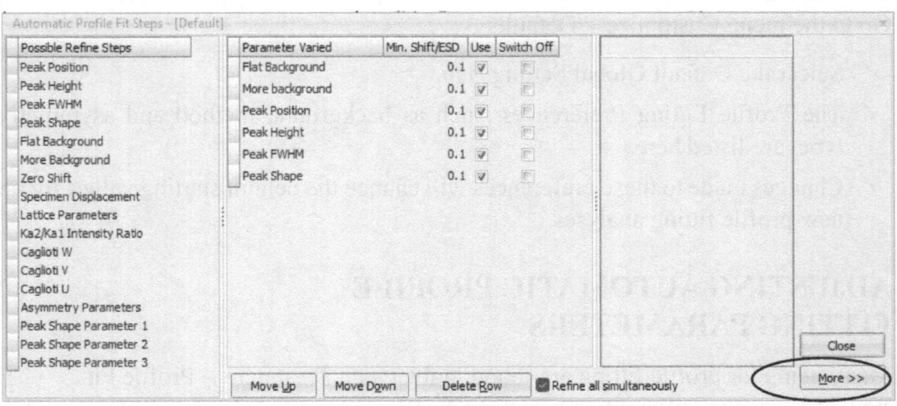

Figure 27. Setting the parameters for the profile fit.

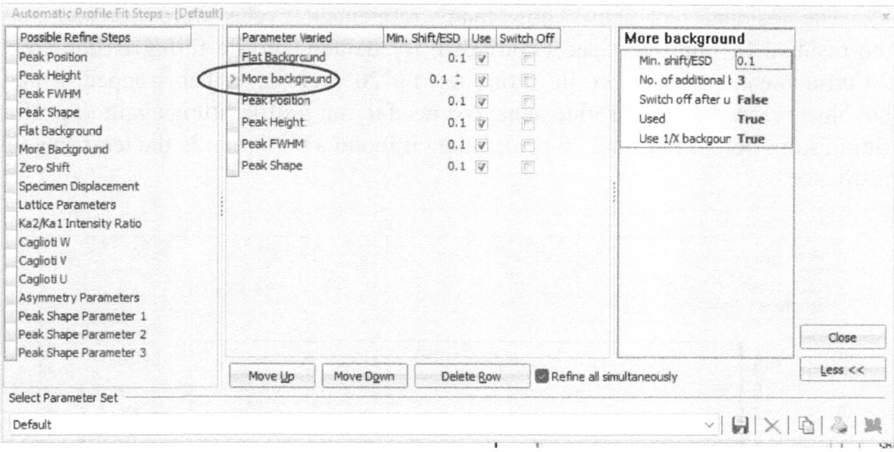

Figure 28. Additional parameters for profile fitting.

When using a background function such as a polynomial, the "Number of additional background terms" will determine the order of the polynomial. In this case, 3 was taken as you can see in the box to the right of Figure 28.

The 1/x background term is effective for modeling the exponentially increasing background at low angles.

TO RUN PROFILE FITTING

To execute the profile adjustment, click on the little triangle enclosed in the circle in Figure 29.

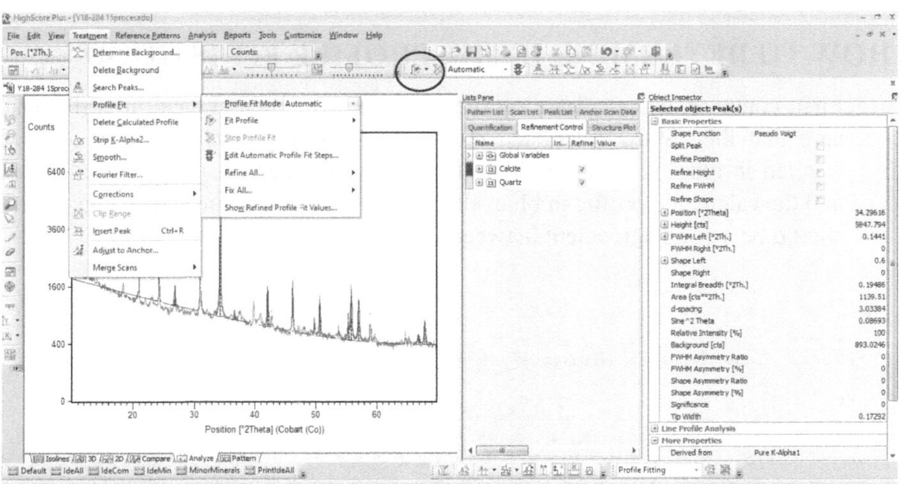

Figure 29. Running profile fitting.

The residual plot shows how many refinement cycles were run and how the residual fit improved (see Figure 30). By default, profile fitting is limited to 20 refinement cycles. If profile fitting ran for 20 cycles and then stopped, it may not have reached optimal refinement. You need to run profile fitting again. If profile fitting stops before reaching 20 cycles, then it found a minimum in the least squares resilience.

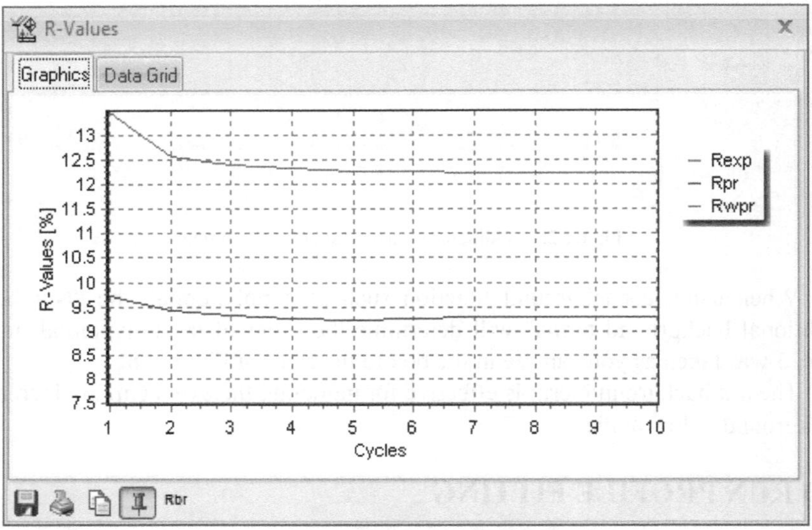

Figure 30. Residual graph of profile fitting.

3.4 Step 4. Evaluate the quality of the profiler fit

HOW TO DETERMINE IF THE PROFILE FIT IS GOOD

1. First, compare the calculated profile with the experimental data. In HSP, make sure that the "Show calculated profile in Analysis view" option is selected (circled in black) (see Figure 31). Experimental data is usually plotted in red and the calculated profile in blue, although this color scheme can change. There should be a good agreement between the two graphs.

Figure 31. Profile calculated in Analysis view.

2. Use the difference plot to determine if the fit is good

   The difference plot is shown in the Additional Graphics area (see Figure 32). Right-click in the Additional Graphics area. Select the menu Show Graphics > Difference Plot

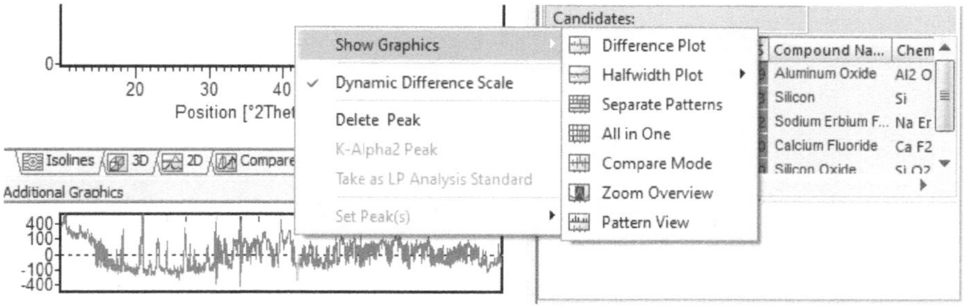

Figure 32. Difference plot to determine if the fit is good.

3. Agreement Indices quantify the difference between the experimental and calculated profiles

   To see the Agreement Indices

   ✓ Select the Refinement Control tab in the Lists Pane.

   ✓ Left-click on the entry Global Variables.

   ✓ Look in the Object Inspector pane.

   ✓ Expand the area "Agreement Indices".

   - $R_{expected}$ is a statistical evaluation of the noise of the data. A smaller $R_{expected}$ indicates better quality data.
   - $R_{profile}$ is the residual difference between observed and calculated plots.
     - Weighted $R_{profile}$ (R_{wp}) uses a weighting function to place more emphasis on good fit between high-intensity data points (such as intense peaks) and less emphasis on low-intensity data points (such as background).
     - In a good fit, R_{wp} should be less than 10.
   - Goodness of Fit is *Rwp2/Rexp2*. In a good fit, the GOF should be less than 4. But even less than 10 could be considered a good fit in particular and complex cases, such as cementitious materials, refractory ceramics, clays, among others, which in addition to crystalline phases also have a significant percentage of non-crystalline phases.

4. Check the peak list for anomalous values, such as high FHWM values

   ✓ For each phase of a sample, values such as FHWM and Shape should be similar, with slight variations as a 2theta function.

   ✓ If a peak in a cluster that is very close has a larger FHWM (see Figure 33), it could mean that there are two overlapping peaks that are fit with a single profile function.

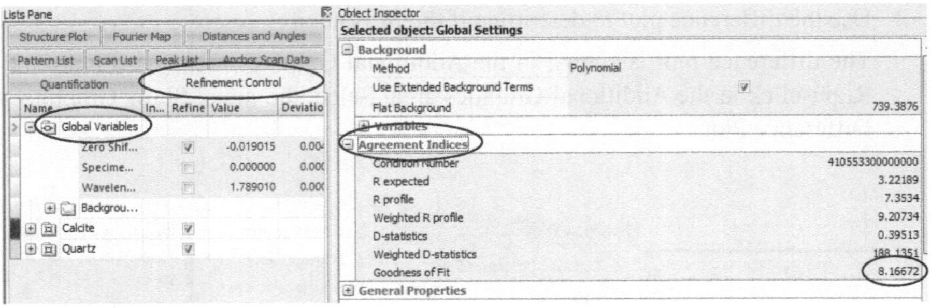

Figure 33. Agreement Indices.

✓ In the list of peaks on the right, the typical FWHM is 0.16 to 0.4 and the typical shape is 0.5 to 1.

✓ Peaks with a width of 0 or 3 are outliers and should be inspected.

✓ Peaks with a shape of 0 are outliers.

To do this, first label the numbers with their peak number as shown in Figure 34.

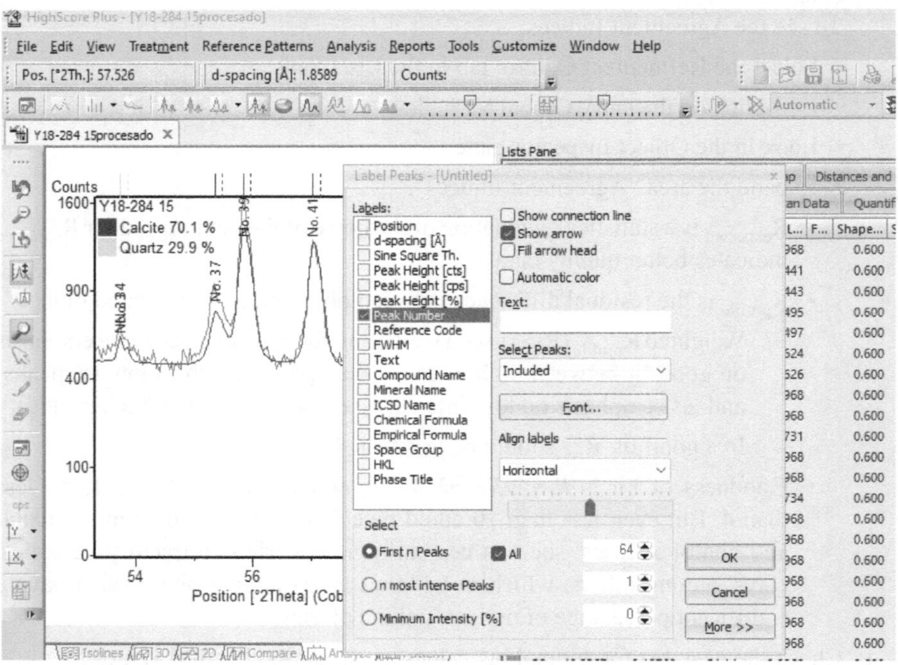

Figure 34. Peak number label.

Then review the data associated with each peak—in this case peak number 37 is being reviewed—and verify the previous criteria (see Figure 35).

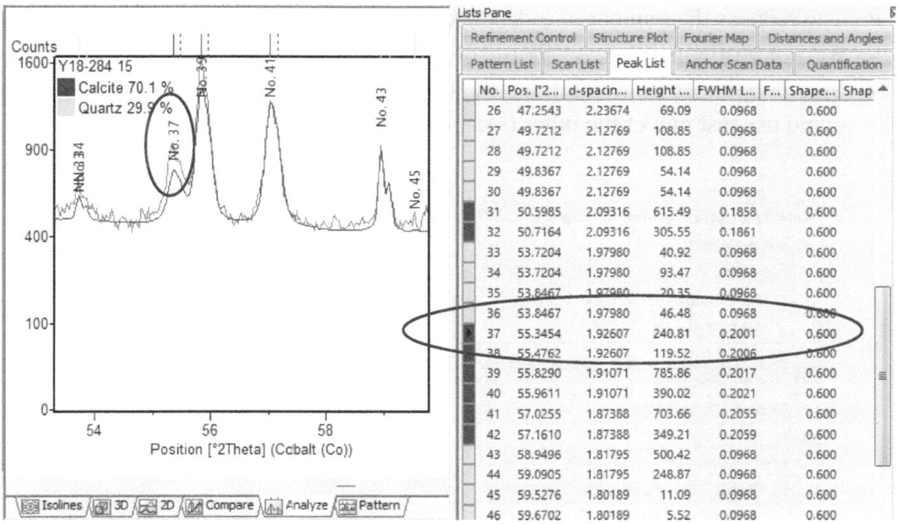

Figure 35. Checking the peak data.

OTHER PROFILE FIT METHODS

Individual peak fitting: Each peak is treated independently. The Position, Height, FWHM, and Shape of each peak are independent variables.

Constrained peak fitting: The position and height of each peak are independent parameters. However, the width, shape and asymmetry of all peaks are described by an equation that slowly varies as a function of 2theta. This constraint of the width and shape of peaks is useful when peaks are highly overlapping, or data are noisy. This method may not work well in a mixture, since the different phases may have different peak shapes and widths.

Pawley fit: Peak positions are generated and constrained by a unit cell and symmetry information. The unit cell lattice parameter is refined rather than individual peak positions (zero shift or specimen displacement may also be refined). Peak heights are fit independently. The width, shape, and asymmetry of peaks are described by a function for each phase in the sample.

Constrained Peak Fitting: The Caglioti Fit

Select the Refinement Control tab in the Lists Pane

- ✓ Left-click on the phrase Global Variables in the Refinement Control pane (circled in red)
- ✓ Look in the Object Inspector pane for the Global Settings.

- ✓ In the Profile Fit Settings area, check the boxes for "Use Caglioti Function" and "Use Shape Function".
- ✓ All of the peak widths and shapes will be fit with simplified equations.
- ✓ The reduces the number of independent variables and will improve the stability when a profile is fitting a pattern with many overlapping peaks.
- ✓ You do not have to use both the Caglioti function and the Shape function—you could use just one or the other (see Figure 36).

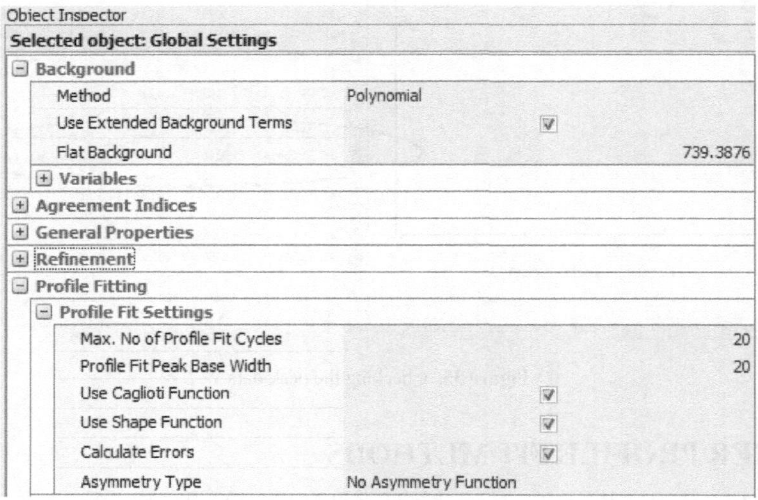

Figure 36. Caglioti Fit.

Edit the profile fit parameters for the Caglioti fit

- ✓ Go to the menu Treatment > Profile Fit > Edit Automatic Profile Fit Steps.
- ✓ Select the Default parameter set.
- ✓ Modify is the More Background term.
  - You can change the No. of Additional Background Terms.
  - You can choose whether or not to use the 1/x background term.
- ✓ Select which Caglioti parameters U, V, and W you want to use
  - Always Refine W. Most typical would be to refine U, V, and W.
- ✓ Select the number of Peak Shape Parameters.
  - Typical is 1. A very strong pattern may be fit with 2.
  - If you changed the parameter set, you will need to save it as a new parameter set

THE PAWLEY FITTING

For Pawley fitting, a unit cell is used to generate the initial peak list. Peak positions are adjusted by refining the lattice parameter of the unit cell rather than the individual peak positions. Peak heights are refined individually. Peak width and shape are constrained with functions such as the Cagliotti equation, as described for constrained peak fitting (see Figure 37).

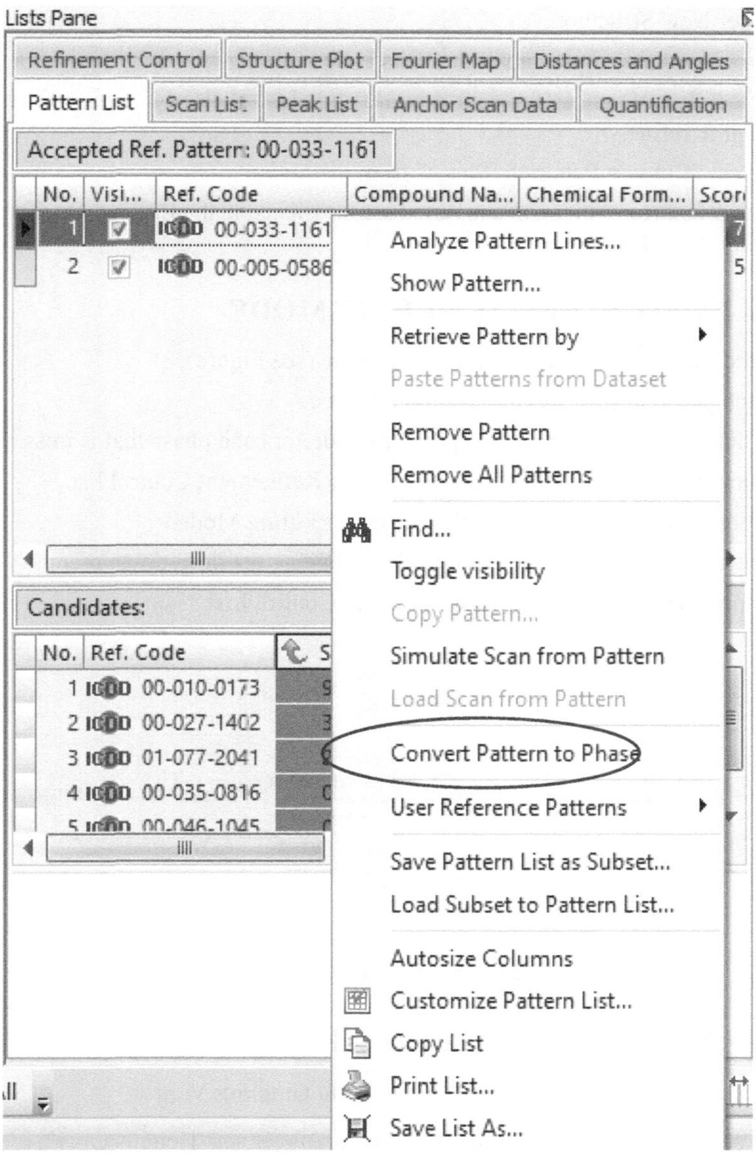

Figure 37. Pawley Fit – convert pattern to phase.

STEPS FOR PAWLEY FIT: LOAD THE REFERENCE PATTERNS

First, manually fit the background as done previously

- ✓ Instead of a Peak Search.
- ✓ Retrieve the reference cards for the phase(s) in your sample.
- ✓ You may also manually input the unit cell from the menu Analysis > Rietveld > Enter New Structure.
- ✓ Go to the Pattern List tab in the Lists Pane.
- ✓ Select the references for your mixture and right-click on the phase(s) that you want to refine.
- ✓ Select "Convert Pattern to Phase" from the menu.
- ✓ The reference(s) that you converted are now listed in the Refinement Control tab of the Lists Pane.

PAWLEY FIT STEPS: SET UP FIT MODE

Go to the Refinement Control in the Lists Pane (see Figure 38)

- ✓ Each phase in your mixture should be listed.
- ✓ If not, repeat the steps on the previous slide for each phase that is missing.
- ✓ Click on the name of the first phase in the Refinement Control list.
- ✓ In the Object Inspector, find the entry for "Fitting Mode".
- ✓ Change the "Fitting Mode" to Pawley Fit using the drop-down menu.
- ✓ Repeat for each phase in the Refinement Control List.

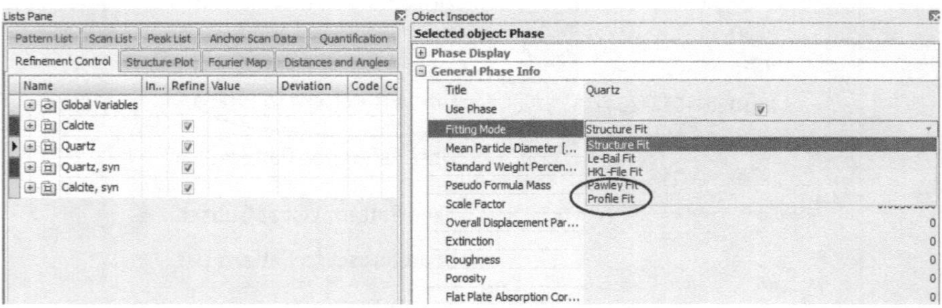

Figure 38. Pawley Fit – Fitting mode.

STEPS TO PAWLEY FIT: CREATE AN INITIAL PEAK LIST

Make sure that you are zoomed out in the Main Graphics View.

- ✓ Execute a Profile Fit using the built-in parameter set "Heights Only".
- ✓ You can select this fitting mode from the menu Treatment > Profile Fit > Fit Profile > Heights Only or by using the Profile Fit toolbar.

✓ This will generate a peak list based on the unit cell(s) of the phase(s) that you loaded into the Refinement Control list.

✓ In case some peaks are created by the reference card that are not observed in your data, you may need to delete those peaks.

STEPS FOR PAWLEY FIT: REFINE THE PROFILE FIT

You may refine the profile fit by using the built-in Pawley Fit parameter set or by creating your own parameter set

✓ The parameter set should include:

- Lattice parameters,
- Background functions,
- Peak Height,
- Caglioti W, V, and maybe U,
- Peak Shape 1, possibly Peak shape 2,
- Zero Shift OR Specimen Displacement,
- Asymmetry Parameters.

✓ To refine asymmetry parameters, you must set the Asymmetry Type to "Finger, Cox, Jephcoat"

✓ The Asymmetry Type is accessed in the Object Inspector: Global Settings, just like the Split Width and Split Shape options

PAWLEY FITTING

✓ When the Pawley Fit is complete:

- The unit cell lattice parameters for each phase will have been refined.
- The peak width and peak shape functions for each phase will have been individually refined.
- These can be found in the Refinement Control List by expanding the entry for a Phase Name and then expanding the Unit Cell and/or Profile Variable entries.
- The Zero Shift or Specimen Displacement error will have been refined.

✓ This can be found in the Refinement Control List by expanding the Global Variables entry.

✓ If you click on a phase name in the Refinement Control List, the peaks for that phase will be highlighted in the Main Graphics as long as the "Show Selected Phase Profile" option is enabled.

✓ You can enable this option from the menu View > Display Mode > Show Selected Phase Profile or from the View toolbar.

3.5 Step 5. Calculations based on peak list:

UNIT CELL REFINEMENT ANALYSIS

Most often, unit cell lattice parameters are refined based on a known phase. Much of the lattice parameter refinement is used to analyze modifications of known materials. For example, to quantify the change in the material when doped. The first steps for this type of analysis are to load the reference card for the material and fit the data profile. The reference card must be converted before it can be used as an initial basis for unit cell refinement.

After you have loaded the reference card(s), go to the Pattern List tab in the Lists Pane:

- ✓ Right-click on the phase(s) that you want to refine.
- ✓ Select "Convert Pattern to Phase" from the menu.
- ✓ The reference(s) that you converted are now listed in the Refinement Control tab of the Lists Pane.

Now you can refine the unit cell lattice parameters:

- ✓ Go to the menu Analysis > Crystallography > Refine Unit Cell… .
- ✓ Select the entry for the phase that you want to refine first.

The following screen then appears to Modify the parameter set for unit cell refinement.

The parameter set is not found by clicking on the "More" button, like it is in every other dialogue window:

- ✓ Instead, the parameter set is at the end of the left column.
- ✓ Select a saved parameter set or click on the "…" button to open the window to edit it.

In the Refine Unit Cell Settings, you can set the Refine entry to (see Figure 39):

– Cell only

– Cell + zero shift

– Cell + displacement

- ✓ This controls the parameters that will be refined.
- ✓ Zero Shift and Displacement are two common sources of systematic error in the observed peak positions.
- ✓ If not already created in your copy of HSP Plus, it is useful to create three saved parameter sets for Cell only, Cell + Zero Shift, and Cell + Displacement.

Specimen displacement and zero shift are two common sources of systematic error in observed peak positions. Specimen displacement is the most common and most significant source of error in modern diffractometers and is created when the

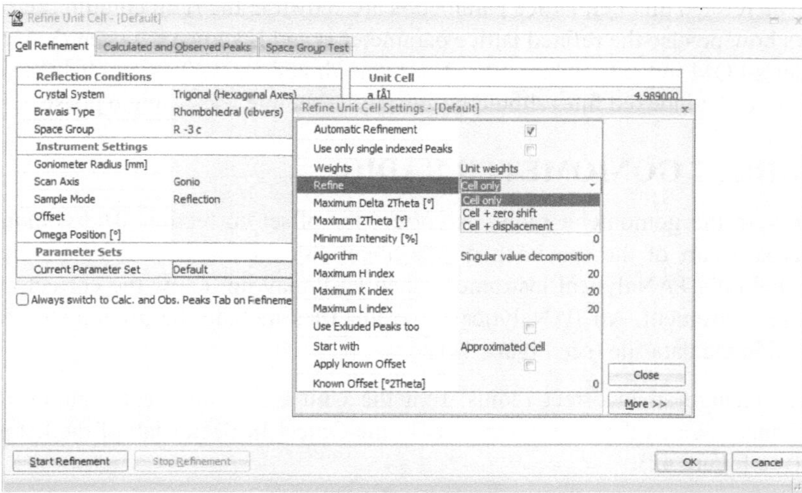

Figure 39. Refine unit cell settings.

sample is not properly positioned on the focusing circle of the goniometer. Specimen displacement varies as cos θ, according to the equation:

$$\Delta 2\theta = -2s\cos\theta / R$$

Zero shift is more common in older diffractometers that do not have optical encoders and is produced when the detector arm of the goniometer is not properly aligned at zero. Zero shift is better to use when refining data is collected over a small angular range, even if the source of error is actual specimen displacement.

REFINE THE UNIT CELL USING THE PARAMETER SET AND EVALUATE THE RESULTS

In the Refine Unit Cell window, click the button "Start Refinement" (see Figure 40):

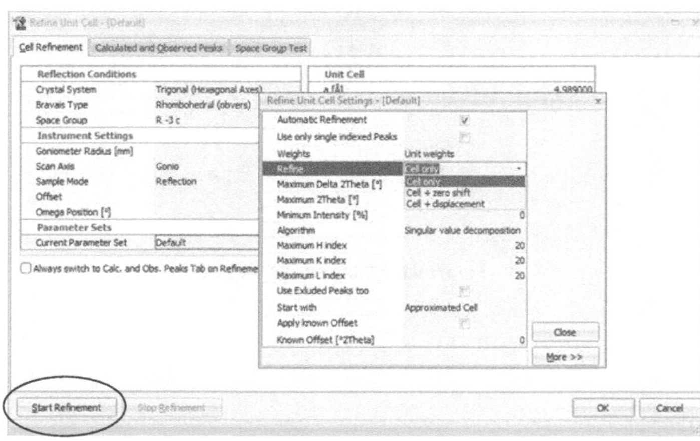

Figure 40. Start refinement.

The refined unit cell lattice parameters are shown in the right column. The ESD shows how precise the refined lattice parameter is and is shown between (). A larger Snyder's FOM indicates a better fit. The FOM should be > 20 for publishable data. The No. of Unindexed lines should be zero if the sample was a single phase.

CORRECT GONIOMETER RADIO

Make sure the goniometer radius is correct for offset correction. The goniometer radius is a part of the specimen displacement calculation. If your data were not collected on a PANalytical instrument, then HSP may not know the correct radius of your instrument. All PANalytical *. xrdml files include the goniometer radius defined in the data file (see Figure 41).

- ✓ To change an incorrect radius, Exit the Unit Cell Alignment window (click Cancel). Open the scan parameters in the Object Inspector tab of the list panel (see Figure 42).
- ✓ An easy way to do this is to double-click the scan name in the scan list. Scroll down to the bottom of the scan information in the Object Inspector. Find the Goniometer Radius entry in the Instrument Settings section. Type the correct goniometer radius (mm). Press the Enter key to record the change. Return to the unit cell refinement procedure.

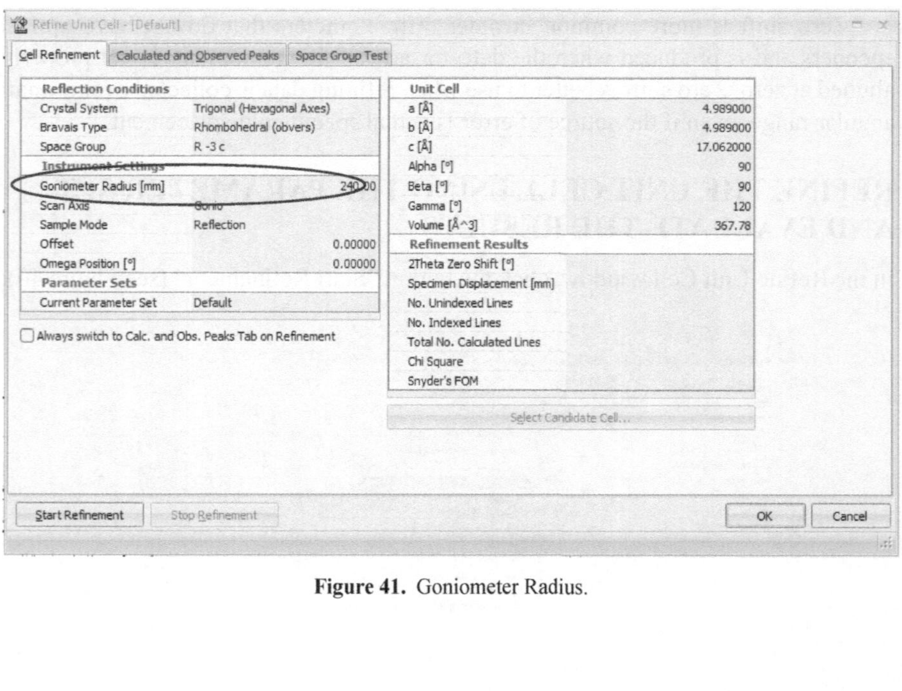

Figure 41. Goniometer Radius.

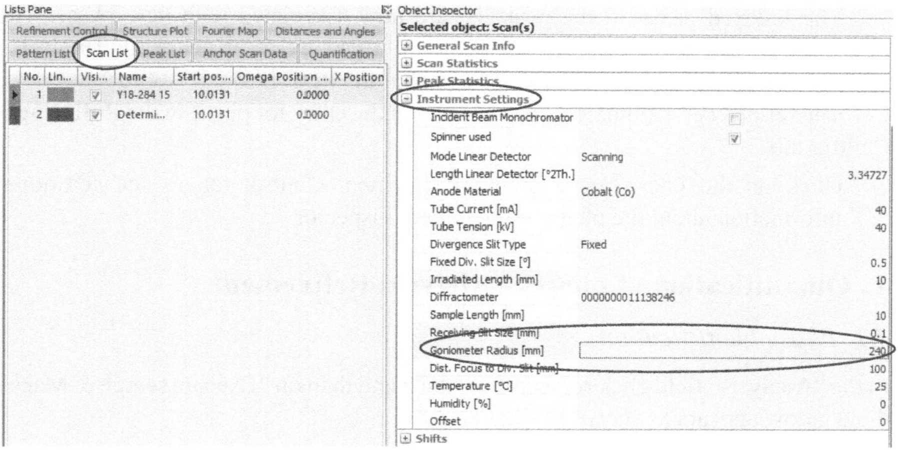

Figure 42. Instrument setting - Goniometer Radius.

The Delta 2Theta chart (see Figure 43) will reveal systematic errors in refinement:

- ✓ Click on the Calculated and Observed Peaks tab
- ✓ The table and graph show the difference between the experimental peak positions and the calculated peak positions based on the refined unit cell
- ✓ In a good refinement, Delta 2Theta in the graph will be randomly distributed around zero.
- ✓ A systematic trend may reveal a sample shift error (delta 2Theta varies with 2Theta) or a zero-shift error (constant delta 2Theta as a function of 2Theta)
- ✓ Once the refinement is acceptable, click OK

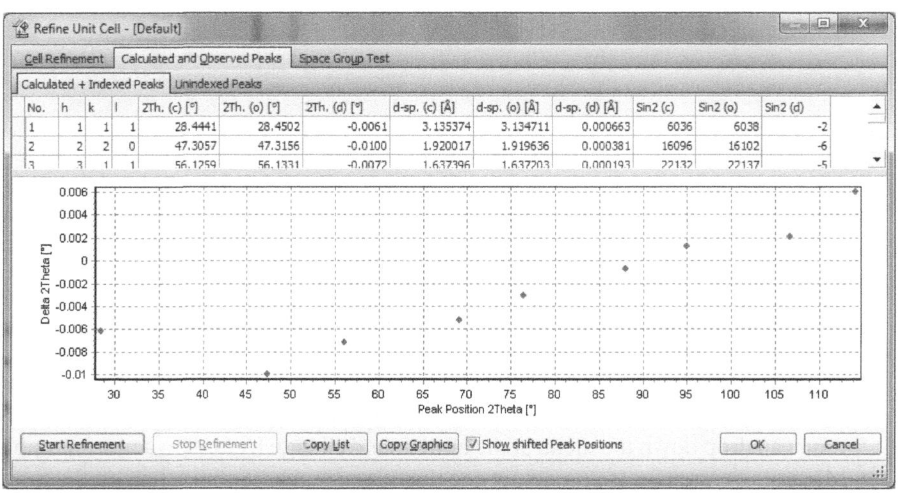

Figure 43. Delta 2Theta Graphic.

The indexed peaks in the peak list are now associated with new phase. The peak profiles can only be further refined with a Pawley fit or phase fit. The peaks associated with the phase will be color coded in the Peak List

The refined cell parameters can be found in the entry for phase in the Refinement Control tab

✓ Click on the phase name in the Refinement Control tab to see additional information about the phase in the Object Inspector

4. Quantification of phases - Rietveld Refinement

4.1 Phase identification

In the "Analysis" field click on search & Mach and then on "Execute search & Mach" the window appears as shown in Figure 44.

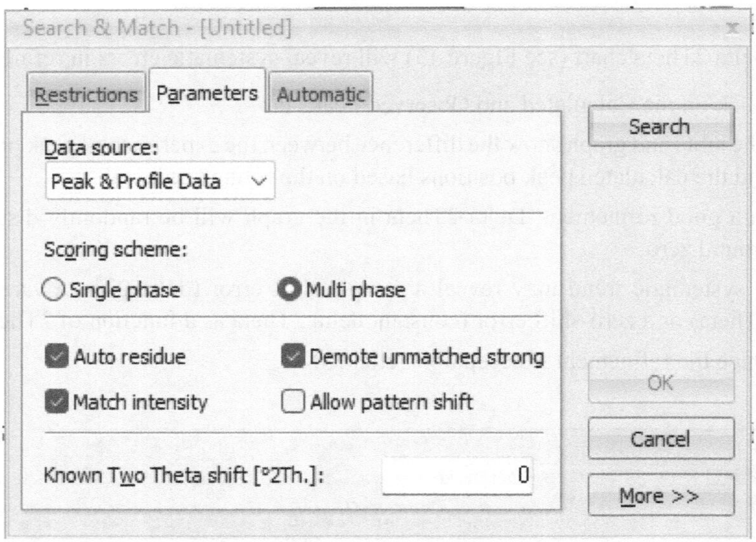

Figure 44. Peak – Profile Data.

We click on the Restrictions tab and then click Restriction set (see Figure 45), you can display the menu in Select restriction set. Select the restriction set that best fits your material; it can be Inorganic, Minerals, or other.

Click on Edit Restriction Set. In the pop-up box, on the "Subfiles" tab, select the most appropriate option for your sample type, for example: "Mineral".

In the same box, on the "Chemistry" tab, you can type the name of the chemical elements that you want to be identified in the crystalline phases.

You can drop down the menu at the bottom and change Default to a more suitable set, Minerals or another. If you do not see that menu at the bottom, click on the more tab on the right side. You can also click on Periodic Table to set which chemical

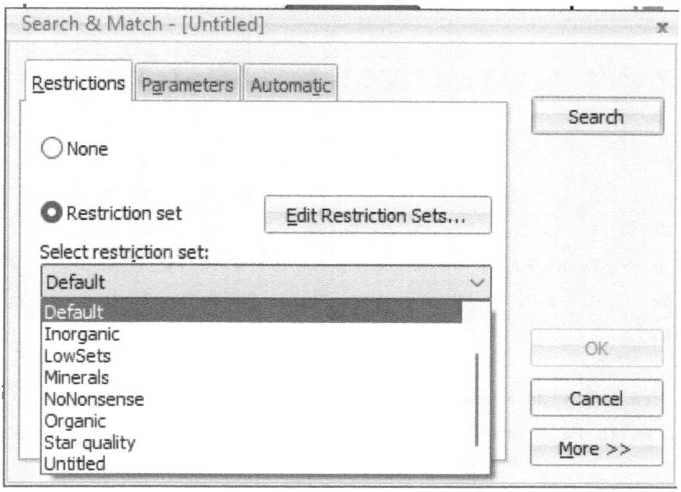

Figure 45. Restriction set.

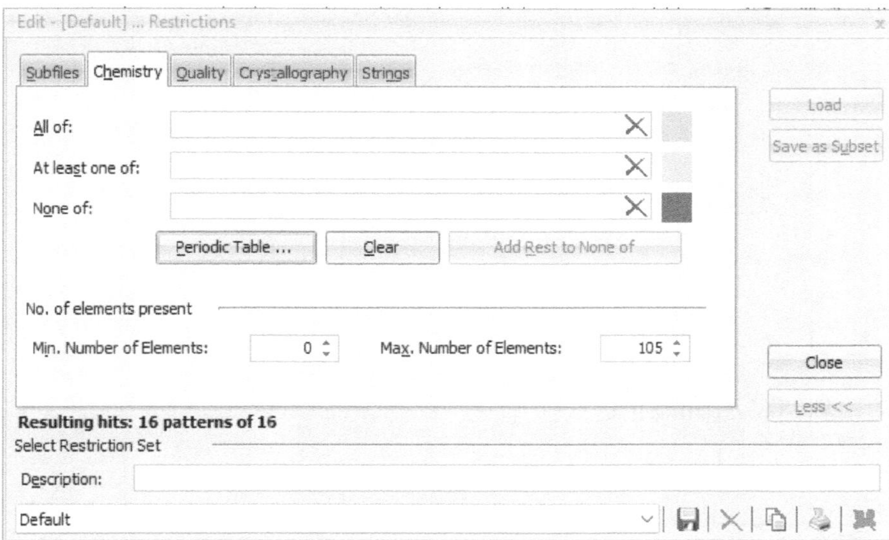

Figure 46. Edit Restriction Set - Chemistry.

(see Figure 46) elements are to be identified and which are not. With these instructions you tell the database which phases it must strictly identify (see Figure 47).

In the "Crystallography" tab, you can enter the phases to be identified manually if you know the detailed information of the unit cell, symmetry group, among others (see Figure 48).

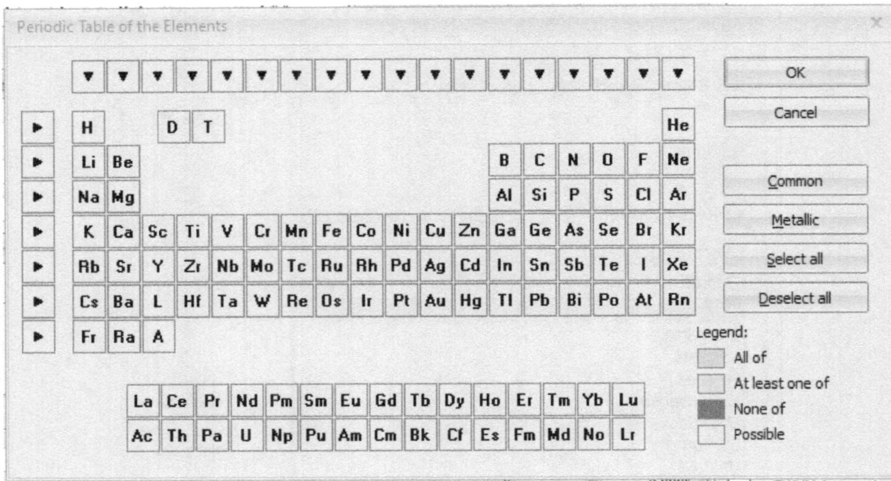

Figure 47. Edit Restriction Set – Periodic Table.

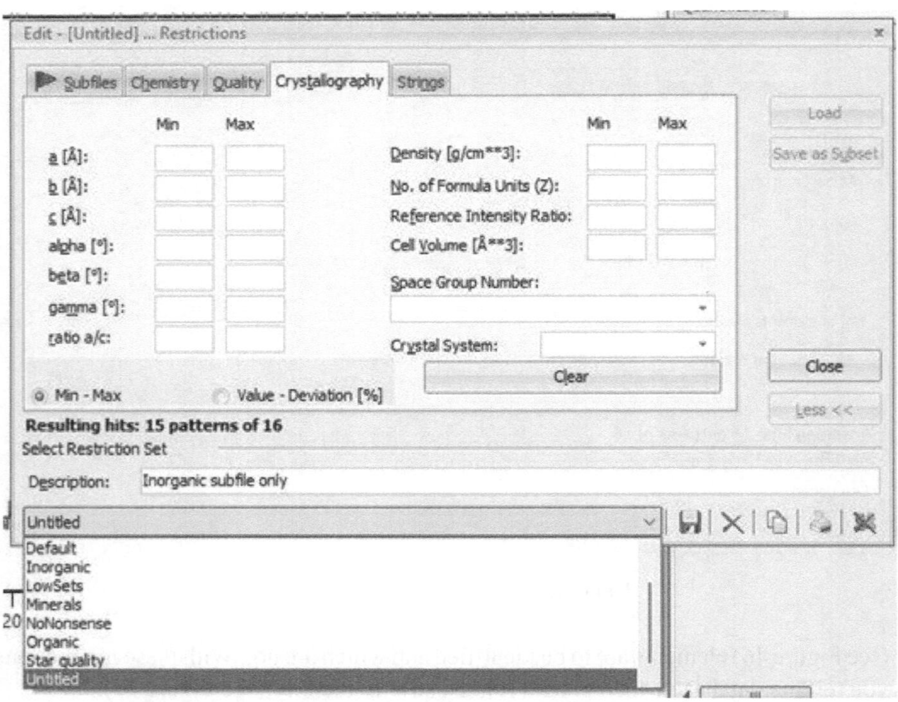

Figure 48. Edit Restriction Set – Crystallography.

After defining all the restrictions, click on Close and then click on Search and Ok. All the possible phases established by the database will appear. But from all these possibilities, the correct phases must be identified (see Figure 49).

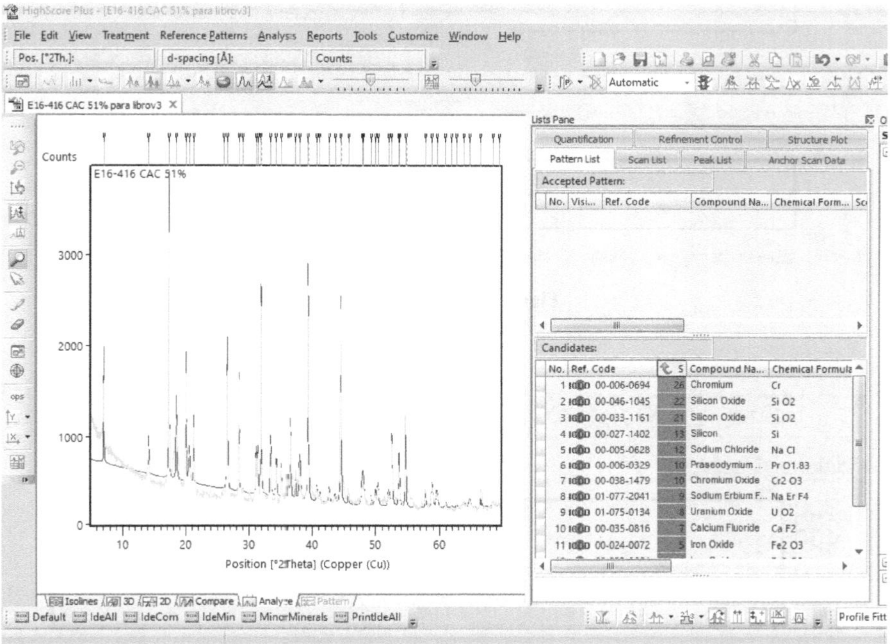

Figure 49. Proposed phases in the databases.

It is possible that the system has suggested some phases to you and they appear in the window at the top right. In this case, it appears in the name of the phase and with a long click drag the phase down,

4.2 Strategies for phase identification

It is possible that the software has selected some compounds, however, it is advisable to return these compounds to the list of possible candidates, for this in the "Lists Pane" box, in the "pattern list" tab at the top select the compound and with a long click, drag it to the bottom, repeat the procedure with each phase until the window is empty as shown in Figure 49.

1. Click on View field >> Panel Default Setting. Then in the bottom box that appeared, look there, and right-click, select Show Graphics >> Difference Plot (see Figure 50)

   Use this option when you have all the peaks identified. If the phases are well identified, this graph should tend to be linear as shown in Figure 51.

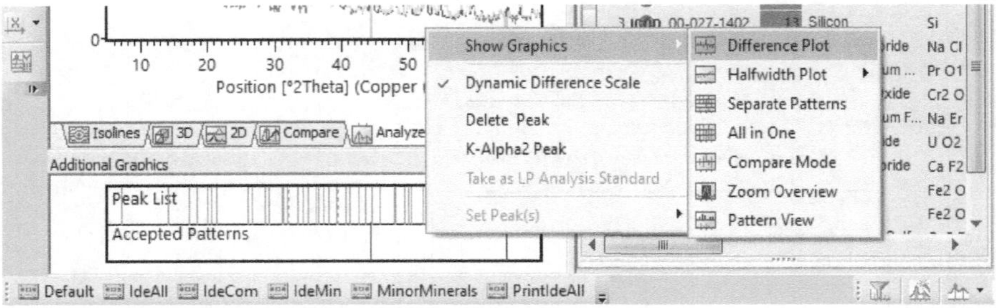

Figure 50. Difference plot.

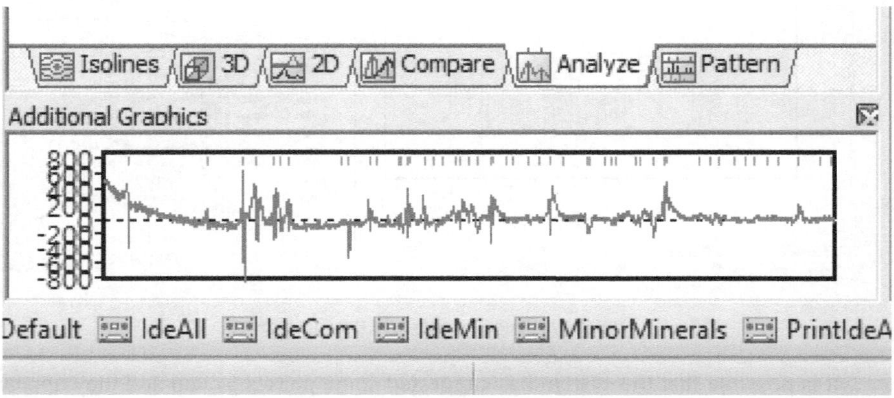

Figure 51. Difference plot – Identified Phases.

2. Double-click on one of the identified phases and a box will appear with the information for that phase as shown in Figure 52.

   Note that peak number 2 has an intensity of 100% at a 2 theta angle of 26.640. Make sure that this peak appears in your experimental diffractogram. One way to do this is to plot the experimental diffractogram in the OriginLab software (see Figure 53) and with the software's experiments, you can accurately locate if this peak is in the experimental diffractogram. This comparison between the experimental and the identified phase can be done with any software you want or in any way you want as long as it is a very precise method.

3. In the Lists Pane window, right click on the first candidate, select "Analysis Pattern lines" (see Figure 54).

   In the pop-up window, verify that the main peaks of the candidate compound match the peaks of the material under study (for this, in the "Matched" row, verify that the √ symbol appears in the box). In a well-identified phase, all or almost all of the cool little boxes should appear (see Figure 55).

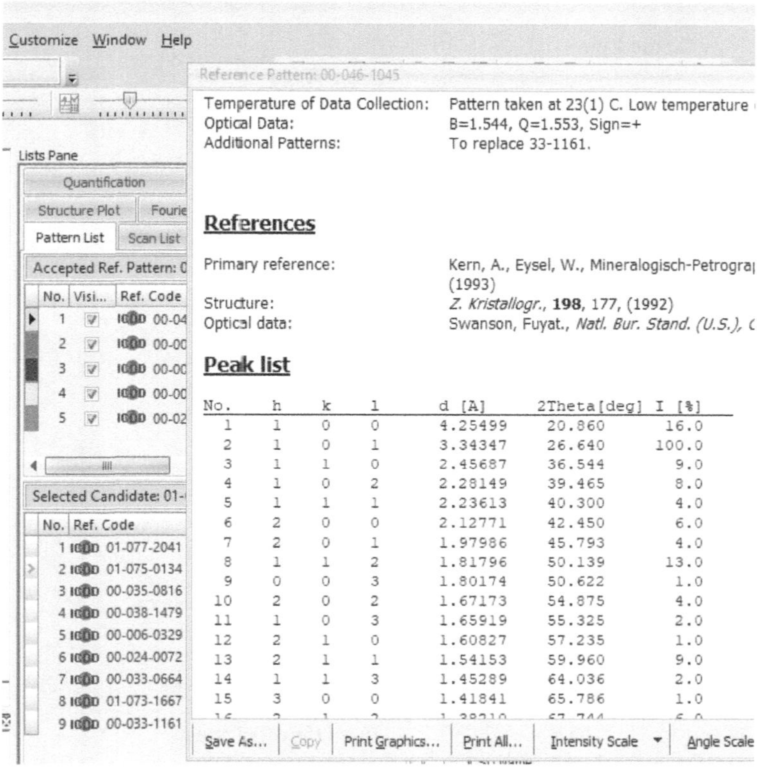

Figure 52. Information of an identified Phase.

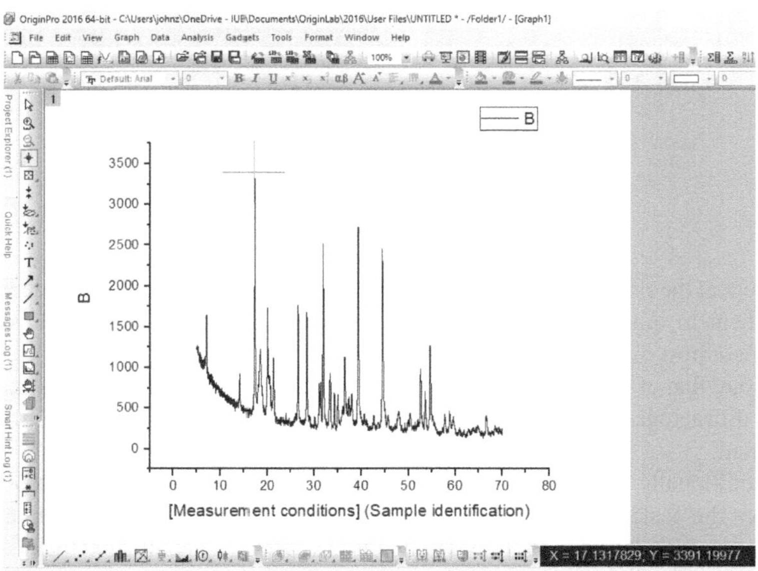

Figure 53. Identifying peaks with OriginLab.

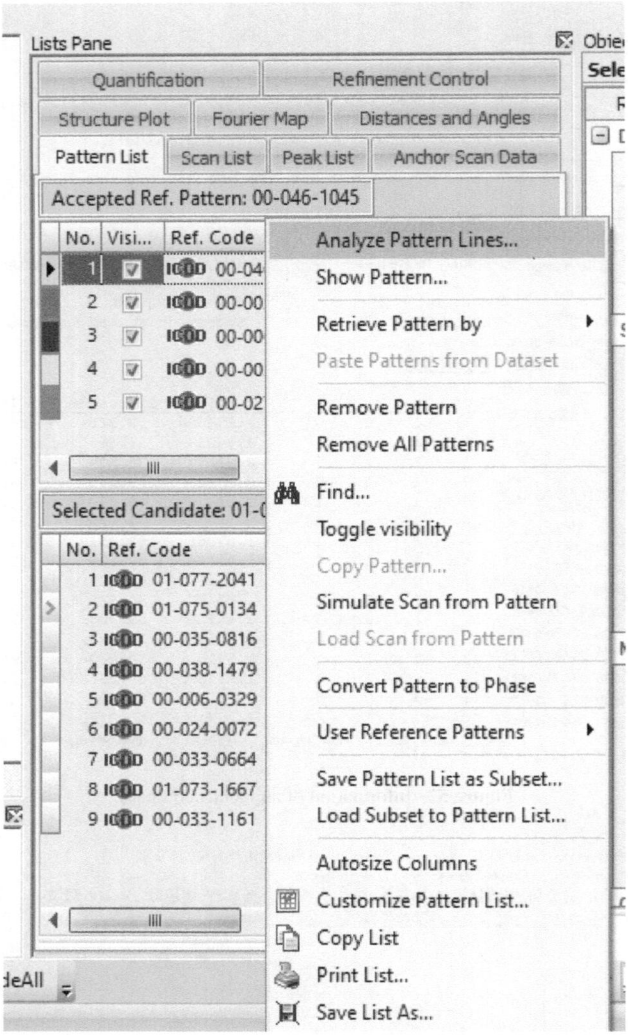

Figure 54. Analyze Pattern Lines.

Repeat the procedure with all candidates. For those with high peak coincidence, drag them to the top. Repeat the analysis with other compounds until you have identified most of the peaks in your sample.

Note, that at this point, most of the triangles above the orange lines at the top of the diffractogram have disappeared, indicating that most of the peaks have been identified.

Additionally, you can zoom in on the graph to check if there is correspondence between the peaks; to do this, click on a starting point on the diffractogram and then drag to select the zoom area; if everything is OK, you have completed the qualitative analysis.

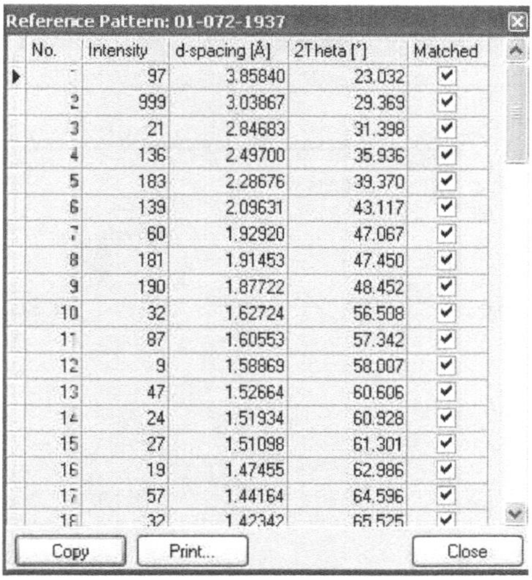

No.	Intensity	d-spacing [Å]	2Theta [°]	Matched
1	97	3.85840	23.032	✓
2	999	3.03867	29.369	✓
3	21	2.84683	31.398	✓
4	136	2.49700	35.936	✓
5	183	2.28676	39.370	✓
6	139	2.09631	43.117	✓
7	60	1.92920	47.067	✓
8	181	1.91453	47.450	✓
9	190	1.87722	48.452	✓
10	32	1.62724	56.508	✓
11	87	1.60553	57.342	✓
12	9	1.58869	58.007	✓
13	47	1.52664	60.606	✓
14	24	1.51934	60.928	✓
15	27	1.51098	61.301	✓
16	19	1.47455	62.986	✓
17	57	1.44164	64.596	✓
18	32	1.42342	65.525	✓

Figure 55. Well-identified phases.

4. Pattern comparison. Another way to check that the phases are well identified is to see the correspondence between the peaks of the experimental diffractogram and the peaks established by the software. To do this, click on the Pattern tab and a window will appear as shown in Figure 56.

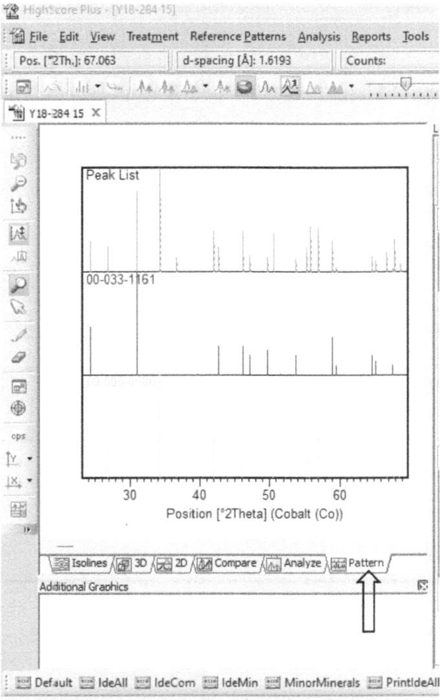

Figure 56. Experimental pattern comparison - software.

Now, you get a Semiquantitative result which needs to be refined see Figure 57.

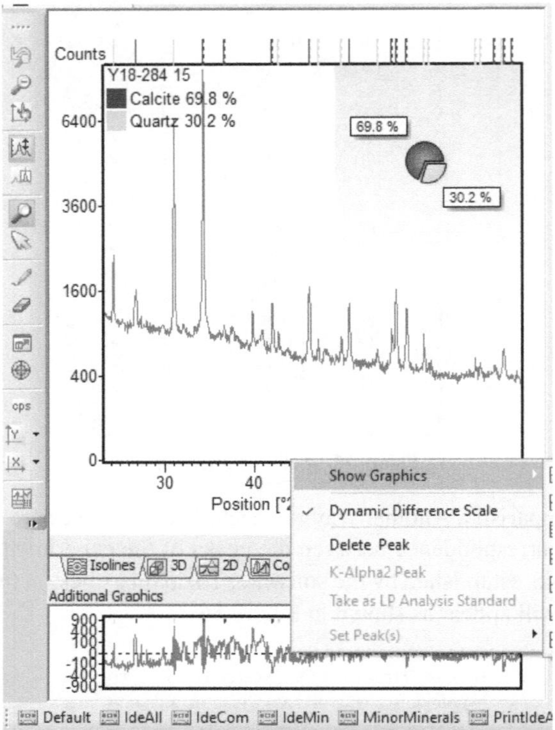

Figure 57. First quantification that needs to be refined.

In the Additional Graphics window, on the bottom left, right-click and select Show graphic >> Separate pattern (see Figure 58).

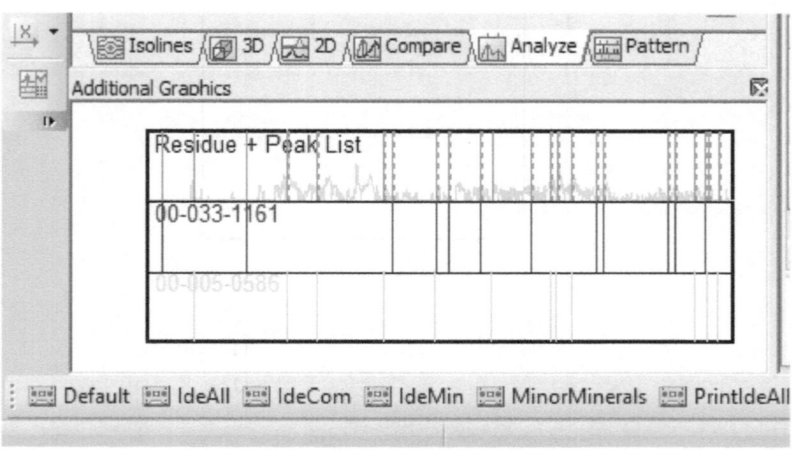

Figure 58. Correspondence between the identified and experimental peaks.

Click on the Refinement control. Play with the buttons at the top of the graph and you will see a clean diffractogram with the identified phases superimposed on the diffractogram peaks. You can click and hold to zoom in on the peaks to see the correspondence (see Figure 59).

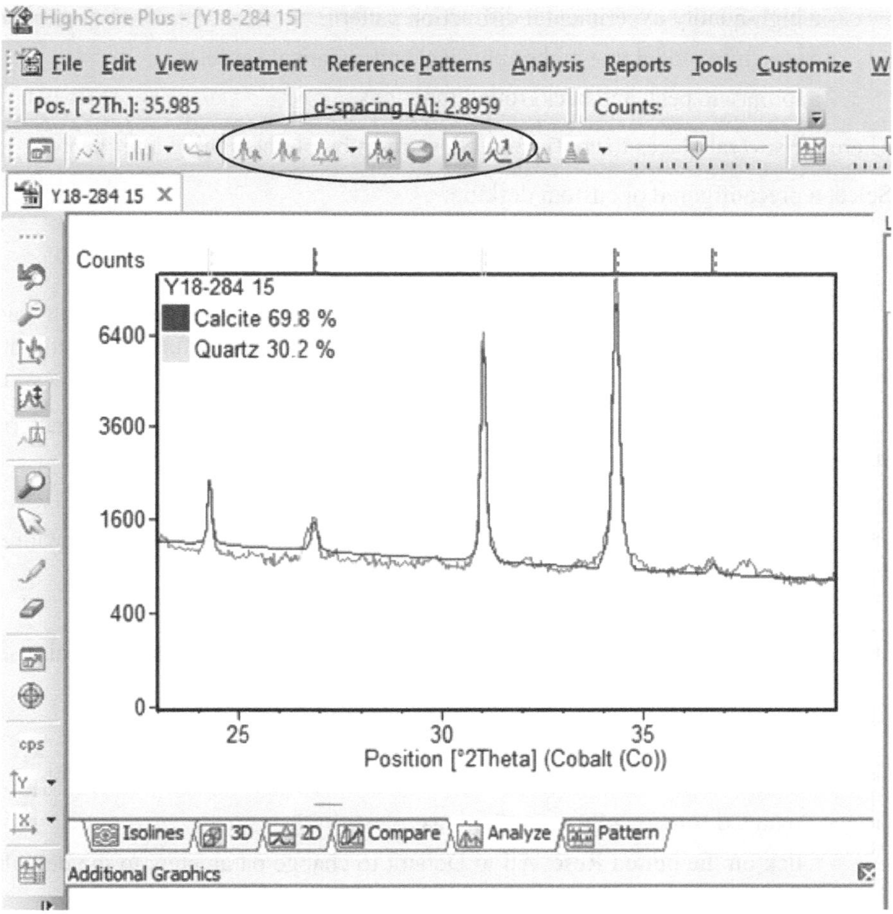

Figure 59. Clean diffractogram.

4.3 *Rietveld refinement*

Refinement by the Rietveld method in general terms is carried out with the following steps:

1. The parameters discussed above for profile fitting are used to create a model.
2. A diffraction pattern is calculated from the model.
3. The calculated pattern is compared to the observed data.

4. Differences between the observed and calculated patterns are minimized by varying the parameters in the model through a least squares process.
5. When the calculated and experimental diffraction patterns match, the model can represent the original sample.

The Rietveld method requires:

– a high-quality experimental diffraction pattern.

– a structure model that makes physical and chemical sense.

– appropriate peak and background functions.

There are several places to configure HSP settings before beginning a refinement.

Select a preconfigured or custom desktop:

– Go to View > Desktop to select a desktop.

– The “Structures” desktop is well suited for Rietveld refinement.

– A quirk of HSP: Imagine that the “Structures” desktop was selected but was then changed. In order to reset to the desktop to the “Structures” layout, you need to select another desktop and then select the “Structures” desktop again.

In Customize > Defaults you can set the starting values for several XRD pattern simulation parameters

– Default Instrument Settings

✓ Change these if you are routinely analyzing data that were not collected with a PANalytical diffractometer

– Default Global Settings

✓ Set the Background Method to the type you most often use (polynomial, use available background)

✓ Set the Profile Base Width (in Refinement) to 8 or more

The program settings contain several important options for Rietveld refinement.

Go the menu Customize > Program Settings

– Click on the button Reset All to Default to change parameters to the default behavior.

– After doing this, you might want to change a couple of settings to be better optimized for Rietveld analysis

✓ In the General tab

– Set Auto-save time and Number of Undo/Redo steps as you like

– Adding more undo steps will slow down the program

✓ In the Rietveld tab

– Check “Show selected Phase Profile”.

– Check “Calculate Correlation Matrix”.

- Check "Keep R-value Graphics Open after Refinement".
- Check "Undo Refinement Control Editing".

Do not process data before Rietveld refinement (Speakman 2010):

- ✓ Do not subtract the background before refinement
  - The background will be modeled during the Rietveld analysis
- ✓ Do not strip K alpha-2 peaks before refinement
  - The Kalpha-1 and Kalpha-2 peak doublet will be modeled by the profile function
- ✓ Do not smooth the data before refinement
  - Rietveld refinement relies on a statistically comparison to your raw data which will be invalid if your data are smoothed
- ✓ The only processing you should consider is converting ADS data to FDS
  - Use this option only if you collected data from an infinitely thick sample using Automatic mode with the Programmable Divergence Slits

PRINCIPLES OF RIETVELD REFINEMENT

The base parameters were:

- ✓ The peak shape,
- ✓ the peak width,
- ✓ and the preferred orientation.

You must define the instrument parameters and correct your data if variable slits were used.

The Divergence Slit Type is reported in the Object Inspector for the Scan

- In the Lists Pane, select the Scan List tab. This will load scan information into the Object Inspector
- In the Object Inspector, scroll down to the Instrument Settings area
- If your data were collected with a PANalytical instrument, these values are already defined.
- If your data were not collected with a PANalytical instrument, you should input the correct Divergence Slit Type, Anode Material, and Goniometer Radius.

- ✓ The Divergence Slit Type is:
  - Fixed for a fixed divergence slit aperture
  - Automatic for a variable divergence slit aperature
- ✓ If your data were collected using variable divergence slits, you must correct for the constant volume assumption
  - Variable slits are the Programmable Divergence Slit (PDS) in "Automatic Mode" on the PANalytical X'Pert Pro

You must correct your data if it was collected using variable divergence slots.

If your sample was 'infinitely' thick, then the irradiated volume was increasing. This can be corrected in HSP.

- Select the menu *Treatment > Corrections > Convert Divergence Slit*
- Load the default parameters
- Click the **ADS to FDS** button
- Click the **Replace** button
- Click the **Close** button

✓ If your sample was thin, then the irradiated volume was constant. In this case, you must change the instrument setting for your data in HSP.

- In the Lists Pane, select the Scan List tab. This will load scan information into the Object Inspector.
- In the Object Inspector, scroll down to the Instrument Settings area.
- In the entry for Divergene Slit Type, change it from Automatic to Fixed.

✓ This will tell HighScore Plus that your data fulfilled the constant volume assumption, even though variable slits were used.

If your sample was thin and you used fixed divergence slits, then the data cannot be analyzed by Rietveld refinement because they do not fulfill the constant volume assumption.

The reference card must be converted before it can be used as an initial basis for unit cell refinement (see Figure 60).

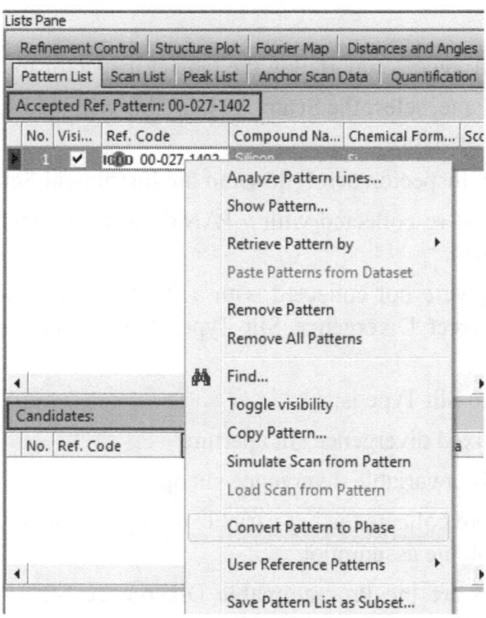

Figure 60. Convert Pattern to Phase.

After you have loaded the reference card(s), go to the Pattern List tab in the Lists Pane

- ✓ Right-click on the phase(s) that you want to refine
- ✓ Select "Convert Pattern to Phase" from the menu
- ✓ The reference(s) that you converted are now listed in the Refinement Control tab of the Lists Pane

STARTING A REFINEMENT

Calculate the diffraction pattern

  – Click the "Start Pattern Simulation" button in the Rietveld toolbar, or

  – Select *Analysis > Rietveld > Start Pattern Simulation*

- ✓ The total intensity will not match because the scale factor is wrong
- ✓ Make sure that peak positions match up approximately
- ✓ The range and step size for the simulation will be automatically set to match the experimental data

Semi-automatic refinement mode gives you control over which parameters are refined.

Set the Refinement to Semi-automatic Mode

  – Use the drop-down menu in the Rietveld toolbar

  – Use the drop-down menu in the menu *Analysis > Rietveld > Refinement Mode*

- ✓ How to flag parameters for refinement

  – In the Lists Pane, select the Refinement Control tab (see Figure 61)

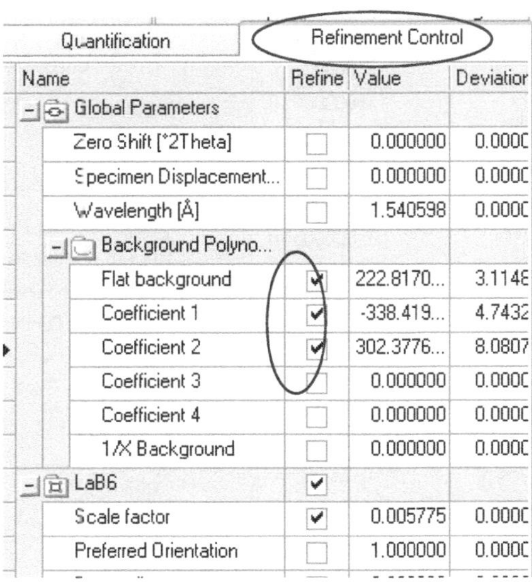

Figure 61. Refinement Control.

- To mark a parameter for refinement, put a check mark next to the parameter in the 'Refine' column
- Then click on the "Start Rietveld Refinement" button

Set refinement to semi-automatic mode

- Use the drop-down menu on the Rietveld toolbar
- Use the drop-down menu on the Analysis > Rietveld > Refinement Mode menu

How to mark parameters for refinement

- In the list pane, select the Refinement Control tab.
- To mark a parameter for refinement, place a checkmark next to the parameter in the "Refine" column
- Then click the "Start Rietveld refinement" button.

✓ Define how the background will be fit (see Figure 62).

- You must decide how you are going to fit the background.
- You could manually fit the background and use that during the refinement.
- You could fit the background using a polynomial or other equation and allowing the background parameters to refine during the Rietveld refinement.

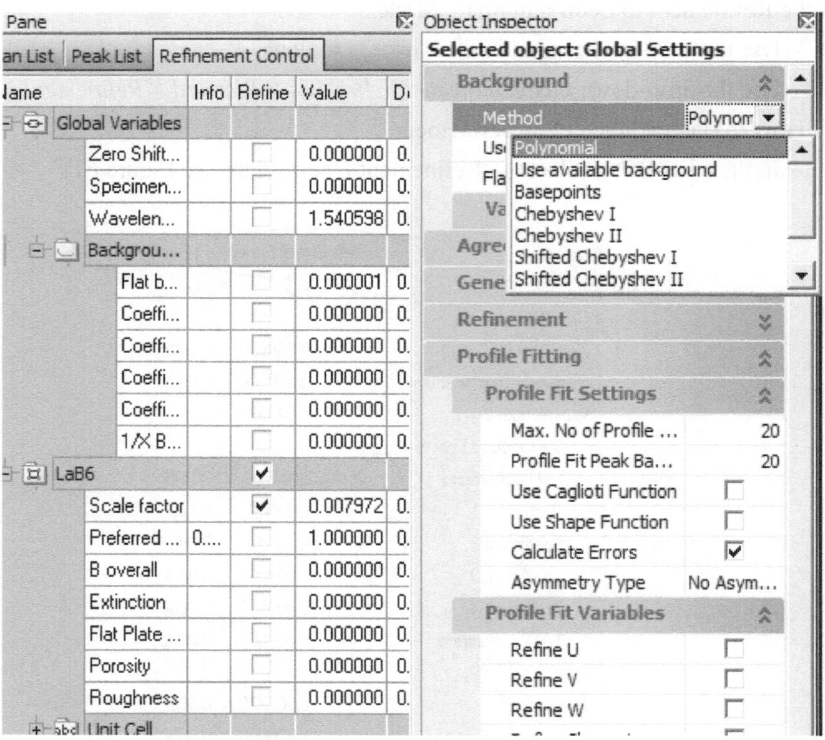

Figure 62. Setting the Background.

✓ To change the background method:
   – In the Refinement Control Lists Pane, select Global Variables.
   – In the Object Inspector, select an option from the drop-down menu for the Background Method.

✓ This menu is near the top of the list in the Object Inspector.

When you start a refinement, don't refine too many parameters at the same time. When you start a refinement, you refine only a few parameters at a time. If you refine too many parameters at the same time, the refinement will diverge, that is, fail. The computer only cares about minimizing the least squares residual.

THE STEPS FOR RIETVELD REFINEMENT

1. First, refine only the scale factor. Place a checkmark next to the scale factor. Click the Start Refinement button (see Figure 63). The calculated peak intensities should now match the experimental data. The calculated peak intensities may be different if there is a substantial peak shift, such that the calculated peak positions do not match the observed peak positions.

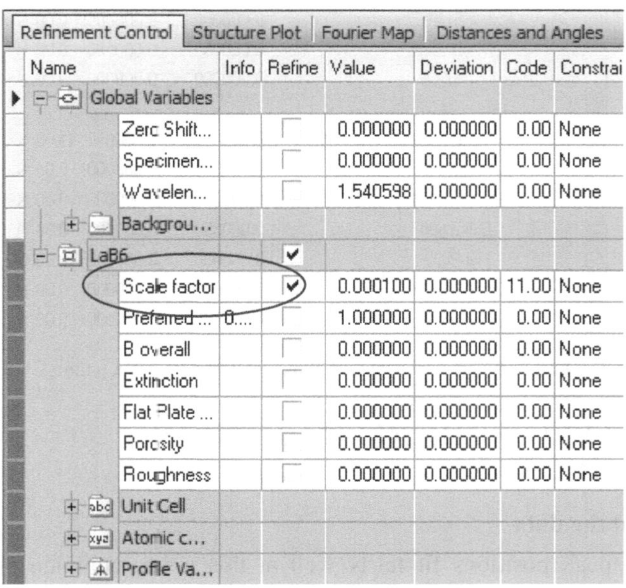

Figure 63. Parameters to refine one at a time.

Refine the background, adding one additional term at a time

2. Scale Factor + Flat Background

   sometimes it helps to change the y-axis to square root scale to better evaluate the background fit

3. Scale Factor + Flat Background + Coefficient 1

4. Scale Factor + Flat Background + Coefficient 1 + Coefficient 2
5. Scale Factor + Flat Background + Coefficient 1 + Coefficient 2 + 1/X Background

The increase in background at low angles of 2theta can be well modeled using the 1/X Background term

That is, I refine with one coefficient, then I go back and refine with the other coefficient, and do the same with the third coefficient (see Figure 64).

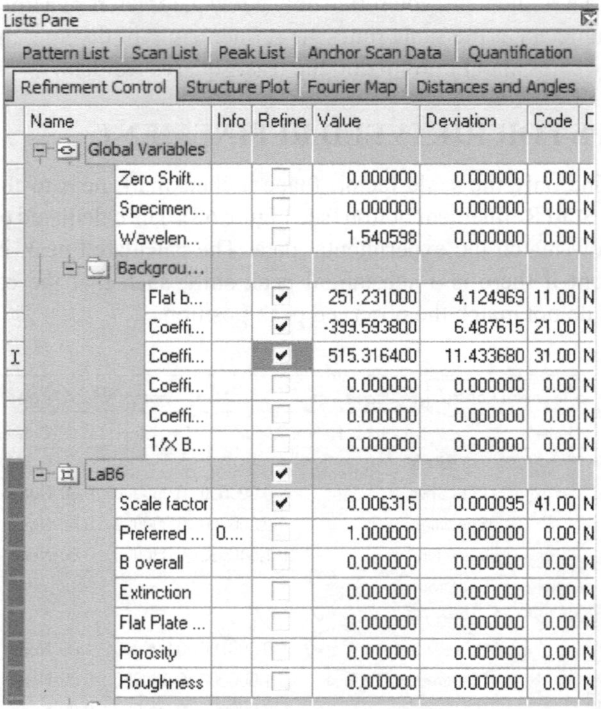

Figure 64. Refinement with three coefficients, one at a time.

HOW TO TELL IF THE REFINEMENT IS PROCEEDING WELL

First, look at the data

– If the peak positions fit fairly well at this point, continue improving the background fit.

– If the peak positions don't fit well at this point, refine peak positions starting with specimen displacement.

Set Additional Graphics to show 'Difference Plot'

– Go to *View > Additional Graphics > Difference Plot.*

– This plot will show the difference between the observed and calculated diffraction patterns.

Also Use Agreement Indices to Evaluating the Refinement (Speakman 2011)

1. The Agreement Indices quantify how well the calculated XRD pattern fits the experimental data:
   - Click on "Global Parameters" in the Refinement Control list.
   - The Agreement Indices are shown in the Object Inspector.
2. Rexpected shows the best possible value for the Residual:
   - A lower Rexpected value indicates higher quality data.
3. R profile and Weighted R profile show how well the calculated pattern matches the experimental data":
   - We want the Weighted R profile to be less than 10% and as close to the R expected value as possible.
   - R expected is an estimation of the best possible R profile based on the statistical noise of the experimental diffraction pattern.
4. GOF (goodness of fit) is Rwp/Rexp:
   - This should approach 1.

GOF IS MORE AND MORE IRRELEVANT BECAUSE MODERN DETECTORS MESS UP THE STATISTICAL VALIDITY

We need to refine the maximum position

5. Previous parameters + Specimen Displacement

✓ displacement-peak shift varies as *cos θ*

$$\Delta 2\theta = -\frac{2s\cos\theta}{R}(in\,radians)$$

✓ zero shift-peak shift constant vs θ

NEVER, NEVER, NEVER REFINE BOTH AT THE SAME TIME!!!

✓ Usually, you should refine lattice parameter next, but we can see that the peak width is the more significant error

With the peak and background positions fine-tuned, begin refining the peak profiles.

✓ Keeping all previous parameters checked, start refining profile parameters:
   - We keep refining the peak position parameters because as the peak width and shape changes the peak position may change in response.
   - Initially, we only refine 1 profile parameter at a time.

✓ If we try refining too many too soon, the refinement will diverge.

6. Start refining Cagliotti parameters one at a time:
   - W only
   - V only
   - U only
7. Refine Peak Shape 1 (not refining any Cagliotti parameters).
8. Look at the lowest angle peaks to determine if you need to refine the peak profile asymmetry:
   - In this case, we do need to refine the asymmetry parameter.
   - Repeat steps 6–8, refining parameters one at a time, a few times.

Look at the high angle peaks_notice how both Ka1 and Ka2 are modeled. If the Ka2 ratio is out of sync, you may need to change the polarization correction. To find the POL values, check the help (search for the keyword polarization). The values are entered as follows: Click Global Parameters (in the Refinement Control list). In the Object Inspector, look for Polarization Correction Coefficient under General Properties. In this case, it is correctly set to 1. Look at a low angle peak – notice if the peak profile is truncated. Change the values: Click Global Parameters (in the Refinement Control list). In the Object Inspector, look for Profile Base Width under Refinement. Change it to 8, run the refinement again. You can change it "permanently" in Customize > Defaults.

In addition to the agreement indices, we can look at the estimated standard deviation to assess the accuracy of the refined parameters. Look at the columns: value, deviation, maximum, minimum, use min/max, constraint, lastchange. I usually rearrange the columns in the order shown below (just drag and drop the column headers), a small deviation value for the network parameter indicates that the refinement is fairly stable and we are getting close to a good solution.

Profile Base Width: Limits the 2theta range over which the intensity contribution of a peak (*hkl*) is calculated, assuming that the intensity contribution of a peak is zero beyond this range, base width values restrict the range to multiples of the FWHM.

Maximum asymmetry correction angle: Peaks below this angle are corrected for asymmetry. Peaks above this angle are assumed to be symmetrical.

Intensity limit: Refinement does not calculate or refine using peaks with a calculated intensity less than the intensity limit.

- The profile function used to calculate the shape of all diffraction peaks.
- The weighting scheme puts greater emphasis on Iobs or Icalc in the refinement residual.
- Maximum number of least squares cycles.

✓ Maximum number of times parameters are refined before refinement stops (assuming other criteria do not cause refinement to stop).

✓ If this number is set too low, refinement may cease before the parameters are truly optimized.

✓ If this number is set too high, refinement may take much longer than necessary or may be unstable.

Before we continue improving the profile fit, lets remind ourselves what is currently being refined:

✓ Right-click on any value in the Refinement Control list.

✓ Choose "Show Refined Values/Constraints ... "

✓ The table shows us what parameters are currently checked to be refined.

To finish the refinement, we need to begin refining multiple profile parameters at the same time:

✓ Peak width and profile parameters are going to affect each other.

✓ For a complex refinement, we would go through several iterations of refining one parameter at a time.

✓ Refine W only, then V only, then U only, then Shape 1 only, then W only, then V only, then U only, then Shape 1 only, then asymmetry only, then W only, then V only, then U only, then Shape 1 only, then asymmetry only ... .

✓ Then begin refining some parameters together at the same time.

✓ Because this is a high-quality experimental pattern of a simple and well crystallized sample, we can begin refining multiple profile parameters at the same time.

1. W
2. W+V
3. W+V+U

Notice large changes in these parameters as they influence each others:

4. W+V+U + Peak Shape 1
5. W+V+U + Peak Shape 1 + Asymmetry

REFINING THERMAL PARAMETERS

✓ Stop refining all profile parameters

✓ Right-click to access shortcuts from seeing all parameters currently being refined. Turning multiple parameters on (refine) or off (fix) at the same time.

– Refine the B isotropic, that is, B = 0 or negative value.

✓ Max/min is automatically set to allow only a minimum value of 0.

– If B keeps going to zero, it is probably really trying to go negative but the minimum constraint does not let it.

SAVE ANALYZED DATA IN THE HPF FORMAT

- ✓ Save the result as a *.hpf file.s
- ✓ This file format bundles together the original data, the results of all analyses done on the data, and the history of the data analysis.
- ✓ You can view the history of data analysis in *File > Properties.*

Limitations of Rietveld Refinement and the Case of Portland Cement

Quantitative phase analysis using the Rietveld method has the following limitations:

1. Only phases for which the crystalline structure is available can be quantified.
2. The method is useful for simple systems of phase mixtures, when the number of phases is very high, it becomes very complex.
3. During the refinement all the phases, presents must be included, since in the quantitative analysis the sum is a constant value (1 or 100).
4. The presence of a non-crystalline part in the sample in relatively small quantities makes quantification very difficult because the error produced in the calculation can be of the same order of magnitude.
5. The different linear absorption coefficients of the phases with respect to the sample must be considered when evaluating the results.

The Rietveld method encounters the difficulty of having a large multiplicity of phases in the study of Portland cement. Commercial software packages typically only support up to 16 different phases in a model, due to the computational cost of including a larger number.

This quantity may be insufficient, especially if the phases of partially hydrated cement are considered. Although the four main phases of clinker are the majority in the samples, an incorrect simulation of other phases present can produce important variations in the mineralogical quantification. The application of the Rietveld method for the study of cements has shown significant advances in recent years, generally showing good conditions of repeatability and reproducibility (León-Reina et al. 2009, Le Saoût et al. 2011, De la Torre and Aranda 2003).

Furthermore, the application of the method requires a high level of qualification of the operator in the methodology. The greatest limitations are related to the ambiguities related to accidental or systematic overlapping in the XRD of the powder. Therefore, its application requires a thorough knowledge of crystallography to be able to overcome these obstacles.

For the construction of the structural model, it is necessary to define the spatial arrangement of each compound, the atomic positions, the cell parameters and the factors of thermal agitation, anisotropy and occupation. These phases present structural differences according to the different methods and processes of obtaining the minerals. In commercial cements, logically there are alite (A) (in polymorphic variants), belite (B) (in polymorphic variants), ferrite (F) and aluminate (in polymorphic variants, Ac, A and Am), but also lime (L), pericas (Pe), gypsum (Y), hemihydrate (H), anhydrite (N), portlandite (P), calcite (C), dolomite (D), quartz (Q), arcanite (Ar), aphthitalite (Af) and syngenite (S).

In the added cements, amorphous phases may also appear, which present bulges and not characteristic peaks in the DRX, and in which case it is necessary to incorporate a known quantity of an internal standard to avoid over-evaluation of the other phases (Scrivener et al. 2004).

There are many factors that make quantification by the Rietveld method difficult. Firstly, sample preparation plays an important role: a cement sample with large grain sizes results in wider peaks that overlap with other peaks, making refinement difficult. For anhydrous cement, grain sizes smaller than 10 μm are recommended (Stutzman 2010). This contributes to phase reproducibility, reducing the effect of preferential orientation and reducing the absorption of radiation by ferrite. On the other hand, the reduction of the effects of preferential orientation can also be achieved with a good arrangement of the powder in the diffractometer, which implies that the conformation of the pellet when it is pressed is conveniently done vertically.

Another aspect to take into account is the polymorphism of some phases. For example, alite is normally monoclinic, but, depending on the clinker manufacturing process and the SO_3 and MgO contents present in the raw material and in the fuels used, it can be present in triclinic form.

(Polymorphs known as T1, T2 and T3), in monoclinic form with different atomic arrangements (M1, M2 and M3), or as a combination of various forms (usually M1 and M2) (De Noirfontaine et al. 2006). The same occurs with tricalcium aluminate, which depending on the NaO content in the clinker can be in cubic form (Ac) (the most common), in orthorhombic form (Ao), or in monoclinic form (Am) (Gobbo et al. 2004). Belite can also be in monoclinic or orthorhombic form depending on the cooling process of the clinker (Barbier and Hyde 1985).

If all possible polymorphs of all phases are included separately in the structural model, it would become a very computationally expensive process. Another possible effect lies in the variation of the occupancy factor (which defines the level of substitution in the structure by other minor elements) of the Al and Fe atoms in tetracalcium ferroaluminate, and although the refinement of these parameters can improve the model fit, it has little impact on the quantification (Le Saoût et al. 2011).

Microabsorption of radiation by some phases can also affect quantification. Studies carried out with clinker samples under standard diffractometers and compared with synchrotron (De la Torre and Aranda 2003) have shown that ferrite is normally under-quantified in weight due to its higher microabsorption of radiation than the rest of the phases. Finally, the preferential orientation appears as a factor that directly influences the quantification by the Rietveld method, since if not considered it can distort the coupling of the peaks. It is therefore important to take special care with the phases and planes that are known to present preferential orientation: alite, calcite, gypsum, basanite, anhydrite and dolomite (Le Saoût et al. 2011).

Based on all the factors described above, the outcome of the refining process can be significantly affected by the operator and the procedure he performs.

Simulations do not necessarily represent the data your diffractometer will produce. There are more factors that contribute to the diffraction pattern you collect with an instrument. Some factors are calculated from fundamental equations.

Simulations automatically take into account:

- ✓ Multiplicity
- ✓ Kα2 radiation
- ✓ Lorentz polarization factor
- ✓ Some factors are approximated by empirical formulas
  - – The width and shape of the diffraction peak are a product of the diffractometer optics, sample preparation, and sample microstructure.
  - – The widths and shapes of the diffraction peaks in these simulations have been calculated using predetermined parameters in an empirical formula.
- ✓ Some programs attempt to model the instrument contribution to peak width and shape using "Fundamental Parameters". This is not yet implemented in HSP.

Other things that affect intensity

$$I_K = SM_K L_K |F_K|^2 P_K A_K E_K$$

- ✓ S is an arbitrary scale factor—used to adjust the relative contribution of individual phases to the overall diffraction pattern
- ✓ M is the reflection multiplicity—considers the fact that some observed diffraction peaks are actually the product of multiple equivalent planes diffracting at the same 2theta position (e.g. (001) (100) (010) etc. in cubic) automatically calculated based on the crystal structure
- ✓ L is the Lorentz polarization factor.
- ✓ P is the intensity modification due to preferred orientation.
- ✓ A is the absorption correction.
- ✓ E is the extinction correction.
- ✓ F is the structure factor, which is the amplitude of scattered light due to the crystal structure (already discussed).
- ✓ The Lorentz polarization factor corrects for various effects

$$L_p = \frac{1 + POL\cos^2 2\theta}{\sin^2\theta\cos\theta}$$

Additional intensity factors

- ✓ Absorption correction

  considers the transmission and absorption of X-rays through the irradiated volume of the sample, depends on the geometry of the sample: significant for cylindrical or transmission samples with a high sample absorption; also significant for a flat plate sample with a low sample absorption.

- ✓ Extinction correction

  secondary diffraction of scattered X-rays off a large perfect crystal can decrease the observed intensity of the most intense peaks scattering by higher grains

limits the penetration depth of most X-rays, resulting in a smaller irradiated volume "ideally imperfect" crystals, grains with sufficient mosaicity to limit extinction, are desired; it is best to reduce extinction by grinding the powder to a finer size.

These two corrections are rarely used.

We now know how to calculate the diffraction peak intensity, but there a couple of more factors involved in simulating real diffraction data.

✓ The intensity, Y_{ic}, of each individual data point i is calculated using the equation:

$$Y_{ic} = Y_{ib} + \sum_{k=k1}^{k1} G_{ik} I_k$$

We already know how to calculate I_K, the intensity of th e Bragg diffraction peak k:

$$I_K = SM_K L_K |F_K|^2 P_K A_K E_K$$

✓ Y_{ib} is the intensity of the background at point i in the pattern.

✓ $k1 - k2$ are the reflections contributing to data point i in the pattern.

- Sometimes multiple Bragg diffraction peaks overlap, resulting in multiple contributions to the observed intensity at a single data point.

✓ G_{ik} is the peak profile function

- This describes how the intensity of the diffraction peak is distributed over a range of 2theta rather than at a single point.
- This profile is due to instrument broadening, sample broadening, etc.

Diffraction peaks have profiles that must be empirically modeled

- The intensity equation I_K predicts that diffraction peak intensity occurs at a single point 2theta but in reality, the diffraction peak intensity is spread out over a range of 2theta. The total area of the diffraction peak profile is the predicted intensity I_K.
- Diffractometers contribute a characteristic broadening and shape to the diffraction peak based on their optics and geometry. Each separate combination of divergence slit apertures, Soller slits, beta filter or monochromators, and detectors will have their own characteristic.

SAMPLE ATTRIBUTES ALSO CONTRIBUTE TO THE DIFFRACTION PEAK SHAPE AND WIDTH

✓ Mass absorption coefficient (sample transparency)

✓ Sample thickness

✓ Crystallite size

✓ Microstrain

✓ Defect concentration

Diffraction peaks have profiles that must be modeled empirically. The intensity equation I_K predicts that the intensity of the diffraction peak occurs at a single 2theta point, but the intensity of the diffraction peak is spread over a 2theta range. The total area of the diffraction peak profile is the predicted intensity I_K.

Diffractometers contribute a characteristic broadening and shape to the diffraction peak depending on their optics and geometry, each separate combination of divergence slit apertures, Soller slits, beta filters or monochromators and detectors will have its own characteristic instrument profile.

Sample attributes also contribute to the shape and width of the diffraction peak:

- mass absorption coefficient (sample transparency)
- sample thickness
- crystal size
- micro strain
- concentration of defects

PROFILE PARAMETERS

Rietveld refinement assumes that peak profiles vary systematically as a function of 2theta. The peak profile can also vary according to the crystallographic direction [*hkl*]. The Cagliotti equation describes how the peak width varies with 2theta

$$H_k = (U tan^2 \theta)^{1/2} + V \tan \theta + W$$

H_k is the Cagliotti function where U, V and W are refinable parameters

- ✓ HSP treats the Cagliotti equation as a convolution between instrument and sample broadening functions
- ✓ The change in U and W between an instrument profile standard (e.g., NIST 640c) and your sample can indicate the amount of nanocrystallite size and microstrain broadening

SYSTEMATIC VARIATION OF PEAK WIDTH WITH 2THETA

For a standard Bragg-Brentano diffractometer U and W will be positive and V will be negative. This trend may vary for data collected with parallel beam, GIXD and capillary geometries. Diffraction Peaks from laboratory diffractometers are a mixture of Gaussian and Lorentzian profiles.

Several different functions can be used to describe the mixing of Gaussian and Lorentzian contributions

- ✓ The profile function is set in the Object Inspector for Global Parameters (found in Refinement Control list)
  - the same profile function is used for all phases in a single pattern
  - Pseudo Voigt and Pearson VII are the most used
  - Pseudo Voigt

– Pseudo Voigt with FJC Asymmetry correction
– Pearson VII
– Voigt

The Pseudo Voigt equation uses up to 3 parameters to describe Gaussian and Lorentzian components

$$G_{jk} = \gamma \frac{2}{H_k \pi}(1+4X_{jk}^2)^{-1} + (1-\gamma)\frac{C_1^{1/2}}{H_k \pi^{1/2}}\exp(-C_1 X_{jk}^2)$$

$$\gamma = \gamma_1 + \gamma_2\, 2\theta + \gamma_3\, (2\theta)^2$$

The parameters γ_1, γ_2, and γ_3 are the refinable shape parameters, listed as Peak Shape 1, Peak Shape 2, and Peak Shape 3 in the Refinement Control list. Usually, Peak Shape 1 is the only one used

✓ H_k is the Cagliotti peak width function
✓ C_1 is $4\, ln\, 2$
✓ X_{kj} is

$$X_{jk} = \frac{(2\theta_i - 2\theta_k)}{H_k}$$

Background can be empirically fit with an equation, manually fit, or evaluated with a blank sample (see Figure 65)

✓ refinement of a polynomial function
✓ refinement using a type I or II Chebyshev polynomial function
✓ By amorphous sinc function
✓ linear interpolation between base points
✓ independent data collection run without the sample in place
  – It is difficult to use this technique effectively
  – It requires careful evaluation of the validity of the background fit

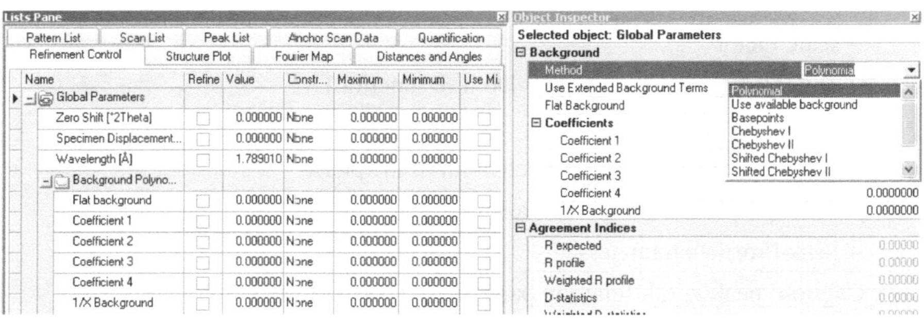

Figure 65. Background fit.

Systematic error in diffraction peak positions can also be calculated (see Figure 66)

- ✓ The systematic error functions are found in Global Parameters
- ✓ They apply to all phases in the sample
  - – Zero shift
  - – Specimen displacement
  - – Wavelength

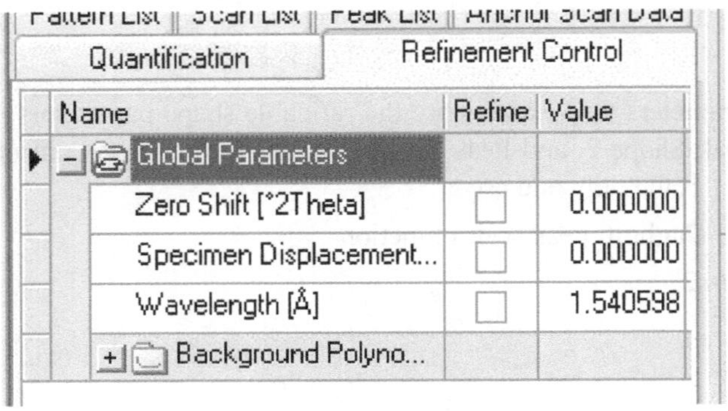

Figure 66. Zero Shift – Specimen Displacement.

THE PARAMETERS OF A PATTERN SIMULATION

- ✓ Global parameters affect the entire sample rather than individual phases
  - – Background function
  - – Zero Shift
  - – Specimen Displacement
  - – POL (Lorentz polarization factor)
  - – Sample Geometry
  - – selecting the Profile Function
- ✓ Other parameters are set individually for each phase
  - – scale factor
  - – preferred orientation (March-Dollase function)
  - – unit cell
  - – atomic parameters are for each atom in the phase
- ✓ position, SOF, thermal parameter
  - – Peak Profile parameters
- ✓ Cagliotti peak width function parameters U, V, and W
- ✓ Profile shape parameters

- ✓ peak asymmetry profile
- ✓ anisotropic broadening function

USING A TEMPLATE TO SIMPLIFY ANALYSIS OF SIMILAR SAMPLES

A template can be used as a starting point for the analysis of multiple datasets that are similar

- ✓ For the template approach, you will refine one dataset that is representative of the group that you want to analyze
  - You might use Cluster analysis to sort and identify similar datasets and the most representative of those datasets
- ✓ After refining the first dataset, you will create a template file to serve as the starting point for all additional refinements
  - A template file is an empty HPF document that contains several settings, such as:
- ✓ The Instrument Profile Calibration Curve (known as a Size-Strain Standard in HSP)
- ✓ Reference patterns
- ✓ Peaks in the peak list
- ✓ Phases for refinement

CREATING A TEMPLATE

- ✓ The template file should be empty of all information that may change significantly between datasets or that you want to recreate from scratch for each analysis step
  - Peak List
- ✓ You may keep the peak list if you will use this as the starting point of profile fitting for samples that contain the same phases
- ✓ You might delete this if you will run a peak search for each dataset
  - Go to the Peak List tab in the Lists Pane
  - Right-click in the Peak List and select the menu option *Delete > Included Peaks*
  - Refinement Control
- ✓ Global Variables such as background and specimen displacement might change between datasets since this depend on individual sample preparation
  - If sample preparation is well controlled, these might be consistent between datasets
- ✓ To remove Global Variables
  - Go to the Refinement Control tab in the Lists Pane
  - Expand the entries Global Variables and Background

– Set the values of Zero Offset and Specimen Displacement to 0

– In every parameter within Background (Flat Background, Coefficient 1, etc.), set the value to 0

✓ Remove Phases in the Refinement Control if different specimens contained different phases

– Pattern List

✓ The template might contain reference patterns for phases that you expect to see in all/most datasets.

✓ To remove reference patterns

– Go to the Pattern List in the Lists Pane

– Right-click and select "Remove all Patterns" or remove individual patterns by selecting "Remove Pattern"

– Go to the Scan List in the Lists Pane

✓ Delete all experimental scans loaded in the Scan List

✓ Save the document in a *.HPF format with a name like "Data Template.hpf"

References

Barbier, J. and Hyde, B.G. (1985). The structures of the polymorphs of dicalcium silicate, Ca_2SiO_4. Acta Crystallogr. Sect. B Struct. Sci. 41: 383–390.

De la Torre, A.G. and Aranda, M.A.G. (2003). Accuracy in Rietveld quantitative phase analysis of Portland cements. J. Appl. Crystallogr. 36: 1169–1176.

De Noirfontaine, M.-N., Dunstetter, F., Courtial, M., Gasecki, G. and Signes-Frehel, M. (2006). Polymorphism of tricalcium silicate, the major compound of Portland cement clinker: 2. Modelling alite for Rietveld analysis, an industrial challenge. Cem. Concr. Res. 36: 54–64.

Gobbo, L., Sant'Agostino, L. and Garcez, L. (2004). C3A polymorphs related to industrial clinker alkalies content. Cem. Concr. Res. 34: 657–664.

Le Saoût, G., Kocaba, V. and Scrivener, K. (2011). Application of the Rietveld method to the analysis of anhydrous cement. Cem. Concr. Res. 41: 133–148.

León-Reina, L., De la Torre, A.G., Porras-Vázquez, J.M., Cruz, M., Ordonez, L.M., Alcobé, X. et al. (2009). Round robin on Rietveld quantitative phase analysis of Portland cements. J. Appl. Crystallogr. 42: 906–916.

Scrivener, K.L., Füllmann, T., Gallucci, E., Walenta, G. and Bermejo, E. (2004). Quantitative study of Portland cement hydration by X-ray diffraction/Rietveld analysis and independent methods. Cem. Concr. Res. 34: 1541–1547.

Speakman, S.A. (2010). Fundamentals of Rietveld Refinement I. XRD Pattern Calculation and Refinement Parameters. Massachusets Inst Technol 1–54.

Speakman, S.A. (2011). Fundamentals of Rietveld refinement, II. Refinement of a single phase. MIT Cent Mater. Sci. Eng. Available http//prism mit edu/xray/Rietveld_SinglePhase pdf (accessed Novemb 29).

Stutzman, P. (2010). Direct determination of phases in portland cements by quantitative X-ray powder diffraction. Tech Note (NIST TN), Natl Inst Stand Technol Gaithersburg, MD.

APPENDIX 2

Methods for the Elimination of the Kalpha2 Peak

In X-ray diffraction (XRD) analysis, K-alpha (Kα) radiation consists of two spectral lines, Kα1 and Kα2, caused by the transition of an electron from the 2p level to the 1s level in the atom. The Kα2 line has a slightly lower energy and therefore a slightly longer wavelength than the Kα1 line. The presence of these two lines can introduce what is commonly referred to as "peak splitting" in XRD patterns, which can complicate data interpretation, particularly for materials with complex diffraction patterns or where the peaks are widely spaced. "Cutting out" or removing the Kα2 line (often referred to as "K-alpha2 removal") in the data processing step is performed to remove the lower intensity and slightly shifted peak arising from the Kα2 radiation. This can help simplify the diffraction pattern and facilitate identification and quantification of the phases present.

However, whether or not this step is "necessary" may depend on a variety of factors: For simple phase identification, it may not be strictly necessary to remove the Kα2 line. The presence of the Kα2 line does not change the positions of the peaks corresponding to the Kα1 line, which are typically used for phase identification. For more complex analysis, such as accurately determining lattice parameters, quantifying phase ratios, or determining crystallite size or strain from peak broadening, removing the Kα2 line may be beneficial. Therefore, if your analysis is straightforward and you have no problems interpreting your data due to the presence of Kα2, you may not need to perform Kα2 removal. However, it is typically recommended for more complex analysis or more accurate results.

In X-ray diffraction, the Rachinger correction is a method to account for the effect of an unwanted K-alpha 2 peak in the energy spectrum. Ideally, diffraction measurements are made with X-rays of a single wavelength. In practice, the X-rays for a measurement are typically generated in an X-ray tube from the K-alpha line of a metal. This generation creates X-rays at a variety of wavelengths, but most non-K-alpha X-rays can be blocked from reaching the sample by filters. However, the K-alpha line is actually two X-ray lines close together: the stronger K-alpha 1 peak and the weaker K-alpha 2 peak. Compared to other radiation, such as stopping radiation, the K-alpha 2 peak is more difficult to filter out mechanically. The

Rachinger correction is a recursive method suggested by William Albert Rachinger (1927) to remove the disturbing peak Kα 2.

IN HIGHSCORE, THREE ALGORITHMS ARE IMPLEMENTED FOR KA2 STRIPPING

RACHINGER METHOD:

The RACHINGER method uses the exact wavelengths of the Kα1 and Kα2 lines and their intensity ratio. The intensity measured at the first point in the profile is assumed to be entirely stemming from Kα1. The d_{hkl} causing the diffraction at this 2θ angle will, of course, also diffract the Kα2 at the angle prescribed by Bragg's law. Because the intensity ratio of the $K\alpha 1{:}K\alpha 2$ reflections is exactly 2:1, half of the intensity measured at the low angle 2θ can be subtracted from the higher angle intensity, which is diffracted by the Kα2 wavelength.

The Rachinger method is a good first approximation to Kα2 deconvolution; however, it assumes that the profile of Kα1 is identical to the Kα2 profile. In fact, the Kα2 peak occurs at a higher angle and is somewhat broader than the Kα1 peak. This difference in profile shape gets more pronounced as 2θ increases.

MODIFIED RACHINGER METHOD:

According to DELHEZ and MITTEMEIJER, the doublet separation, δR, is calculated as a 2θ dependent parameter as shown in the formulas below:

$$I\alpha 2(2\theta) = R_{int}\, I\alpha 1(2\theta - \Delta R)$$

$R_{int} = 0.5$

$\Delta R = 2arcsin\,(R_{wav}\sin\theta 1) - 2\theta 1$

$R_{wav} = \lambda 2/\lambda 1$

MODIFIED WIDE RACHINGER METHOD:

This method does not distinguish between background and peak intensity. It is optimized for very broad features as encountered in semi-crystalline materials. This method should not be used for ordinary crystalline samples, where the other two methods produce better results.

LADELL METHOD:

PIKE and LADELL 1961 carefully determined the profiles of Kα1 and Kα2 peaks of resolved lines. They also calculated how the difference in peak shape varies with 2θ, using Fourier techniques. Based on this analysis, a table, which is used to subtract a portion of the Kα2 intensity from a series of adjacent 2θ points was created.

This method describes the Kα2 profile as the summation of 3, 5, or 7 sub profiles instead of only 1 sub profile as in other modified Rachinger methods. The sub profiles are indicated with levers (positions as wavelength ratios) and weights (intensity ratios). The algorithm of HSP uses 3 sub profiles.

CHANGING THE MODEL FOR THE ELIMINATION OF THE KALPHA PEAK 2

Once the first Rietveld refinement is done (the removal of the Kα2 peak was done automatically with the Rachinger model). After the first refinement, you can change the model and refine each parameter again one by one with the new chosen model which can be the Ladell or Wide Raching model.

Go to the **Treatment** field at the top of the screen, click to display the menu, then click on **Strip K-Alpha 2** and the screen in Figure 1 will appear.

The values for the wavelengths shown are characteristic for a copper anode and are set by the program. In the Rachinger method, click to display the menu and choose the method to remove the Kα2 peak, which can be Rachinger, Ladell or Wide Raching.

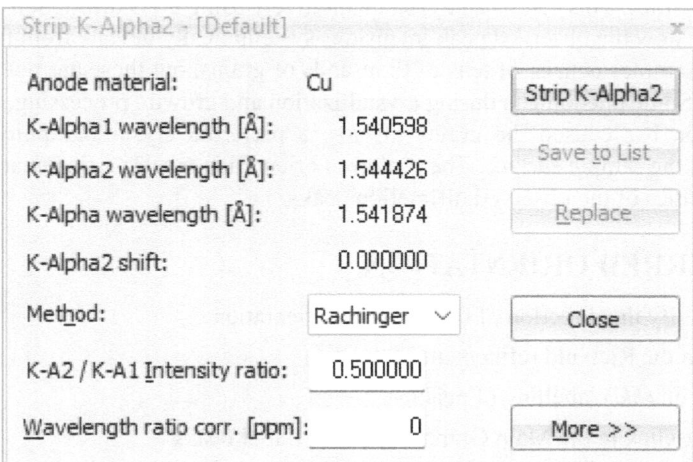

Figure 1. Models for KAlpha2 peak removal.

APPENDIX 3

Preferred Orientation

The preferred orientation of crystallites can create a systematic variation in diffraction peak intensities. This can be analyzed qualitatively using a 1D diffraction pattern by observing how the observed peak intensities systematically deviate from the ideal.

The samples consist of tens of thousands of grains, but these are not randomly oriented. Some phenomena during crystallization and growth, processing, or sample preparation has caused the grains to have a preferred crystallographic direction normal to the sample surface. The preferred orientation creates a systematic error in the intensities of the observed diffraction peaks.

PREFERRED ORIENTATION

First, identify the direction of the preferred orientation:

- ✓ Begin the Rietveld refinement of the data
- ✓ Turn on (*hkl*) labelling of peaks
- ✓ Right-click in the Main Graphics - select Label Peaks
- ✓ In the Label Peaks dialog, select HKL
- ✓ Click OK -The peaks will now be labelled

This only works if the calculated pattern has been simulated or refined (see Figure 1).

Look at the (*hkl*) values to see obvious trends in intensity vs. geometry. Remember that (*hkl*) is parallel to ($n * h\ n * k\ n * l$). For example, (011) and (022) are parallel. In Figure 2, the (011) (higher left) and (022) (higher right) peaks are both stronger than calculated. The other peaks are either not identified or the difference between calculated and experimental is not as dramatic. This indicates a preferred [011] orientation. Sometimes a related set of peaks will be weaker rather than stronger; this corresponds to a needle-like particle shape.

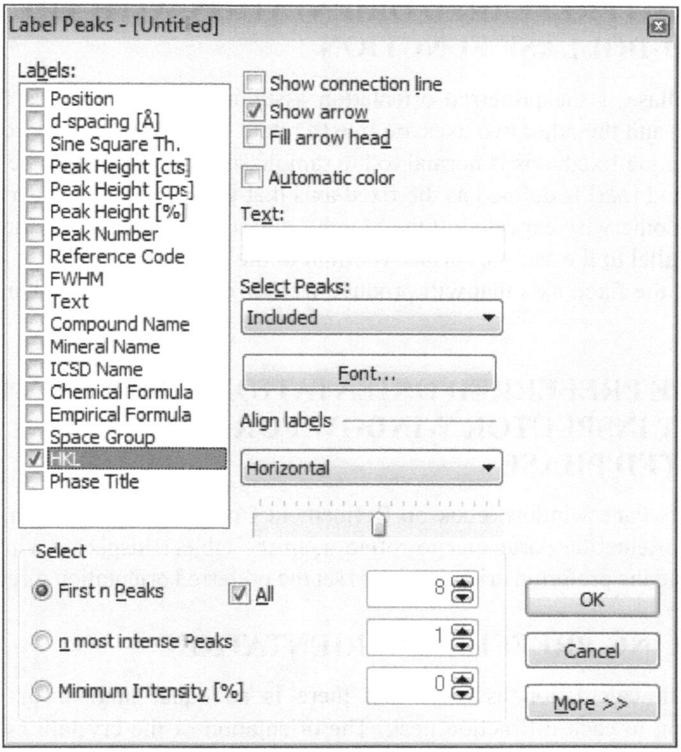

Figure 1. Label peaks with Miller indices.

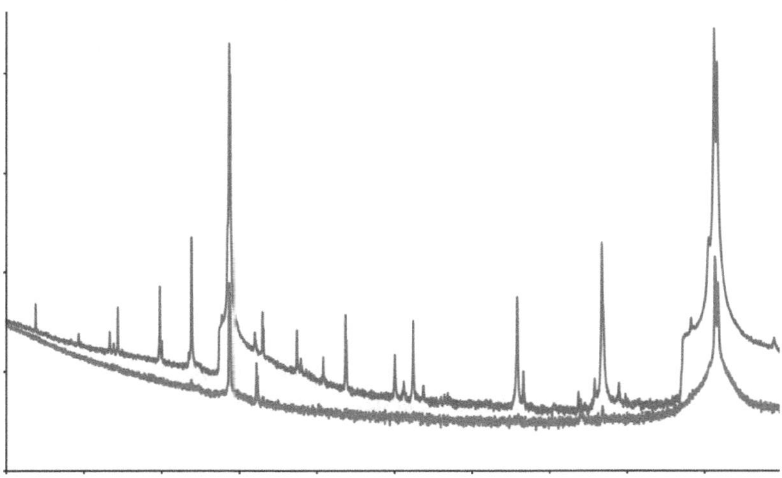

Figure 2. Preferred orientation.

DEFINING PREFERRED ORIENTATION WITH THE MARCH-DOLLASE FUNCTION

March-Dollase is the preferred orientation assuming that one axis is fixed in its orientation and the other two axes are free to rotate. If the March-Dollase parameter is < 1, then the fixed axis is normal to the sample surface (parallel to the diffraction vector s) and [*hkl*] is defined as the fixed axis that will produce stronger diffraction peaks than otherwise expected. If the March-Dollase parameter is > 1, then the fixed axis is parallel to the sample surface (normal to the diffraction vector s) and [*hkl*] is defined as the fixed axis that will produce weaker diffraction peaks than otherwise expected.

SET THE PREFERRED ORIENTATION DIRECTION IN THE OBJECT INSPECTOR WINDOW FOR THE SELECTED PHASE

In the Lists Pane window, click on Refinement Control. At the bottom, check the preferred orientation parameter to refine it. In the Object Inspector window at the bottom, find the preferred orientation and set the preferred orientation direction there.

MODELING PREFERRED ORIENTATION

The default calculation assumes that there is an equal number of crystallites contributing to each diffraction peak. The orientation of the crystallites is random and there is a statistically relevant number of grains. The preferred crystallographic orientation produces a non-random distribution in crystallite orientation which will cause the intensities of the observed diffraction peaks to vary systematically.

The March distribution function empirically models the preferred orientation effect

The March distribution function is a model for fitting preferred orientation of cylinder or needle-shaped crystals.

$$P_k = (r^2 \cos^2 \alpha_k + r^{-1} \sin^2 \alpha_k)^{-3/2}$$

α_k is the angle between the preferred orientation vector and the normal to the planes generating the diffracted peak

r is a refinable parameter in the Rietveld method

There are other models for Preferred Orientation, such as spherical harmonics
The spherical harmonics model will be added in HSP v4

Modeling Preferred Orientation in HSP

In the Refinement Control list, click the phase that has the preferred orientation. In the Object Inspector, expand the Preferred Orientation settings and set the *h*, *k*, and *l* directions to represent the [*hkl*] of the preferred orientation. Set the Preferred Orientation parameter, this parameter is located in the Object Inspector as "Parameter (March/Dollase)", this parameter is also located in the Refinement Control List as "Preferred Orientation", 1 means the [*hkl*] orientation is completely random < 1 means the [*hkl*] direction is the preferred orientation of the crystallites, > 1 means the [*hkl*] direction is preferentially avoided. Recalculate the simulation to observe the change in peak intensities (see Figure 3).

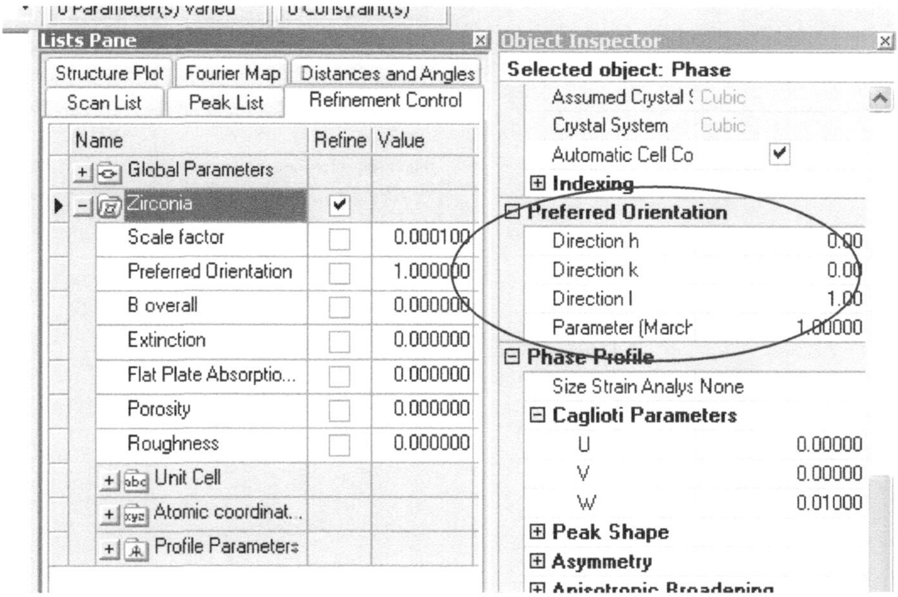

Figure 3. Setting the preferred orientation.

For high symmetry phases, you should enable the "Summed March/Dollase function" (see Figure 4):

- ✓ In the Refinement Control list, click on Global Parameters
- ✓ In the Object Inspector, scroll to the Refinement settings
- ✓ Check the option "Use summed March/Dollase function"

Refine again and observe the best fit of your refinement.

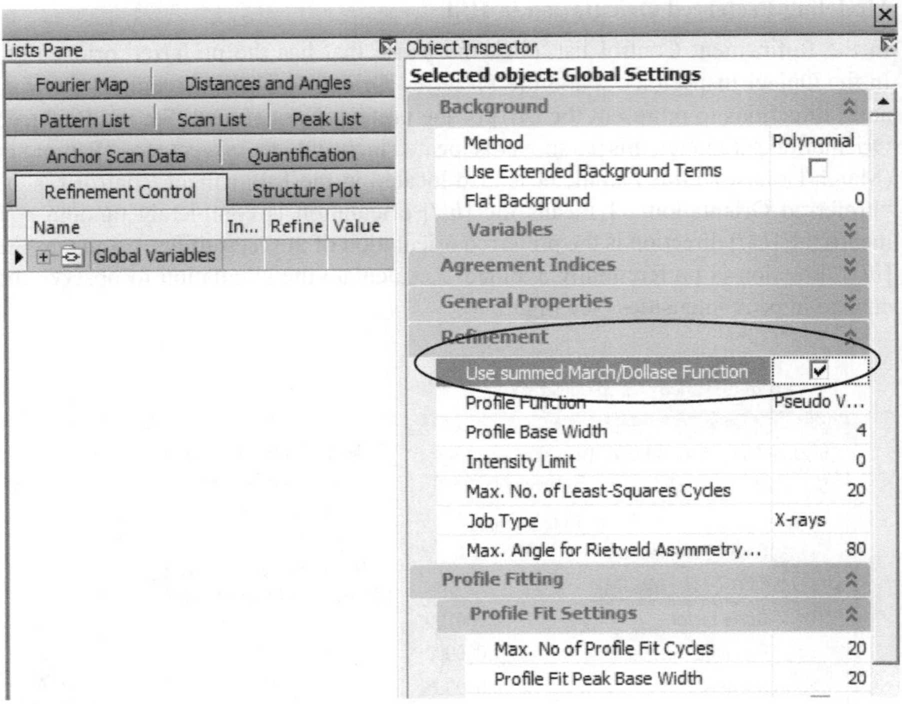

Figure 4. For the case of high symmetry.

APPENDIX 4

Material Properties from the Symmetry Group

In the Lists Pane window, click the Pattern List tab, locate one of the identified phases and double-click. A file with crystallographic information for the phase will appear, as shown in Figure 1.

In this example case, in the crystallographic parameters it is found that the Space group is $P3_221$.

With the techniques given in Chapter 2, it can be established that the group $P3_221$ denotes that the phase has a trigonal primitive cell + ternary helical axis

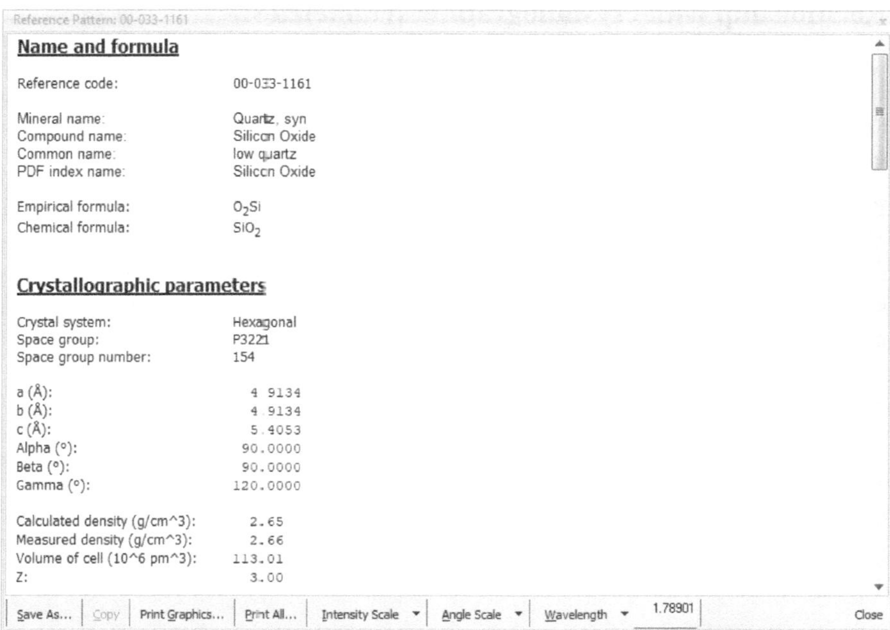
Reference Pattern: 00-033-1161

Name and formula

Reference code: 00-033-1161

Mineral name: Quartz, syn
Compound name: Silicon Oxide
Common name: low quartz
PDF index name: Silicon Oxide

Empirical formula: O_2Si
Chemical formula: SiO_2

Crystallographic parameters

Crystal system: Hexagonal
Space group: P3221
Space group number: 154

a (Å): 4.9134
b (Å): 4.9134
c (Å): 5.4053
Alpha (°): 90.0000
Beta (°): 90.0000
Gamma (°): 120.0000

Calculated density (g/cm^3): 2.65
Measured density (g/cm^3): 2.66
Volume of cell (10^6 pm^3): 113.01
Z: 3.00

Save As... | Copy | Print Graphics... | Print All... | Intensity Scale | Angle Scale | Wavelength | 1.78901 | Close

Figure 1. Crystallographic parameters.

with left-hand rotation + binary axis + monary axis. With the techniques given in Chapters 4 and 5, its properties derived from the symmetries can be established.

Silicon oxide (SiO_2) is a material of great interest in various fields due to its electrical and magnetic properties. These properties are closely related to its crystal structure and the $P3_221$ group, which depends on the way the silicon and oxygen atoms are arranged in its crystal lattice.

Electrical properties

Electrical Insulator: Silicon oxide is an excellent electrical insulator due to the lack of free charge carriers in its structure. This stems from the nature of its strong covalent bonds between silicon and oxygen atoms, which makes the electrons highly localized. Low Dielectric Constant: SiO_2 has a relatively low dielectric constant (~ 3.9 in quartz). This property is related to the rigidity of the crystal lattice, as the trigonal structure prevents significant polarization when subjected to an electric field.

Magnetic properties

Diamagnetism: SiO_2 is a diamagnetic material, meaning that in the presence of an external magnetic field, it generates a small, opposing magnetic field. This property is derived directly from its crystal structure ($P3_221$ group) and the absence of unpaired electrons in the silicon and oxygen atoms. Since its atoms have no intrinsic magnetic moments, it does not exhibit significant magnetism.

Relationship with crystallographic symmetries:

Tetrahedral symmetry: In many forms of SiO_2, such as quartz, the silicon atoms are surrounded by four oxygen atoms, forming a tetrahedral structure in addition to the hexagonal one. This symmetry restricts the movement of electrons and contributes to its behavior as an electrical insulator.

Piezoelectric symmetry: In some forms of SiO_2, such as α-quartz, there is piezoelectric behavior. This means that, under mechanical deformation, the material can generate an electric field. This phenomenon is directly related to the lack of a symmetry center in the crystal structure, which allows the generation of electrical polarization when the lattice is distorted.

Index

G

H

I

K

L

M

N

O

P

Q

R

S